AF388973

The Built Environment in a Changing Climate: Interactions, Challenges and Perspectives

The Built Environment in a Changing Climate: Interactions, Challenges and Perspectives

Editors

Giulia Ulpiani
Michele Zinzi

MDPI • Basel • Beijing • Wuhan • Barcelona • Belgrade • Manchester • Tokyo • Cluj • Tianjin

Editors
Giulia Ulpiani
Climate Change Research Centre
University of New South Wales
Sydney
Australia

Michele Zinzi
Energy Technologies and
Renewable Sources Department
ENEA - Italian National Agency
for New Technologies, Energy
and Sustainable Economic
Development
Rome
Italy

Editorial Office
MDPI
St. Alban-Anlage 66
4052 Basel, Switzerland

This is a reprint of articles from the Special Issue published online in the open access journal *Climate* (ISSN 2225-1154) (available at: www.mdpi.com/journal/climate/special_issues/Climate_Built).

For citation purposes, cite each article independently as indicated on the article page online and as indicated below:

LastName, A.A.; LastName, B.B.; LastName, C.C. Article Title. *Journal Name* **Year**, *Volume Number*, Page Range.

ISBN 978-3-0365-2356-9 (Hbk)
ISBN 978-3-0365-2355-2 (PDF)

Contents

About the Editors

Giulia Ulpiani

Giulia Ulpiani is a Postdoctoral Research Assistant currently working joint with UNSW Climate Change Research Centre and UNSW CRC Lab. She holds a Ph.D. in Industrial Engineering (specialized in Environmental Applied Physics) and a Master's Degree in Mechanical Engineering. Giulia's research, with a focus on climate and energy in the built environment, spans: i) urban heat island mitigation/adaptation technologies and holistic assessment of the impacts on health, energy, and comfort, ii) linkage between urban overheating and pollution, with special emphasis on spatio-temporal heterogeneity and extreme weather events (e.g. heatwaves, bushfires), iii) outdoor/indoor comfort and air quality, iv) advanced cooling technologies based on caloric materials and daytime radiative coolers, and v) smart automation of energy systems.

Michele Zinzi

Michele Zinzi, Ph.D. in Energetics. Full time researcher at ENEA since 2000. Main field of interest and activities: expert in building physics, testing on building envelope and urban materials, development of methodologies for the energy performance of buildings and energy analyses, assessment and mitigation of urban overheating and heat island. Editorial board member of Energy and Buildings (Elsevier) and *Climate* (MDPI). Advisory and scientific committee member of many international conferences. Author of many papers published in scientific journals and in international conferences. ENEA scientific responsible in several national and European Projects. National Executive Committee member in the Energy in Buildings and Communities Programme of IEA.

climate

Editorial

Introducing the Built Environment in a Changing Climate: Interactions, Challenges, and Perspectives

Giulia Ulpiani [1,*] and Michele Zinzi [2]

[1] Climate Change Research Centre, Level 4 Mathews Building, University of New South Wales, Sydney 2052, Australia
[2] Agenzia Nazionale Per le Nuove Tecnologie, L'Energia e lo Sviluppo Economico Sostensibile (ENEA), Via Anguillarese 301, 00123 Rome, Italy; michele.zinzi@enea.it
* Correspondence: g.ulpiani@unsw.edu.au

Citation: Ulpiani, G.; Zinzi, M. Introducing the Built Environment in a Changing Climate: Interactions, Challenges, and Perspectives. *Climate* 2021, 9, 104. https://doi.org/ 10.3390/cli9070104

Received: 15 June 2021
Accepted: 20 June 2021
Published: 23 June 2021

Publisher's Note: MDPI stays neutral with regard to jurisdictional claims in published maps and institutional affiliations.

1. Introduction

Planning for climate change adaptation is among the most complex challenges cities are facing today. Worldwide, unprecedented levels of urban overheating have been recently recorded. Hosting more than half of the world population, cities are major receptors and drivers for climate change with disruptive and mutually empowering impacts on both the natural and the socio-economic environment [1,2]. Global and local dimensions of a changing climate are especially worrisome in cities because the rate of change in the patterns of human settlement, energy use, transportation, and industry is escalating much faster than elsewhere, causing a complex network of feedback and amplifying loops. Furthermore, cities are fertile ground for the interaction between heat and pollution hot spots to the point that where an urban heat island (UHI) exists, likely an urban pollution island (UPI) co-exists [3–5], causing compound effects on human health. On top of that, extreme events, such as heatwaves, floods, bushfires, and cold spells, are becoming more frequent, severe, and longer lasting, which is projected to double-to-triple the concerted effect on mortality rates [6,7]. Indeed, climatic alterations put a strain on (i) energy needs for cooling and release of anthropogenic heat, (ii) mortality and morbidity due to overheating and air pollution, (iii) productivity and wellbeing, and (iv) accessibility to public spaces and social prosperity. Recognizing the potentially negative consequences of these events is key to take active and bold actions, to stem the tide of change, and move towards adaptation and resilience [8].

Therefore, some critical questions arise:

- What is the future of the urban realm in a changing climate?
- What is the role of a growing population with expanding patterns of urbanization and consumption?
- How can we mitigate buildings' and cities' burden on local/global environmental change?
- How can we design to provide adequate housing and outdoor spaces to the vulnerable population?

It is now very well documented that urban overheating causes the cooling energy consumption of buildings to double [9,10], forces the construction of new power plants to compensate for increasing peak electricity demand [11], and empowers social injustice [12]. Several mitigation and adaptation technologies have been proposed [13], and many successful examples have been recently demonstrated [2,14,15], yet a full understanding of how cities are supposed to reach enough dynamicity to cope with climate change is still unveiled.

2. Aims and Scope

The Special Issue on "The Built Environment in a Changing Climate: Interactions, Challenges, and Perspectives" is aimed to collect answers to the above questions on a worldwide scale. Submissions were encouraged that contribute to:

- collecting criteria and methods to develop meteorological datasets including climate changes;
- establishing innovative monitoring systems to capture the multifarious impacts of an evolving climate on the built environment;
- defining the energy and comfort metrics in future buildings;
- estimating the impacts in terms of air quality and heat-related mortality and morbidity rates;
- investigating the interaction between global and local climate changes;
- defining governance models, legal frameworks, and agenda-setting methods to prioritize climate policies; and
- defining criteria and targets for urban- and building-integrated design in a warmer world.

3. Presentation of the Published Papers

This Special Issue collects 11 studies from all continents around the world (see Figure 1). Four macro topics are covered and discussed in detail in the following subsections: (1) future-proof design criteria, (2) urban heat island (UHI) mitigation, (3) urban health in a changing climate, and (4) development of new methodological frameworks. Most papers (36.4%) deal with topic 1, 27.3% with topic 2, and 18.2% with topic 3 or 4. The same macro topic could be addressed by a variety of perspectives and scales ranging from individual building level to city level. Each paper is identified with a different marker on the map in Figure 1: the colour code discriminates across macro topics, whereas the marker shape denotes the scale of investigation.

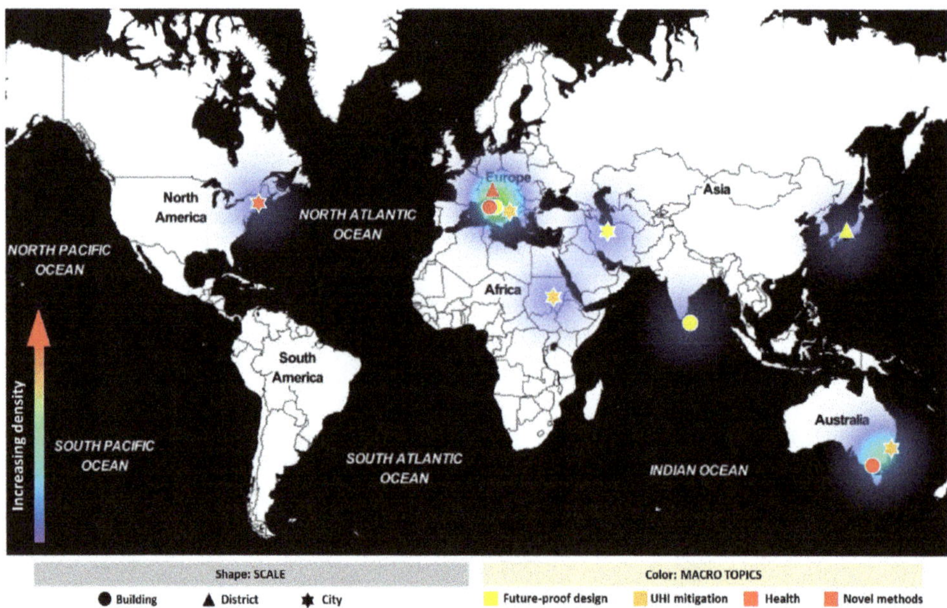

Figure 1. Density heatmap of the studies collected in the Special Issue across the globe. The shade goes from violet to red for increasing number of papers in the same geographic areas, whereas markers identify the cities where the studies were conducted: the shape defines the scale; the colour defines the macro topic. For multiple-city studies and for continent-wide studies, as those conducted in Australia, we used the most populous city or the city of affiliation.

3.1. Future-Proof Design Criteria

Long-lasting, comfort-oriented redesign measures are the core topic of four papers in this collection. In [16], the indoor heat stress in modern, air-conditioned, multi-level office buildings is investigated in the tropics (Colombo, Sri Lanka). Design interventions on plan layout, orientation, sectional layout, and envelope characteristics are assessed in terms of heat-gain risk and associated energy consumption in 12 case studies. Plan form is identified as the key driver for heat vulnerability. Deep plans allow better control and homogeneity of the heat transfer to the indoors, as compared to shallow and covered layouts. The peripheral zones are most vulnerable to heat extremes, exceeding 40 °C when the air conditioning is inactive. Shading and insulation are found to mitigate the local heat stress together with a careful design of the building sectional layout, especially in terms of night ventilation. The study overlooks factors like space lighting systems and occupancy patterns, as these varied only marginally across the investigated buildings.

Climate change may cause premature obsolescence of todays' energy efficient paradigms, too. This is addressed in [17], where the authors propose a methodology to evaluate the change in energy performance for near-zero energy buildings (nZEBs) by comparing present and future demand under the hypothesis of unchanged nZEB legislative requirements. Two future scenarios from the IPCC Fifth Assessment Report (AR5) [18] are considered that incorporate not just the effects of global warming but also the change in radiative forcing. Indeed, both temperature and solar gains are critical variables in the energy balance of current nZEB definitions. A case study in Rome is used to perform hourly dynamic simulations, following the calculation method in force, described in ISO 52016-1:2017 [19]. The annual power consumption is demonstrated to increase by almost 20% in the future, largely caused by longer-lasting use of air conditioning and intensified peak demand. Comfort is jeopardized for 5–6% more time during the year. However, newly-developed, long-term comfort metrics based on the statistical temperature distribution reveal a milder penalty, since diurnal and annual swings are strongly levelled out in future scenarios.

Thermal environmental design is key to ensure (and preserve) not just indoor but also outdoor liveability. It requires a comfort-oriented selection and allocation of built elements (e.g., buildings, pavements, roads), vegetated areas (e.g., parks, tree-lined avenues, gardens, green roofs), and water features (e.g., fountains, pools, sprinklers, ponds). In [20], a redevelopment plan for Central Osaka Station is proposed by combining different heat island countermeasures and by applying computational fluid dynamics, surface heat budget equation, and GIS to track wind distribution, surface temperatures, and mean radiant temperatures. The results are compared in terms of standard new effective temperature (SET*) on a typical summer day. Solar shading is the most impactful redesign strategy in terms of SET* reduction at peak hours (6 to 8 °C), followed by surface material change (0 to 2.5 °C) and ventilation (0 to 1.5 °C). A general conclusion is that climate-resilient outdoors can be effectively achieved by adjusting the shade provided by buildings and that provided by trees and by selecting appropriate surface materials.

A conclusive critical point on future-proof design discussed in the Special Issue is the need for re-inventing even the way we set the right conditions for fighting climate change, especially at the local scale. Some cities in the world act as flagships in implementing and testing the effects of climate-resilient urban plans. This is, for instance, the case of Tehran in Iran, where the impacts of climate upheavals have been especially worrisome in recent years [21]. In this collection, Ghasemzadeh and Sharifi [22] investigate the barriers that are hindering climate change adaptation, notably in low-resilience districts, to appreciate their significance, interconnectedness, and hierarchy. The analysis is approached through a mixture of qualitative and semi-qualitative methods, including focus group discussion, questionnaire-survey, interpretive structural modelling, and confirmatory factor analysis. The most critical barrier is found right in the "structure and culture of research", which deals with the absence of a centralized research body that coordinates and supervises studies and pilot projects for climate adaptation, policy and decision-making processes, and academic efforts towards winning strategies and experiences. The second greatest independent barrier is imposed by "laws and regulations", which are untargeted, incongruous, and loopholed when it comes to climate change and performance verification. Last, but not least is the barrier in "Planning", triggered by a lack of local policy-making bodies and authorities to build up cohesive land use plans, adaptation models, assessment schemes, and integration mechanisms. The study lays the basis for restructuring the way climate adaptation is typically tackled by stressing the need for identifying and removing/softening the barriers before attempting any uncoordinated and unsupported urban plans.

3.2. Urban Heat Island Mitigation

The Urban Heat Island intensity can be determined based on air temperature (atmospheric UHI or AUHI) or surface temperature (SUHI). In [23], a micro-scale experimental investigation of the AUHI intensity and spatial distribution in Bari, Italy is conducted on a daily, monthly, seasonal, and annual basis across five years (2014–2018). The study elucidates how sea breeze and urban attributes modulate local hot spots. The UHI intensifies during the summer and typically overnight. On a 24h basis, it gets exacerbated in high-density local climate zones (LCZ = 2 according to the classification by Stewart and Oke [24]) reaching 4.0 °C. However, the daytime maxima (nearly 5 °C) are measured in lower-density areas (LCZ = 5), countervailed by reduced night-time intensity (less than 3 °C). Future scenarios are evaluated at selected locations by downscaling the trends depicted by the IPCC A2 scenario [25]. A 2 to 4 °C maximum increase in urban air temperature is expected between 2071 and 2100, thus posing further strain on urban liveability.

Coastal cities are especially cumbersome when it comes to UHI mitigation design. This is thoroughly expressed in [26] by Yenneti et al., who review the impacts of urban overheating on health, energy, labour productivity, and social behaviour and quantify the urban cooling potential in Australian cities by collecting evidence from several heat-mitigation strategies. The dualistic effect of sea breeze and hot desert winds in most Australian cities establishes a highly dynamic interaction with the local heat island circulation, which results in extreme temporal and spatial UHI heterogeneity. Despite the complex mechanisms at stake, the use of reflective or green materials and the implementation of water-based technologies proved to considerably alleviate the thermal unbalance with respect to rural and suburban areas. The average maximum mitigation potential of individual heat mitigation strategies is quantified at 1.0 °C for urban greenery (e.g., trees, hedges), 0.1–0.2 °C for building greenery (e.g., green roofs, vertical gardens), 0.3 °C for reflective surfaces (e.g., roofs, pavements), and 1.0–2.0 °C for water features (e.g., misting systems, sprinklers, fountains). The combination of multiple techniques on account of local specificities is the winning strategy in most studies, with average maximum UHI mitigation in the order of 1.5 °C.

However, the key drivers for UHI generation and intensification may be profoundly different in developed, developing, and undeveloped countries, which ultimately defines the set of best practises in terms of mitigation and adaptation. In [27], the authors present the unique case of East Africa, where UHI is escalating fast, caused by the collective effect of climate change and rapid urban population growth, but very little is the limelight in international literature. The study focusses on the five most populated cities in different climatic zones, namely Khartoum in Sudan, Dar es Salaam in Tanzania, Nairobi in Kenya, Addis Ababa in Ethiopia, and Kampala in Uganda. The comparison is based on annual daytime and night-time SUHI intensities across time (2003 versus 2017) and space. SUHI drivers are found in climate conditions, urban development patterns, and informal settlement growth. Blue and green features are identified as essential means of urban heat mitigation and comfort enhancement to be carefully planned not just in view of microclimatological constraints but with due attention to potential gentrification, social unrest, and spread of disease.

3.3. Urban Health in a Changing Climate

In [28], the authors evaluate the heat-related mortality at home and outside of the home on hot days in Boston, MA, between 2000 and 2015. Subject and neighbourhood-scale attributes are analysed to interpret the degree of associativity between hot days and mortality at three temperature thresholds, while geography-weighted regression is used to further scrutinize the spatial heterogeneity. It is found that at-home mortality is triggered by both social and environmental vulnerabilities. Among social factors, low-to-no income and limited English proficiency are conducive to higher mortality. On the bright side, even small-scale or individual heat mitigators, such as street trees and enhanced energy efficiency, are associated with significantly reduced death risk, which encourages the implementation of a wide range of adaptation solutions.

However, the exacerbation of temperature extremes is only one manifestation of a strongly multi-faceted, changing climate that has disruptive effects on the healthcare system. Again, in Australia, bushfires demonstrated their increasing destructive potential during the Black Summer in 2019/2020. In this Special Issue, Rajagopalan and Goodman deal with the need to design future-proof buildings not just to preserve a pleasant thermal environment, but also to protect the indoors from the acute and chronic health effects of air pollution [29]. This becomes an imperative during extended bushfire events when people seek shelter in residential buildings. In actuality, the current building stock in Australia offers inadequate air tightness and filtration. The manuscript explores the potential benefits of re-designing the building envelope and the filtration system and informs on the use of portable air cleaners as coping mechanisms during extreme pollution episodes. Long-lasting air quality preservation calls for coupling of reduced smoke infiltration and improved ventilation if not extraction. Further, analysis of current filtration technologies demonstrates the need for new technological developments to be effective against gaseous pollutants and particulate matter.

3.4. Novel Methods

As the dynamics at building, district, and city scale change, so does the sophistication of methodological frameworks. Two crucial questions arise: (1) How can we plan for the future if our predictions are inaccurate or limited? (2) What is the scale of analysis, action, and prediction? Is it inter-urban, intra-urban, individual-building, or somewhere in-between?

To answer the first question, in [30], the authors focus on the need for reliable and robust future weather estimators and weather files generators by comparing three commonly used tools based on statistical downscaling (WeatherShift, Meteonorm, and CCWorldWeatherGen) against the typical meteorological year prediction obtained by high-quality regional climate modelling and dynamic downscaling. The energy consumption of a residential house and an apartment in Rome (Italy) was simulated by forcing the boundary conditions in accordance with the four generated future datasets. Interestingly, the differences between the two families of weather estimators were not only driven by the different forecasting approach but showed sensitivity to the building type, too. This demonstrates that overlooking regional and local specificities will contribute to forecasting inaccuracies and uncertainties. As such, the smaller scale will become vital in planning based on future urban energy budgets.

The benefits of working against a scale that is smaller than the city but bigger that the individual building become critical when concepts like energy flexibility and smart grid come into play. In an ever more connected and intercommunicating network of buildings that consume and produce energy at the same time (prosumers) and that may exacerbate as well as mitigate local hot spots, there is a growing need for building energy models that take into account the mutual interactions. A dedicated module has been recently introduced as part of the EUReCA (Energy Urban Resistance Capacitance Approach) platform, described in [31]. The authors test the module in Padua (Italy) during the cooling season. The urban energy demand is predicted through a bottom-up approach that integrates mutual shading, heat island effects, and single-building energy demand. The analysis demonstrates the spiral of feedback loops between local heat island and degraded HVAC's efficiency, which results in increased waste-heat rejected from cooling systems. Quantifying these feedbacks is pivotal not just to return accurate estimates of district-level energy consumption, but also to fully appreciate the positive impacts of heat-mitigation strategies.

4. Conclusions

The papers included in this Special Issue tackle multiple aspects of how cities, districts, and buildings could evolve along with climate change and how this would impact our way of conceiving and applying design criteria, policies, and urban plans. Despite the multidisciplinary nature of the collection, some transversal take-home messages emerge:

- Today's energy-efficient paradigms may lose their virtuosity in the future unless accurate estimates of future scenarios are used to design modelling platforms and to inform legislative frameworks;
- Acting at the local scale is key. Future climate change adaptation will be implemented at the local level. Overlooking regional and local specificities will contribute to inaccurate and inefficient action plans. As such, the smaller scale will become vital in predicting future urban metabolic rates and corresponding comfort-driven strategies;
- Energy poverty, heat vulnerability, and social injustice are emerging as critical factors for planning and acting for future-proof cities on par of micro- and meso-climatological factors;
- given that the impacts of climate change will persist for many years, adaptation to this phenomenon should be prioritized by removing any prominent barrier and by enabling combinations of different mitigation technologies.

These topics will receive a global reach in few decades, since also developing and underdeveloped countries are starting their fight against local climate change, with cities at the forefront.

Funding: This research received no external funding.

Institutional Review Board Statement: Not applicable.

Informed Consent Statement: Not applicable.

Conflicts of Interest: The authors declare no conflict of interest.

References

1. Wilby, R.L. Constructing climate change scenarios of urban heat island intensity and air quality. *Environ. Plan. B Plan. Des.* **2008**, *35*, 902–919. [CrossRef]
2. Haddad, S.; Paolini, R.; Ulpiani, G.; Synnefa, A.; Hatvani-Kovacs, G.; Garshasbi, S.; Fox, J.; Vasilakopoulou, K.; Nield, L.; Santamouris, M. Holistic approach to assess co-benefits of local climate mitigation in a hot humid region of Australia. *Sci. Rep.* **2020**, *10*, 1–17. [CrossRef] [PubMed]
3. Ulpiani, G. On the linkage between urban heat island and urban pollution island: Three-decade literature review towards a conceptual framework. *Sci. Total Environ.* **2020**, *751*, 141727. [CrossRef] [PubMed]
4. Ulpiani, G.; Ranzi, G.; Santamouris, M. Local synergies and antagonisms between meteorological factors and air pollution: A 15-year comprehensive study in the Sydney region. *Sci. Total Environ.* **2021**, *788*, 147783. [CrossRef] [PubMed]
5. Li, H.; Meier, F.; Lee, X.; Chakraborty, T.; Liu, J.; Schaap, M.; Sodoudi, S. Interaction between urban heat island and urban pollution island during summer in Berlin. *Sci. Total Environ.* **2018**, *636*, 818–828. [CrossRef] [PubMed]
6. Kong, Q.; Guerreiro, S.B.; Blenkinsop, S.; Li, X.-F.; Fowler, H.J. Increases in summertime concurrent drought and heatwave in Eastern China. *Weather Clim. Extrem.* **2020**, *28*, 100242. [CrossRef]
7. Baklanov, A.; Molina, L.T.; Gauss, M. Megacities, air quality and climate. *Atmos. Environ.* **2016**, *126*, 235–249. [CrossRef]
8. Araos, M.; Berrang-Ford, L.; Ford, J.; Austin, S.E.; Biesbroek, R.; Lesnikowski, A. Climate change adaptation planning in large cities: A systematic global assessment. *Environ. Sci. Policy* **2016**, *66*, 375–382. [CrossRef]
9. Santamouris, M. *Minimizing Energy Consumption, Energy Poverty and Global and Local Climate Change in the Built Environment: Innovating to Zero: Causalities and Impacts in a Zero Concept World*; Elsevier: Amsterdam, The Netherlands, 2018.

10. Santamouris, M. On the energy impact of urban heat island and global warming on buildings. *Energy Build.* **2014**, *82*, 100–113. [CrossRef]

11. Santamouris, M.; Cartalis, C.; Synnefa, A.; Kolokotsa, D. On the impact of urban heat island and global warming on the power demand and electricity consumption of buildings—A review. *Energy Build.* **2015**, *98*, 119–124. [CrossRef]

12. Grove, M.; Ogden, L.; Pickett, S.; Boone, C.; Buckley, G.; Locke, D.H.; Lord, C.; Hall, B. The Legacy Effect: Understanding How Segregation and Environmental Injustice Unfold over Time in Baltimore. *Ann. Am. Assoc. Geogr.* **2018**, *108*, 524–537. [CrossRef]

13. Santamouris, M. Regulating the damaged thermostat of the cities—Status, impacts and mitigation challenges. *Energy Build.* **2015**, *91*, 43–56. [CrossRef]

14. Bartesaghi-Koc, C.; Haddad, S.; Pignatta, G.; Paolini, R.; Prasad, D.; Santamouris, M. Can urban heat be mitigated in a single urban street? Monitoring, strategies, and performance results from a real scale redevelopment project. *Sol. Energy* **2021**, *216*, 564–588. [CrossRef]

15. Cheela, V.; John, M.; Biswas, W.; Sarker, P. Combating Urban Heat Island Effect—A Review of Reflective Pavements and Tree Shading Strategies. *Buildings* **2021**, *11*, 93. [CrossRef]

16. Rajapaksha, U. Environmental Heat Stress on Indoor Environments in Shallow, Deep and Covered Atrium Plan Form Office Buildings in Tropics. *Climate* **2020**, *8*, 36. [CrossRef]

17. Summa, S.; Tarabelli, L.; Ulpiani, G.; Di Perna, C. Impact of Climate Change on the Energy and Comfort Performance of nZEB: A Case Study in Italy. *Climate* **2020**, *8*, 125. [CrossRef]

18. IPCC. *Climate Change—The Synthesis Report*; 2014; Available online: https://www.ipcc.ch/report/ar5/syr/ (accessed on 4 June 2021).

19. International Organization for Standardization. *Energy Performance of Buildings—Energy Needs for Heating and Cooling, Internal Temperatures and Sensible and Latent Heat Loads—Part 1: Calculation Procedures*; ISO: Geneva, Switzerland, 2017.

20. Takebayashi, H. Thermal Environment Design of Outdoor Spaces by Examining Redevelopment Buildings Opposite Central Osaka Station. *Climate* **2019**, *7*, 143. [CrossRef]

21. Ahmadi, M.; Vaziri, B.M.; Ahmadi, H.; Moeini, A.; Zehtabiyan, G.R. Assessment of climate change impact on surface runoff, statistical downscaling and hydrological modeling. *Phys. Chem. Earth Parts A/B/C* **2019**, *114*, 102800. [CrossRef]

22. Ghasemzadeh, B.; Sharifi, A. Modeling and Analysis of Barriers to Climate Change Adaptation in Tehran. *Climate* **2020**, *8*, 104. [CrossRef]

23. Martinelli, A.; Kolokotsa, D.-D.; Fiorito, F. Urban Heat Island in Mediterranean Coastal Cities: The Case of Bari (Italy). *Climate* **2020**, *8*, 79. [CrossRef]

24. Stewart, I.D.; Oke, T.R. Local Climate Zones for Urban Temperature Studies. *Bull. Am. Meteorol. Soc.* **2012**, *93*, 1879–1900. [CrossRef]

25. IPCC. *Climate Change 2014: Synthesis Report. Contribution of Working Groups I, II and III to the Fifth Assessment Report of the Intergovernmental Panel on Climate Change*; IPCC: Geneva, Switzerland, 2014.

26. Yenneti, K.; Ding, L.; Prasad, D.; Ulpiani, G.; Paolini, R.; Haddad, S.; Santamouris, M. Urban Overheating and Cooling Potential in Australia: An Evidence-Based Review. *Climate* **2020**, *8*, 126. [CrossRef]

27. Li, X.; Stringer, L.; Dallimer, M. The Spatial and Temporal Characteristics of Urban Heat Island Intensity: Implications for East Africa's Urban Development. *Climate* **2021**, *9*, 51. [CrossRef]

28. Williams, A.A.; Allen, J.G.; Catalano, P.J.; Spengler, J.D. The Role of Individual and Small-Area Social and Environmental Factors on Heat Vulnerability to Mortality within and Outside of the Home in Boston, MA. *Climate* **2020**, *8*, 29. [CrossRef]

29. Rajagopalan, P.; Goodman, N. Improving the Indoor Air Quality of Residential Buildings during Bushfire Smoke Events. *Climate* **2021**, *9*, 32. [CrossRef]

30. Tootkaboni, M.P.; Ballarini, I.; Zinzi, M.; Corrado, V. A Comparative Analysis of Different Future Weather Data for Building Energy Performance Simulation. *Climate* **2021**, *9*, 37. [CrossRef]

31. Romano, P.; Prataviera, E.; Carnieletto, L.; Vivian, J.; Zinzi, M.; Zarrella, A. Assessment of the Urban Heat Island Impact on Building Energy Performance at District Level with the EUReCA Platform. *Climate* **2021**, *9*, 48. [CrossRef]

Article

Environmental Heat Stress on Indoor Environments in Shallow, Deep and Covered Atrium Plan Form Office Buildings in Tropics

Upendra Rajapaksha

University of Moratuwa, Department of Architecture in University of Moratuwa, Moratuwa 10400, Sri Lanka; upendra@uom.lk

Received: 4 January 2020; Accepted: 19 February 2020; Published: 22 February 2020

Abstract: Environmental heat stress on buildings through façades contributes to indoor overheating and thus increases demand for energy consumption. The study analyzed the problem, heat gain risk, of modern air-conditioned multi-level office buildings in tropics, for example Colombo. *Plan form, orientation, sectional form and envelope* were identified and theorized to understand design interventions to reduce the risk of getting heat stress on indoor environments. On-site thermal performance investigations in multi zones of identified three typical built forms, namely; *shallow, deep and covered atrium plan forms,* quantified the heat stress. Reaching the daytime indoor and surface temperature in peripheral zones of multi-story office buildings during air conditioning "off-mode" up to 38 °C–42 °C was seen as a critical heat stress situation to be addressed through building design. Shading or insulation on façades to control environmental heat gain and manipulation of building section for night ventilation to remove internal heat developed during the daytime are discussed. However, the significance of the plan form depth was found to be a main contributor in dealing with heat transfer to indoor space. Deep plan form was found to be more effective in controlling environmental heat transfer to indoor space across the plan depth.

Keywords: heat stress from outside; indoor environments; tropics; multi-level office buildings

1. Introduction

Thermal comfort experienced by the building occupants plays a vital role in enhancing climate responsive design. Indoor overheating, in this context, is identified as one major problem. Indoor overheating is a condition where indoor air temperature moves above the upper limits of comfort zone. This can result in thermal discomfort while reducing the productivity of the occupants in buildings with free floating conditions. Further, overheated indoors demand extensive use of mechanical systems for cooling and thus increase in operational energy.

Controlling environmental solar heat gains into buildings due to high levels of ambient air temperature, internal heat generation from occupants and equipment together with enhancing heat escape from indoors determines the thermal balance in buildings [1,2]. Depending on the climate type and usage pattern of buildings, excess heat gain could contribute to indoor overheating [3].

Reducing indoor overheating potential of the building design contributes to decrease the demand for cooling energy [4]. To prevent indoor overheating, passive and resilient design interventions may be integrated with the building design. In tropics, with the presence of high levels of diffuse radiation due to cloud cover all year around, the need to overcome indoor overheating is significantly important in lowering the demand for cooling energy and as a prerequisite for passive cooling in lowering indoor air temperature levels than the corresponding ambient levels.

On the other hand, addressing environmental heat stress on air conditioned buildings is an essential phenomenon to study. In such conditioned buildings, environmental heat stress through façades to indoors may not be visible and sensible to the occupants but contribute to increase the energy use in air conditioning, exacerbating the emissions and warming problem [5]. Thus, in-depth understanding of pattern of heat stress on air-conditioned buildings becomes imperative and performance improvements of these buildings in tropics are yet to be achieved. There is lack of consensus surrounding measurement and reporting of heat stress in buildings in tropics. Specific aim of this paper is to develop an empirically tested prediction method to ascertain environmental heat stress pattern on façades and thus indoors of three building typologies i.e., shallow, deep and closed atrium. Research questions the plan depths of these building typologies in relation to heat stress on façades and distribution of indoor air temperature pattern across spaces. This will enable the understanding of plan depth-specific factors of indoor overheating patterns. Objectives of the work were to quantify the extent to which heat stress occurs in these buildings and to analyze indoor air temperature distribution across plan depths with a focus on learning the significance of plan form depth on indoor air temperature distribution. The work used a comprehensive field investigation on thermal behavior of selected multi-level office buildings in Colombo.

2. Background, State-of-the-Art

Addressing indoor overheating risks in buildings in cooling-dominated climates has been seen as a growing research interest even under current climate scenarios. This has been further driven by current climate change projections and unintended consequences of poorly integrated interventions in building design as well. Review of literature, presented below, reflects a mapping of overheating risks due to environmental heat stress and identification of state-of-the-art prioritization areas in order to address these risks in office buildings in both free floating and air-conditioned modes.

2.1. Indoor Overheating and Energy Use

Peer reviewed data on climate change based on main sources such as the Synthesis Report of Intergovernmental Panel on Climate Change [6], suggests an increase of GHG emissions and annual temperatures over the years. IPCC's Synthesis Report shows that increase of GHG emissions would result in further warming in the global climate system [6]. Warming climates could directly enhance indoor overheating potential in buildings [7]. Previous studies have shown strong evidence to justify that increase of average external air temperature and indoor overheating could considerably increase cooling load [6,8–10] and thus emissions.

Research evidence predicts more warm days in sub-tropics and tropics thus limiting the efforts of emissions reduction strategies in building sector. Computer simulations have predicted that contemporary office buildings that are designed in current sub-tropical climatic conditions are subjected to the risk of further overheating with increasing warming weather [10]. Hence the need to understand the thermal behavior patterns of buildings is vital if design strategies need to be integrated more meaningfully to address indoor overheating.

Climate data from Department of Meteorology, Sri Lanka, indicates an increase of monthly mean temperatures over the past decades (Figure 1). Nevertheless, in recent years, new design styles with more glass façades are being added to the building population in Sri Lanka purely on aesthetic purpose and environmental and climatic dimensions are in question. These contemporary buildings add new corporate images for respective client organizations and enhance class one material comfort for the occupants but serious concerns over the thermal and energy performance remain significant.

In such a context, an examination of existing buildings in relation to the global problem of warming and prevailing building practice could easily be useful in quantifying the problem and, then to suggest appropriate solutions in future practices. Further, the research aims at highlighting why heat gain risks in tropics is severe and building–climate interplay is in need of a new direction.

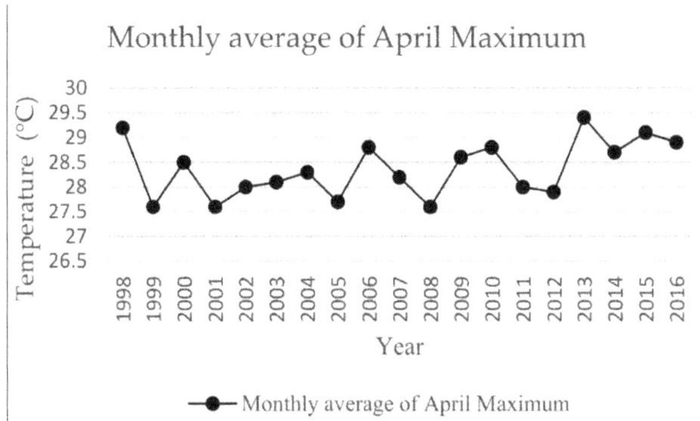

Figure 1. Mean monthly air temperature (for example March) since 1998 in Colombo is slightly on the increasing side (Data source Met Department, Colombo).

2.2. Interplay of Building and Climate in the Tropical Context

Climate is a catalyst that affects building–climate interplay and thus indoor climate in the operational stage of a building and thus need for energy. Microclimate around the building, plan depth and sectional profile of the building form, thermo physical characteristics and design of the building envelope are significant areas that need to be addressed though interventions.

Indoor thermal load profiles in buildings due to internal or environmental heat stress regulate the internal air temperatures of buildings and thus energy use [11]. This profile can either be of internal load dominant or environmental load dominant due to a diversity of heat gain sources.

In warm humid tropical climates, environmental loads play a major role in the problem of making naturally ventilated buildings overheated. The risk of heat stress on the façade of a building makes the process of maintaining the average indoor air temperature close to the required comfort zone more challenging. Due to the complexities of climatic interactions, the building components may tend to play a dual role in climate response. The best example is the building envelop, which can promote day light and ventilation but brings heat into the building indoor environment the same time in addition to the conduction heat gain. Hence, in building–climate interplay, addressing contradicting behaviors like this within a holistic manner becomes vital.

Several interventions in buildings, to promote indoor comfort and reduce the energy consumption for cooling have been investigated and assessed [12]. Givoni [13] highlights that in warm climates, natural ventilation is desirable when the outdoor air is at a lower temperature than the indoor air or when it can prevent indoor overheating caused by direct or indirect solar gain. A pioneering work on thermal mass [14] showed that indoor air temperature in a building with closed windows during day and night ventilation can be stabilized despite outdoor peaks.

Similar experiment method was found applied to a study in Galle, Sri Lanka by Rajapaksha U et al. [15]. This experimental work on the site showed how building envelope with high thermal mass and night ventilation contributed to reduce indoor air temperature during the day by 2.5–4° C lower than the ambient. Potential of thermal mass for passive cooling was seen with the support of diurnal range between 7–8 degrees C for warm humid climates. A previous study of Rajapaksha I, [16] had justified the same potential of thermal mass. The evidence of these researches can be used to justify the efficacy of building–climate interplay for indoor thermal comfort. However, day time ventilation has a relatively small effect on the indoor air temperature of buildings protected from solar radiation and its main function is to directly enhance the comfort of occupants [13].

2.3. Limitations of Airflow in Tropics

The textbook on Introduction to Architectural Science by Szykolay [3] discusses in detail how the physics of heat flow behave with building components and elements of climate in affecting indoor thermal environments of buildings. Ventilation helps to improve thermal comfort, indoor air quality and remove toxic mold grow in buildings [17] but has limitations. Research evidence challenges the linear function of the cooling effect of air velocity. Accordingly, cooling effect of air flow works up to a certain limit and the effect diminishes with elevated velocities after 1 m/s [18].

Work of Szokoloy [19] illustrates that indoor air movement at 1m/s can extend the upper limit of comfort zone, 28 °C, up to 31.7 °C in warm humid climates. Further, an air velocity of 1.5 m/s can extend the comfort zone by 5 degrees K up to 33 °C. An air velocity higher than this speed could cause thermal discomfort. However, it is a known fact that day time dry bulb temperature in these climates (Colombo for example) moves around 34 °C or higher on a typical summer day. Therefore, it is quite difficult to experience thermal comfort given the daytime relative humidity levels around 80 percent and the limitation of the air velocity up to 1 m/s, in addition to internal loads from occupants and heat stress from outside. Therefore, there is a need for lowering the daytime indoor air temperature below the corresponding ambient levels in free floating conditions.

2.4. Multi-Level Air-Conditioned Buildings

Façade area of a multi-level building is greater as compared to a low rise building and plays a significant role in the building–climate interplay. A more climate responsive façade is able to control environmental heat gain and access for daylight and ventilation without heat gain [14]. Even if the building is air-conditioned, control of environmental heat gain through the envelope remains a requirement. The ideal façade architecture in air-conditioned buildings integrates interventions of U-Value technologies [11] which are able to control the elevation of indoor air temperature above the ambient levels through insulation, both resistive and capacitive, [3] and solar shading to avoid heat transfer from outside and thus maximum opportunity for adaptation to outside hot climate and reduce cooling energy loads.

An experimental study on design priorities to restrict dynamics of external heat loads with an average ratio of glazing to façade area around 43 percent, external sun shading devices, insulation, solar control glass and to remove internal heat loads with night ventilation and slab cooling has shown that the primary energy use of office buildings can be reduced to about one-third of the average building stock and kept within a limit of 100 KW h per net floor area per year for moderate climates. This includes energy for heating, ventilation, cooling and lighting as well as auxiliary energy and energy dissipation by conversion from primary energy to end energy [19]. Lam's [20] study using a DOE (Department of Energy) simulation based on a generic model of common characteristics of 146 existing high rise commercial buildings in sub-tropical climates suggests that envelope as a major design aspect that affects building cooling load and air conditioning as the single largest electricity end user.

Shading has extensively been used for protection against solar heat gain from outside. The entertainment of a shaded air layer inside a double skin façade using a shutter with a heat resistant of 1.0 m^2 K/W can give a potential reduction of heat transmission of about 50 percent when compared to the standard choice of a 2-layer low energy glazing, meaning reducing heat development outside the glazing areas can reduce heat gain risks [21]. Similar studies are indicated by others for naturally ventilated environments highlighting the cavity space and the height of the double skin façades are crucial design factors for buoyancy effect of airflow [22] and more studies are available for air-conditioned environments as well. Decrease of solar insolation on façades in sub-tropics and increase of energy saving (while balancing day light and visibility) are linked to self-shading of façades due to orientation, azimuth angles and locations of buildings [23]. High reflectance on the surface finish of the outer façade can lower the heat transfer and thus indoor air temperature and cooling loads significantly [24].

Studies on heat gain risks and required solutions for tropics are least available and most studies available on cooling and façades of office buildings are on free floating conditions and use software simulations as the primary tool of research method [25]. Knowledge gap visible is the need for performance driven on site field investigations on architectural integration in real buildings, specially taking air-conditioned buildings and heat gain risks on them. Removal of internal heat gain from occupancy, lighting and equipment to a heat sink is another significant condition to be addressed. Halawa et al. [26], reviewing on energy conscious designs of building façades, argues that there is a research gap and a lack of a systematic and comprehensive analysis of the available literature regarding the energy and thermal performance of building façades based on the various possible design and technical configurations, especially in hot and humid climates. A recent research to quantitatively analyze the impacts of building envelope design factors upon cooling and heating loads in US cities of different climate zones suggests the importance of having variations in optimal sets of design factors in different climate zones [27].

A review of the effect of building envelops of high-rise buildings in hot-humid climate on the thermal comfort and energy efficiency focusing the Malaysian tropical climate highlights passive design method as one of the most potential strategies applied on building envelope in hot–humid tropical regions and based on the research findings in the same context establishes recommendations for envelop design strategies to be used by the designers for high-rise buildings [28]. Similarly, a number of researches have been carried out worldwide that indicate the importance of the building façade design on thermal performance of a particular building and on energy efficiency [29–32].

The growing trend of modern glass façades buildings in tropical Sri Lankan context signifies the urge of establishing a framework for building design applications along with climate responsive interventions. The common practice in the existing context is based on the aesthetical appearance where no concern to overcome barriers of overheating and high energy demand is given. On the other-hand, the research evidence on such aspects are lacking thus the need for conducting investigations to identify the barriers and opportunities related to building designs and thermal performance is highly required.

2.5. Reducing Heat Stress—An Insight into Theoretical Aspects

Studies indicate that overheating risk due to heat gain in buildings is already a problem in many building types across warm climates around the world. The literature offers a range of long term monitoring and simulation studies on overheating aspect of buildings with free floating conditions. Overheating due to heat gain by conduction through the outer façades and accumulation of internal loads has been discussed. Establishment of level of indoor overheating and the criteria to assess this are typically developed based on expectations of the occupants. Acceptable indoor air temperature levels for occupants are established using adaptive method [33] and acknowledged in international standards [34]. These recent comfort theories have acknowledged the interaction between people and their surrounding environment. It suggests that people who live in warmer climates can tolerate higher levels of air temperatures than the people who live in colder or moderate climates. It also points out that danger of overheating can be assessed in these buildings in which a relationship between the indoor comfort temperatures is derived from the mean of outdoor air temperature [35]. The range of acceptable indoor air temperature can be wider for occupants in both residential and non-domestic buildings with free running spaces where cooling is achieved by behavioral changes or options for user controlled ventilation strategies. The proportion of occupied hours with temperatures above an extended threshold gives an indication about overheating [36]. Overheating criteria has extensively been defined for residential buildings based on comfort expectations of occupants in many European climates and warm humid climates as well. Overheating is found to occur in vulnerable homes in European homes with poor ventilation even in the absence of heat waves [37] attributing to the poor air quality and occupancy behavior.

Overheating potential of buildings cannot be visible in conditioned buildings due to air temperature control by active means inside and little is known about heat gain stress on air conditioned environments/buildings in severe tropics. Addressing heat gain risks would allow a more climate sensitive façade design performance and would lead to avoid heat stress on façades. A current limitation with understanding environmental heat gain is that the complexity of defining the full impact of varied types of surroundings on the building. This is at least partially a result of the behavior of anthropogenic heat from traffic, albedos and geometries of urban canyons, climate type and orientation of the built mass to solar and wind access. This behavior is interconnected and becomes more complex with the dependency on a variety of design factors of the building concerned as well. Further overcrowding which leads to smaller building plot sizes results in diminishing the ability to cool buildings due to the thermal interference of neighboring structures [38]. However, there is currently no simple matrix for quantifying integrated heat risks on buildings in tropics and further the performance of building façade cannot be looked in isolation. Thus, the body of research knowledge on façades and the knowledge on the bioclimatic approach to building–climate interplay need to be brought together.

Bioclimatic design [39] approach is seen as an appropriate basis, which involves a way buildings filter and modify the external climate for occupants' comfort, to deal with energy efficiency opportunities. The need to have bioclimatic design in practice has been discussed as a good human adaptation of free running buildings in warming climates [40,41]. Bioclimatic influence can be effective in the manipulation of environmental loads and internal loads as well. This is easily applicable in skin dependent inclusive modes [42] of buildings because of the potential interaction between climate, building design and occupants. The involvement of building design between climate and occupants for thermal comfort in tropics is based on the integration of the microclimate enhancement, form and envelope of the building [15].

Microclimatic enhancement can be effective in air-conditioned and mixed mode buildings as well. Microclimate enhancement and calibrating the form and envelope design can be considered as effective input measures to reduce heat gain risks of externalities such as climate on the building (and its façade). Manipulating form (both plan and sectional) and envelope can remove internal heat gain from the building interior through buoyancy effect of sectional form and heat sink effect of thermal mass and air. For this, night ventilation is beneficial and tropics are now beginning to appear with diurnal ranges around 6–8 degrees C. Microclimate around a single building is diverse. Research has indicated that different microclimatic effects with regards to air temperature, solar radiation, shading and wind effects can occur around a single building [43]. This non-uniformity can be attributed to the building's plan form, sectional form, its orientation and design components of the microclimate.

Benefit of climatically linking inside and immediate outside of a building with substantial shading has been highlighted for indoor comfort improvements [38]. The research evidences suggest manipulation of plan form, sectional form and orientation can create shading on the external building façades against direct solar access creating shaded building microclimates that would be central to reduce heat transfer through the façade. If the building is ventilated in the night, substantial removal of internal heat and improving heat sink effect of the envelope for the following day is possible based on the buoyancy effect attributable for the sectional form and plan form [15]. Similar work is indicated by others for warm humid climates [44,45].

Manipulation of building geometries for night ventilation is applicable for air conditioned buildings as well suggesting differentiation of thermal performance standards—a new thinking towards mixed mode buildings in the tropics with night ventilation. A critical need is to work out a sensor technology and time management frame to open the stack flow in the night and early morning hours. More research is required in this respect.

A simulation study carried out to identify the performance of different building façades in moderating the indoor temperatures of building interiors and to reduce the external gains established evidence that glass façades in tropical Sri Lankan context influence external gains while increasing cooling load requirement. Onsite investigations conducted in sequence to the same study signified that provision of night ventilation reduces the indoor air temperature of the built spaces the following day [46]. In contrary to this practical scenario the trend is heading to glass façades as stated earlier thus indicating the knowledge gap of building designers and their less concern.

A recent study [47] conducted in Colombo context to identify temperature behavior around building façades in different levels of multi-story buildings implies that the temperature behavior around the façades differ in each side of the building as well as in different floor levels of the building. The findings suggest that the building façade, rather than being monotonous, should be designed considering the orientation of each façade as well as the vertical micro-climatic diversity. As a result of the investigation, it was visible depending on the building plan form the thermal behavior differentiated. However, this is an area to be further investigated in practical applications.

Nevertheless, in tropical Sri Lankan context though theories related to building forms and the thermal performance is adapted in few practices, the practical situation and the actual performance of such theoretical interventions are not investigated in relation to the climatic context of Sri Lanka. This is identified as a research gap to be investigated. With the aim of fulfilling this objective the research attempts to compare at real scale thermal performance of three distinct plan depth types with similar building heights and façade characteristics. Sampling of 86 plan forms available in Colombo Metropolitan Region (CMR) contributed to identify three main plan depths; shallow, deep and covered atrium types.

3. Research Methodology

The paper develops an evidence based narrative for different plan form typologies using on-site thermal performance investigations in quantifying the heat gain risks on them and discussing means of reducing that on air-conditioned buildings. The initial phase of the study was conducted involving 86 multi-level office buildings in CMR. These buildings with multi-level floors ranging from 4 to 14 are being used for office, bank and other commercial functions. The buildings are located in very dense urban areas on small plot of lands ranging from 100 to 3000 square meters. Building Energy Index (BEI) was calculated for all 86 buildings in order to identify the building population in respect to energy usage and efficiency levels. BEI was simply calculated by dividing the total annual energy usage by the total usable floor area of each building. Annual BEI of these office buildings was found to be within a range from 90–412 kWh/m^2/a. Majority of these buildings are found to be with BEIs higher than 110 kWh/m^2/a, which is an acceptable standard for the energy efficiency building codes [48]. Further, the mean BEI for building stock in CMR is 212 kWh/m^2/a and thus solidifies that the office building stock in CMR is energy obsolete [49]. Regrettably, the typical office building stock in CMR is not designed with an explicit intent to include climate and environmentally responsive design considerations.

The next phase of the study selected 12 free standing office buildings (from the larger stock of 86) primarily having distinct plan shapes. This phase categorized the office building stock of 86 into 12 buildings with significant six types within generalized basic plan forms and composite plan forms. Both the basic plan forms and composite plan forms consist of shallow and deep in addition to a courtyard (atrium) form. The selected twelve buildings included an atrium form having a glass roof on top making the atrium space at the ground level a tall internal lobby space—a common practice with most compact urban atrium buildings in Sri Lanka. Since the atrium is covered at the top level with a glass roof, the enclosed space inside the atrium does not experience any air moment due to stack flow except the air coming from the side corridors and entry points to the atrium in different heights. The purpose of having an atrium for vertical air moment is completely lost in these buildings due to the glass roof cover at the top. This research intended to highlight this drawback with performance evidence.

Façades of these 12 sample buildings are composed of glass windows, brick walls and aluminum cladding. Percentage distribution of these materials in front façades is approximately 30–57% compared to the other façades. Glazed façades are primarily fixed with glass panels with very few operable windows. Glazed façades are orientated to varied directions with no concern to control unwanted direct solar. Building forms have almost similar envelope properties of cement plastered high mass concrete and brick without any insulation to control conductive heat flow from outside. Floor to floor height varies between 3 to 3.5 m in all these buildings, a common practice.

All 12 buildings are located in close proximity in Colombo City. Table 1 depicts the identified classification of 12 buildings and demonstrates their orientations and BEIs with envelope characteristics and plan typologies. Building morphology and performance data were focused on orientation, plan shape (form), construction materials and fenestration characteristics such as window to wall ratio (WWR), aspect ratio (façade length/depth). Monitored technical and operational characteristics included capacity of air-conditioning systems, operational work hours and types and usage of equipment. Occupied hours, occupancy profiles, air-conditioning systems and equipment characteristics were found to be nearly similar across this 12 building population but a clear difference was seen in respect to their BEI which was ranging from 106–400 kWh/m^2/a. Of them three buildings with distinct plan forms are found to be with relatively lower BEIs but reason for that was due to their lower usage of air conditioning in occupied spaces. These three buildings accommodate some spaces with free floating conditions as well.

Table 1. A classification of the selected 12 plan forms of office buildings from a larger population of 86 buildings in Colombo. Most forms are composite shallow or deep plans, some are with atriums.

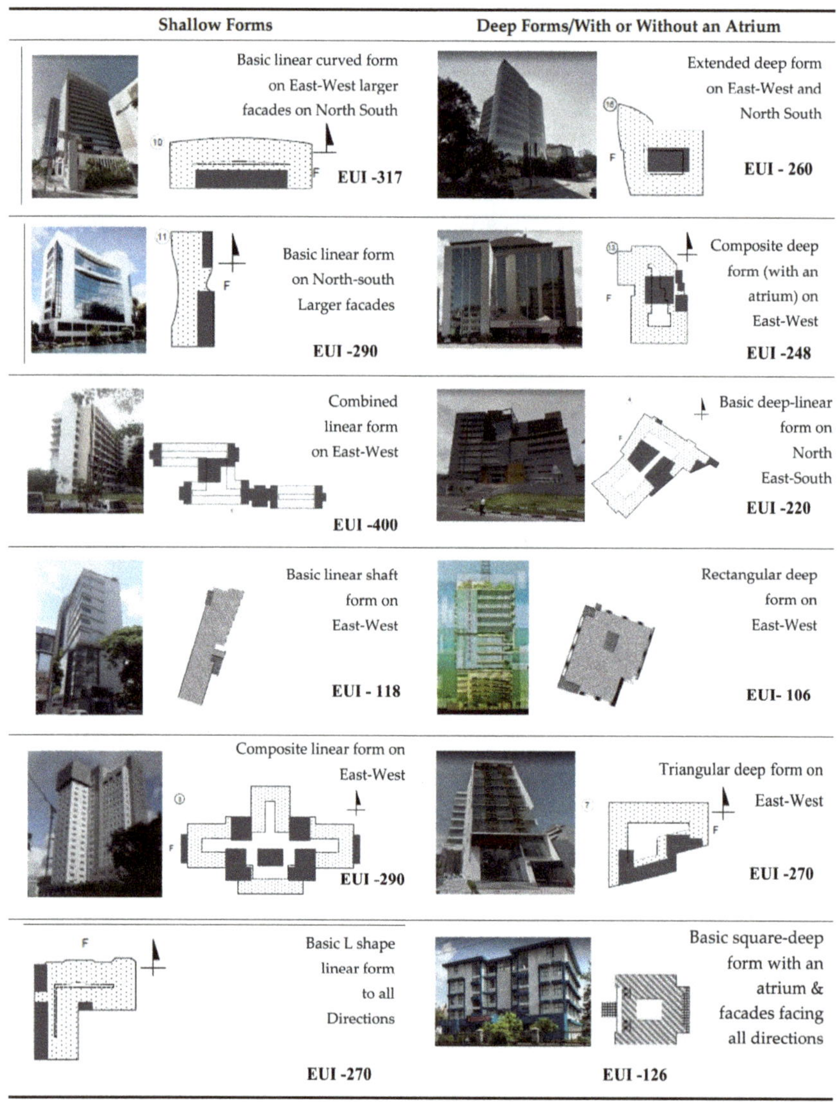

The work measured the levels of elevation of indoor air temperature as compared to the corresponding ambient levels as an indicator of heat stress on indoors. Similar methods have been used by others for research on residential buildings [50]. The buildings were investigated in weekdays during air-conditioned hours and weekends during non-air-conditioned hours as well making the visibility of heat gain risk easier. Free standing buildings provided an opportunity to measure the effects of surrounding climate with no thermal effect from other built structures closer to them.

Figure 2 presents decrement factor—the level of elevation or decrease of indoor air from the corresponding ambient air temperature—for the sample of selected 12 office buildings when air-conditioners are in off-mode in weekends. To obtain an idea about approximate heat stress on buildings, the decrement factor is used as an indicator. The majority of buildings' elevation of indoor air in the peripheral and central zones was remarkably higher than the corresponding ambient levels during AC off-mode with sealed windows, thereby indicating a greater potential for heat stress and indoor overheating in buildings designed for air-conditioned mode. Use of a multi zone method involved an investigation of air temperature behavior in peripheral and central zones giving more accurate picture about temperature distribution across an indoor space than taking a single zone investigation. Readings in four orientations in a peripheral zone were taken individually and averaged to one reference value. Similarly, three readings were taken in center zones of each building.

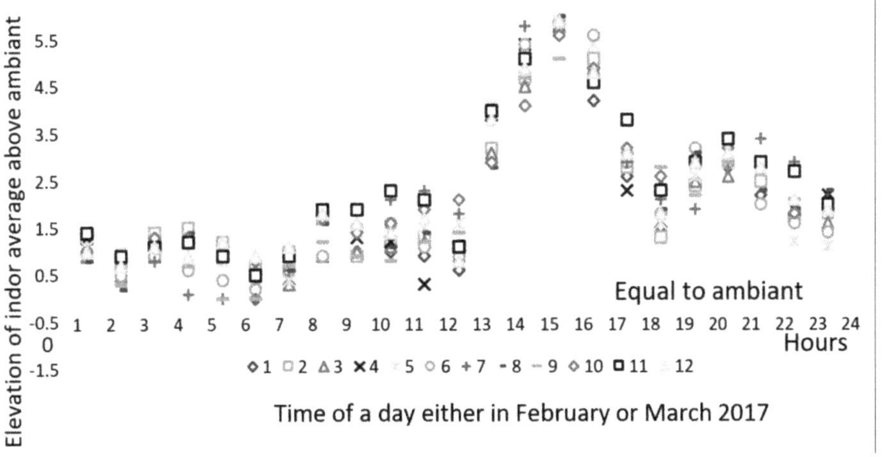

Figure 2. Elevation of indoor air temperature average well above ambient levels in 12 typical building indoors shows environmental heat stress on façades/envelopes.

Buildings are designed to run with conditioned environments and found to be in exclusive mode [42]. The U-value calculations of each case is worked out following the standard equations using the general resistance values of typical construction materials used in Sri Lankan context. It is a commonly accepted factor that in tropical Sri Lankan context the most appropriate building orientation is north-south, in order to avoid direct solar gain into building interiors. However, orientations of selected buildings are not proper, contributing to solar exposure during morning and evening hours. Nearly 95 percent of multi-story office buildings in Colombo are facing major roads without responding to the required orientation. Although the work presented involves the City of Colombo, methodology and performance evidence are relevant to any other tropical climates, building typology and usage. A detailed on-site thermal performance investigation was carried out for three selected building forms shown in Table 2. The three forms, which are in slightly composite and compact are shallow, deep and atrium plans, representing most of the physical characteristics of the larger population of office buildings in Colombo. These three buildings were selected due to their distinct plan form type.

Table 2. Three basic plan form typologies (shallow, deep and covered atrium) selected from a larger population of 86 buildings for the thermal performance investigation.

Building Form	A—Shallow	B—Deep	C—Atrium
Photo of the building			
Sectional profile			
Plan form	Shallow rectangular	Deep plan square	Deep square with an atrium closed on top
Orientation	NE/SW	NE/SW	N/S
Sectional form	Compact stack	Compact stack	Stack with atrium
No of floors	Ground + 7	Ground + 14	Ground + 4
Building height	32m	42m	20m
Net floor area m^2	632	1400	1260
Usage and age of the building	Office (3 years)	Office (3 years)	Office (5 years)
External wall U-Value W/m^2 K	0.22	0.21	0.16
Windows U-Value W/m^2 K	1.10	0.89	1.12
g-value of window	0.63	0.65	0.61
Daylight availability	Nearly 50%	Nearly 50%	Nearly 50%
Energy footprint KWh/m^2/a	118	106	126
Façade's glazing proportion	55%	58%	50%
Solid façade materials	Burned bricks and concrete columns	Burned bricks & concrete columns	Burned bricks and concrete columns
Type of Air-conditioner	Water cooled package	Air-cooled VRV	Water cooled package
Capacity of Air-conditioner	60 BTU/sq. feet	60 BTU/sq. feet	60 BTU/sq. feet
Percentage of floor area that is air-conditioned	60% approx.	61% approx.	55% approx.

3.1. Limitations

The research aims at investigating the heat gain risk of multi-story air-conditioned buildings in tropical context. Air-conditioned office buildings were selected as the building category type for the on-site investigation. Further, Colombo, the existing commercial hub of Sri Lanka with a majority of office buildings and a typical tropical climatic context having an ambient temperature between 28 °C–34 °C during a typical day was selected as the study context. Selected building population is designed with multi levels ranging from 4 to 11 floors in height. Nearly 60 percent of total building population in Colombo falls on to this category [51].

3.2. Instrumentation

Air temperature and wall surface temperatures were measured during April and May, 2016 and 2017, the hottest months of a typical year in these three selected buildings. Calibrated Hobo data loggers (Hobo for air temperature/RH and Hobo thermo couples with four external probe sensors each for surface temperature/RH) were used with temperature measuring range of −20 to +70 degrees C with accuracy ±0.34 K and Relative Humidity range of 5–95% with accuracy ±2.5 K of real scale readings. Data loggers and their sensors were shielded against solar and reflected radiation by placing them in a shade throughout the study.

3.3. Rationale of Longitudinal and Vertical Thermal Measurements

Multi zone temperature readings were taken over a weekend during air-conditioner off-mode and on Mondays with air-conditioner on-mode at 30 s intervals and later averaged to hourly values. Ambient hourly weather data was obtained from the Dept. of Meteorology, Colombo for specific dates, the on-site measurements were taken. The objective of the measurement rationale was to ascertain a number of comparisons as follows;

- Dynamics of air temperature distribution in peripheral and central zones in a typical office floor in order to look at combined effect of façades and plan depth on indoor climate in dealing with environmental heat stress
- Dynamics of indoor air temperature deviation against corresponding ambient levels in order to assess the heat stress from solar gains
- Elevation of indoor air temperature in peripheral and central zones across a typical floor above the set point temperature in air-conditioned mode
- Wall surface temperatures (both external and internal) with indoor air in order to assess heat stress from immediate building or urban microclimate on the façade and heat sink capacity of thermal mass and its impact on indoor climate
- Decrement delays of indoor air and internal wall surface temperatures inside three buildings for assessing any differences of plan form effect in addressing heat stress from outside

Since the atrium form is covered with a glass roof from the top, the atrium space was not monitored for its air flow behavior. It is important to note that monitoring air flow rates and patterns inside buildings was not a subject of the research in the context of buildings designed for air-conditioning. The use of criteria that define heat stress based on the elevation of indoor air temperature in respect to the set point temperature during air-conditioning and corresponding ambient levels during free floating conditions.

Monitoring air temperature distinction between occupied workspaces close to the perimeter façades and central zones away from the façades was expected to indicate the behavior and distribution of environmental heat stress across a façade through to the center of a typical floor plate. Similarly, air temperature distinction on the vertical path in the atrium building indicates the negative effect of glass roof of the atrium space.

4. Results and Discussion

Building A—compact shallow plan form

The typical office floor of shallow plan form (Building A) is shown in Figure 3. The building has a ground floor and 7 upper floors in compact and linear shallow plan form with longer façades facing northwest and southeast orientations. Nearly 80 percent of the total façade area is facing either direct or indirect solar access at least during 10:00 a.m. and 15:00 p.m.—a typical scenario of office buildings in Colombo. 5th and 8th floors were measured for thermal performance from Friday the 22nd April 2016 to Tuesday the 26th April 2016. Building Management System (BMS) indicated 24 °C as the set point temperature of air-conditioned office floors on weekdays from 8:30 a.m. to 17:00 p.m. Air conditioner was on off-mode during the weekend.

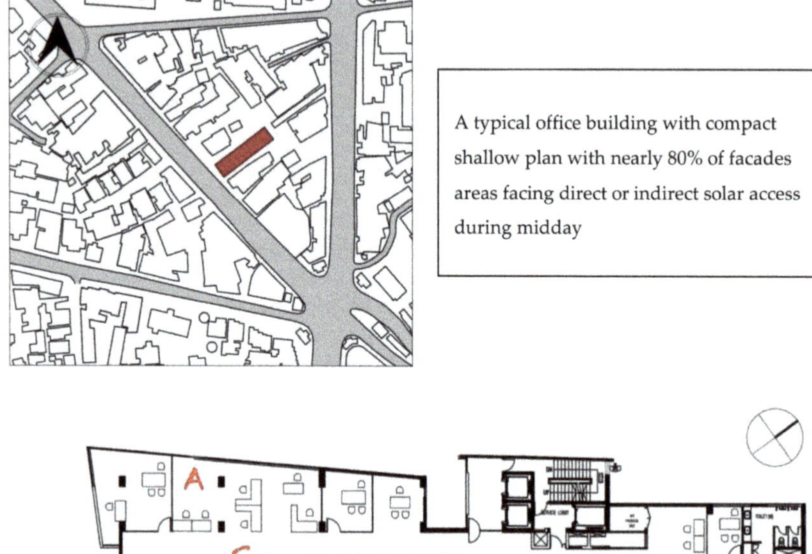

Figure 3. Building A (plan depth—14 m, length—56 m) shallow plan forms with unprotected façades contributes to heat stress from environmental loads.

Indoor air temperature in the peripheral zones A and B moved well above ambient levels during air-condition off-mode on Sunday the 24th April 2016 (Figure 4-Top). Peripheral zone on the southeast (Zone B) reached its maximum of 35 °C by 8:30 a.m. and remained above ambient till 11:00 a.m. while the peripheral zone on the northwest orientation (Zone A) commenced its elevation above ambient by 11:30 a.m., reached its maximum, 42 °C, by 14:30 p.m. and continued to remain higher than the ambient throughout the night. In the shallow plan form, the peripheral zone claims a larger occupied area and results indicate the severity of the heat gain risk due to environmental loads on façades. The central zone which is about 30% of the useable area and 5 m away from the periphery maintained its air temperature below the ambient but remained close to 29 °C throughout the day.

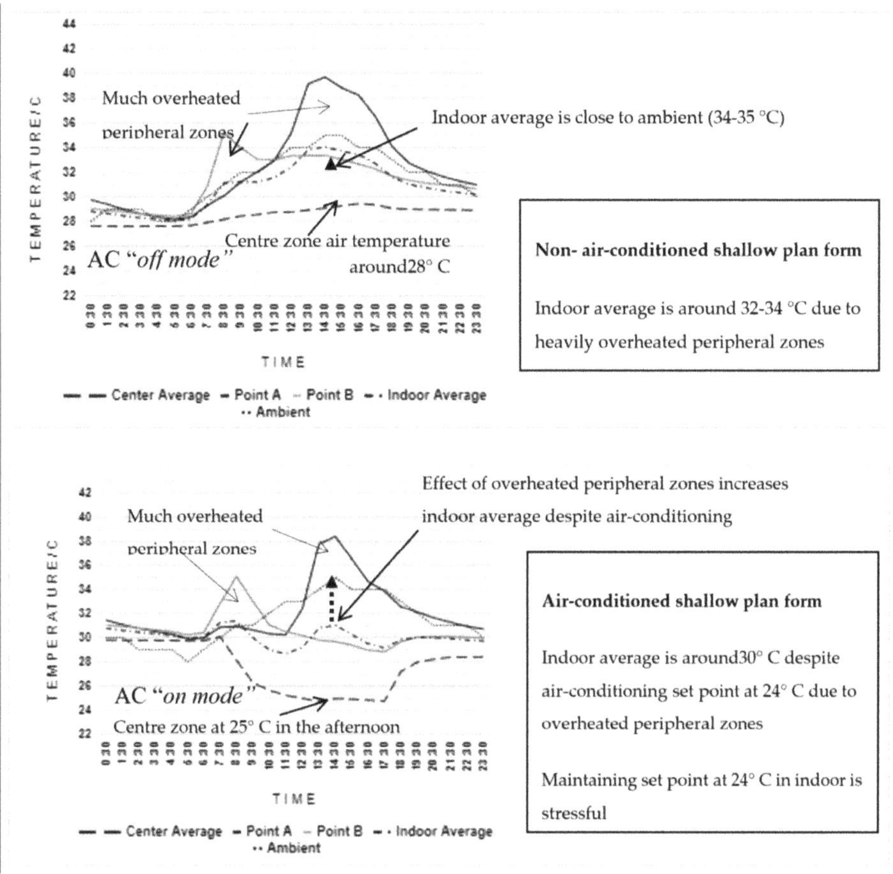

Figure 4. Air temperature inside 5th floor of Building A with shallow plan, Top—AC off-mode and Bottom—AC on-mode, overheated indoors due to environmental heat gain in both situations.

The distribution of occupancy, computers and other equipment were found to be uniform in both zones and investigation was carried out on a Sunday assuring that internal loads are not a contributing factor for this heat gain risk. The complete scenario highlights the severity of the office building population in Colombo and the need to protect the façades from environmental heat gain risks.

Indoor average of all zones closely followed the pattern of corresponding ambient levels but moved well above 33 °C between 12:30 and 17:30 pm indicating heat gain risk on the indoor environment. Figure 4-Bottom shows the indoor thermal behavior of 5th floor when the air conditioner is on on-mode on Monday, the 25th April 2016. Since the building is sealed and walls are insulated, indoor air temperature was recorded as 31 °C, which was about 3 °C higher than the ambient at 6 am. Despite set point temperature at 24 °C, starting at 8 am and during office hours, results indicated that indoor air temperatures in peripheral zones were extensively dynamic throughout the day in air conditioned mode reaching its peak to 35 °C on southeast orientation at 8:30 followed by 38.5 °C on the northwest orientations at 2:30 p.m. Internal air temperature at the center zone moved around 1 °C above the set point temperature and elevated to 28 °C and above just after 17:30 when the air conditioner off, creating a more warmer environment in the night and heat stress on the air conditioner on the following day morning. Indoor average moved almost with the corresponding ambient around 34–35 °C and showed the heat stress on the façade and its impact on the indoor average air temperature.

Figure 5-Top explains the surface temperatures and heat stress risk on the external façade and its impact on the indoor wall surfaces in peripheral zone during AC off-mode. Wall surface temperature behaviors on the air conditioned mode day (25th April 2016) are shown in Figure 5-Bottom. Average of internal wall surfaces of the building façades (tiled floor not included) remained below the ambient around 30 °C during air conditioned mode. Averages of all indoor air in the perimeter zone and outdoor air just outside the external façades were well above the ambient, reaching 38.3 °C at 14:30 p.m. and 41 °C at 15:30, respectively, a situation which cannot be expected in climate responsive design practice.

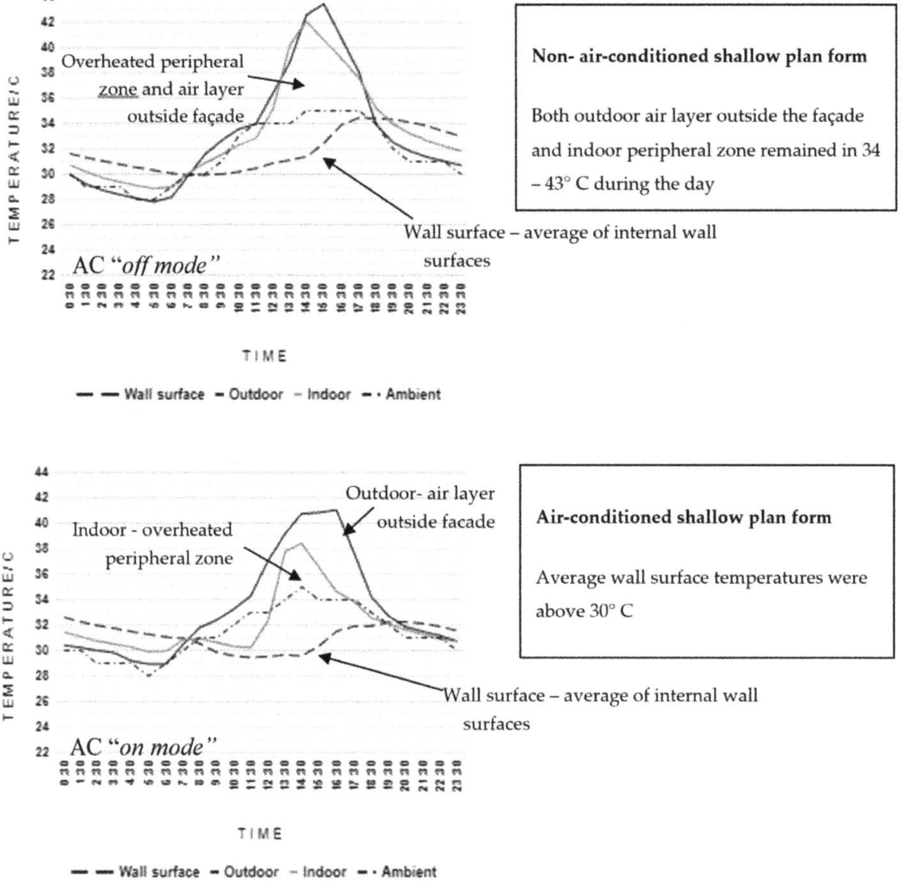

Figure 5. Indoor overheating in Building A with surface temperature behavior of façades and indoor peripheral zones moving around 40 °C in both AC off-mode (Top) AC on-mode (Bottom).

External air layer just outside the building façade moved around 34–44 °C during daytime. During AC off-mode, average of indoor air in the periphery remained high as 32–42 °C during daytime. Comparison of Figures 4 and 5 shows that maximum of indoor air in peripheral zones has come down only by 3.5 °C when the indoor environment was changed from AC off-mode to AC on-mode. Heat sink effect of internal surface of the façade was visible by moving its surface temperature 2–4 °C below the ambient and 2–10 °C below the indoor periphery but has been unable make any impact on indoor air in both AC on- and off-modes due to direct heat transfer through the façade.

Building B—compact deep plan form

The Building B, with a deep plan form, is shown in Figure 6, and has a ground floor and 14 upper floors in compact form. Nearly 80% of the total façade area is facing either direct or indirect solar access during a typical day—a typical scenario of office buildings in Colombo. Parking facilities are given in the 2nd to 5th floors with natural light and ventilation. From 7th floor up to the 14th floor are allocated for typical office functions. Central zone in a typical floor is slightly larger than the peripheral zone which falls within 5 m from the external façade. The 7th and 12th office floors were investigated from Friday the 1st April 2016 to Tuesday the 5th April 2016. Results of 3rd Sunday (AC off-mode) and 4th Monday (AC on-mode) are presented here. Building Management System indicated 24 °C as the set point temperature of air-conditioned office floors on weekdays from 8:00 a.m. to 17:00 p.m. Air conditioner was on off-mode during the weekend.

Figure 6. Deep square plan form of the Building B (plan depth—24 m, length—55 m) Top left—Layout in the urban setting, Images—internal office and equipment setting, Bottom—typical floor plate.

Figure 7-Top shows the thermal performance behavior of the 12th floor on Sunday the 3rd April 2016 when AC was on off-mode. Indoor air in the peripheral zones moved well above ambient in the morning and afternoon reaching its maximum of 42 at 9:30 a.m. and 37 at 16:30. This behavior is a result of heat stress on the façade and its poor performance. However, peripheral zones remained below the ambient during mid-day (10:30 a.m.–15:00 p.m.) due to shading created by overhangs on windows. Solar angles lower than 56° in altitude results in direct solar radiation exposure on the westerly side with Window to Wall Ratio 0.4 (the easterly side employs larger windows WWR 0.6). Due to lower solar angle after 15.00 pm on the westerly façades, peripheral zone on the west showed a sharp increase up to 37 °C by 16:30 creating a risk on the internal zones in the night and following day morning (Figure 7-Top).

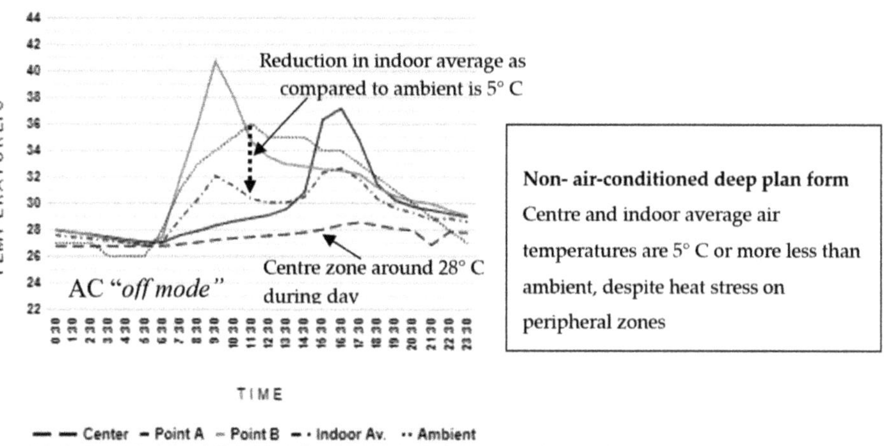

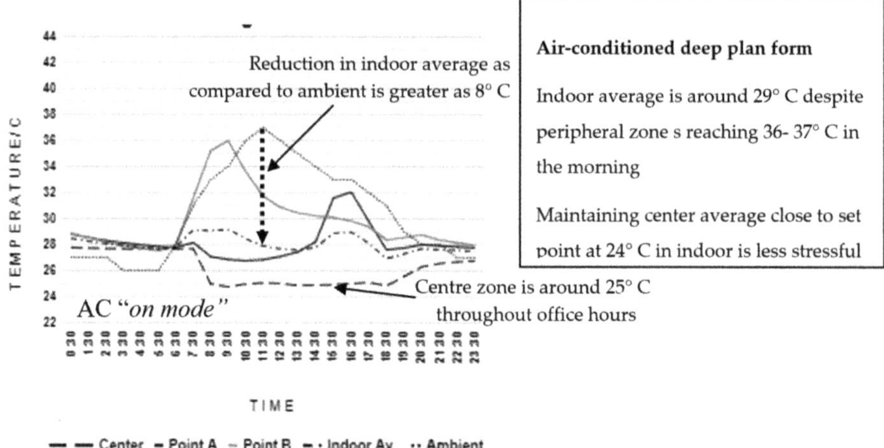

Figure 7. Top—12th floor of Building B in AC off-mode, Bottom—AC on-mode. Despite overheating in peripheral zones, center and average move closer to set point at 24 °C.

Despite peripheral zones reaching extensively higher temperatures, the center zone maintained a consistent level around 28 °C in AC off-mode. Indoor average remained around 2 °C–5 °C lower than the corresponding ambient. Results suggest that elevation of average air temperature in center is minimal in deep plan forms as compared to shallow plans.

Figure 7-Bottom shows thermal performance of the same floor (12th) when the AC was in on-mode during Monday the 4th April 2016. Despite set point at 24 °C, the indoor average was in trouble and moved around 28 °C–29 °C due to heat stress from the peripheral zone B just inside two southeasterly façades in the morning. However, center moved very close to set point temperature and as a result, indoor average was able to show some relief from the heat stress from peripheral zones during the noon. A reduction of 8 °C from the ambient was clear in average temperature inside the spaces indicating an effect from greater percentage of center zones with lower temperatures close to the set point of 24 °C.

These initial results show just how much the monitored intensity and duration of elevated indoor temperatures in the building with a shallow plan form greater than the same in the building with a deep plan form. There remains work to be done on the impact of plan depth on the elevation and distribution of indoor temperature due to heat stress in overheating assessment but this research questions the contemporary climate responsiveness of shallow plan form which is considered to be an ideal solution for daylight and ventilation efficiency. More work involving a larger population of shallow and deep plan forms is yet to come in a future paper.

Figure 8-Top presents surface temperature of the building façades on the 12th floor on Sunday the 3rd April 2016 with off-mode of AC. Internal wall surface temperature followed the ambient pattern and moved around 29 °C–30 °C in the night and 30 °C–32 °C during the daytime showing heat gain. Direct heat gain through windows was visible with indoor peripheral zone moving up to 40 °C by 9:30 a.m. and 37 °C by 16:30 p.m. Thermal mass temperature moved higher than ambient in the night contributing to heat sink to indoor air. Heat sink effect of the internal surface was not visible at all in the absence of a method to remove heat from inside with night ventilation or any other means. Figure 8-Bottom shows that internal wall surface temperature during AC on-mode on Monday the 4th April 2016 remained constant around 29 °C during both day and night. However, indoor air in the peripheral zones showed an increase up to 35.5 °C with AC on-mode irrespective of constant level of internal surface temperature, This suggest lack of heat sink effect of internal wall surfaces but solar gain from the façades.

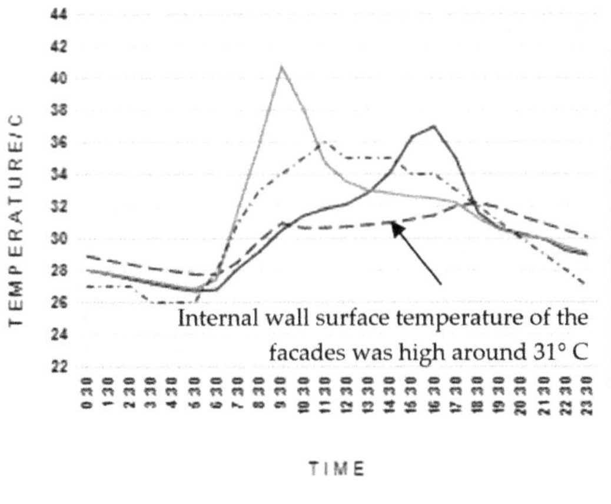

Figure 8. *Cont.*

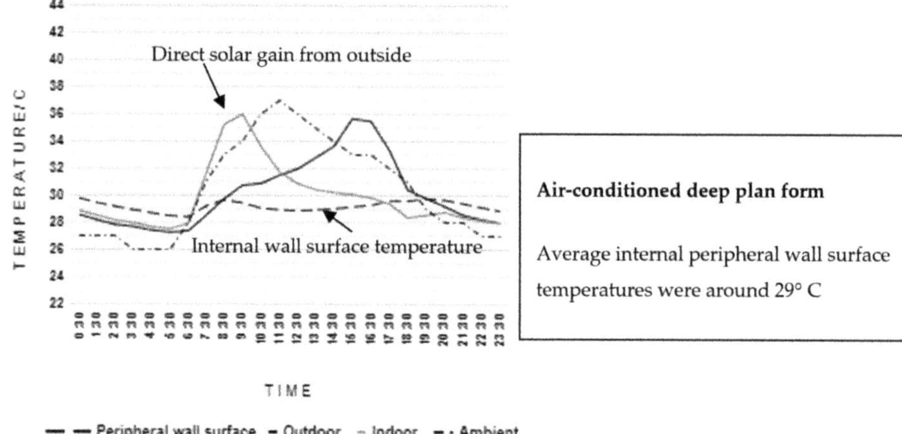

Figure 8. Thermal behavior of peripheral zones with wall surface temperature, Top—during AC off-mode and Bottom—during AC on-mode.

Building C—Atrium plan form covered from top

The building has a ground floor and four upper floors with an atrium (covered at the top) at the center in the deep plan form, which is another common form of practice in urban areas. The size of the atrium is 12 m × 10 m and 15 m in height and is designed to enhance daylight but covered with a glaze roof top, which is in question (Figure 9).

The atrium is open at the ground floor level to outside through the main entrance, which is the only access point placed but on a wind shadow. Wind flow comes from the other side, southwest, of the building, Main office spaces on the east and west, which are air conditioned, have unprotected façades.

Air temperature measured on Thursday the 4th and Friday the 5th May, 2017 at 1.5 m above the ground on four microclimates just outside the building showed a diversified heat gain risk. Microclimate on the North moved around 29 °C–31 °C, while West, South and East microclimates moved to peaks around 34 °C–41.5 °C in the afternoon from 13.30 p.m. onwards. (Figure 10). This diversity can be attributed to the plan form and its orientation. Longer façades (points 1 and 5 on plan) of the main office spaces adjoining the atrium are exposed to direct solar access, a reason mainly attributable to form and orientation. It is interesting to understand that north microclimate was moving above east microclimate where direct solar access was available in the morning. North microclimate is on the wind shadow, close to the traffic and does not get any wind at all.

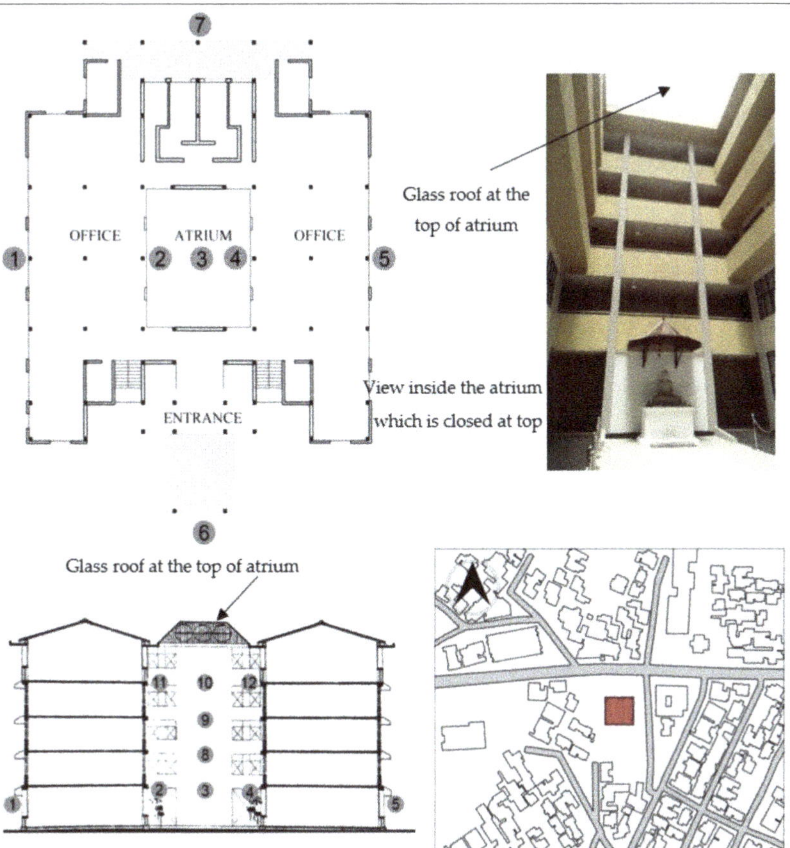

Figure 9. Building C (plan depth—25 m, length—26 m) Top left—Atrium plan form, Top right—Building in semi dense urban context, Bottom left—Section through the building with a central atrium.

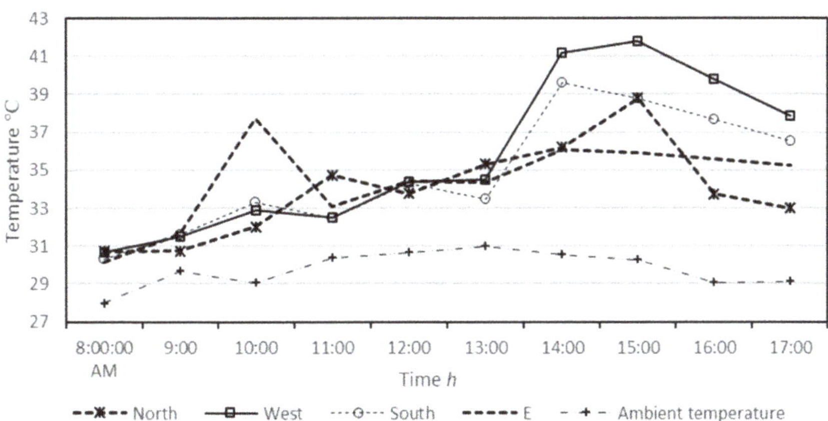

Figure 10. Heat gain risks on microclimates around the building with temperatures reaching 33 °C–38 °C in the morning on east and 35 °C–42 °C in the afternoon on west.

Figure 11 (below) shows a sharp increase of air temperature inside the atrium with its height in center during 09:00 a.m. and 14:00 p.m. Air temperature reached 36 °C at 13.5 m level, 35 °C at 10.5 m, 33 °C at 7.5 m and 4.5 m levels between 13:30–14:30 p.m. indicating an overheated situation of about 3.0 °C–5.5 °C above the ambient level. The air temperature started to decrease after 14:00 p.m. and entire thermal behavior followed the ambient pattern throughout the day and night. A stratification was not visible in the night instead overheating was visible and air temperature at all levels moved close to each other 2 °C–3 °C above the ambient temperature in the night. Results suggest direct heat gain from outside and roof level to the atrium, stagnation of internal heat inside and absence of night ventilation. The atrium is wrongly integrated into the building section hindering the capacity in heat removal, night ventilation and thus heat sink effect of thermal mass. This is a common mistake visible in most of building practices in Colombo.

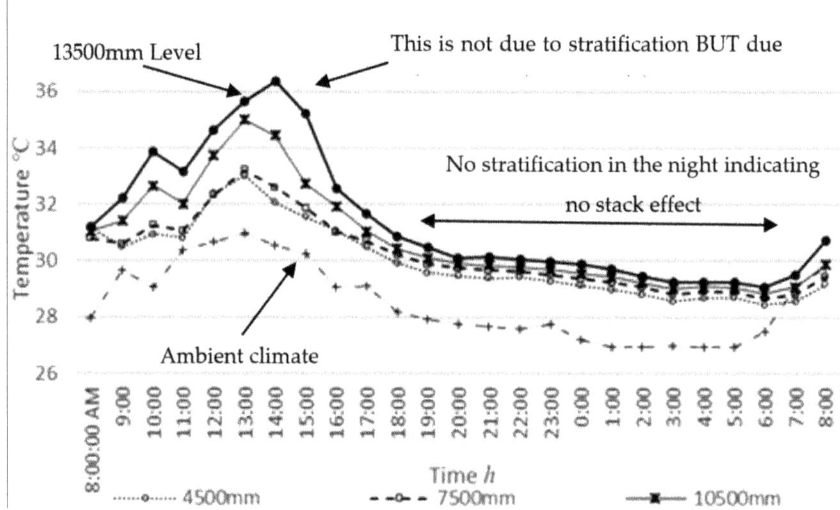

Figure 11. Vertical air temperature diversity in the atrium with its height up to 36.5 °C from 29 °C and indoor overheating condition well above ambient in both day and night.

Air temperature behavior in the perimeter zone of the atrium, adjoining the occupied office spaces, was investigated to quantify the heat gain risk due to atrium of this building (Figure 12). Ground floor readings moved 1.5 °C above the corresponding first floor level indicating higher heat gain risk at the ground level than the first floor level. Second floor and third floor level readings were moved 2 °C–2.5 °C above the corresponding center of the atrium showing heat stress on the envelope dividing office spaces and atrium. Results indicate the risk of getting atriums overheated and their heat gain risks on adjacent occupies spaces.

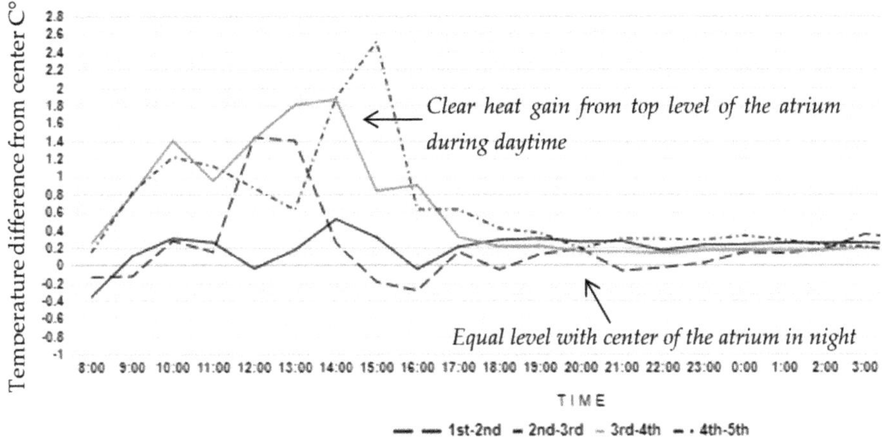

Figure 12. Elevation of air temperature in the perimeter zones of the atrium from its corresponding centers along 1st to 5th floor levels creates a heat gain risk on the adjacent office spaces.

By monitoring the Building C with a covered atrium, an attempt was made to predict a severity due to the heat stress scenario in closed atriums. Since closed atriums are commonplace in urban contexts in Colombo and other cities, future case versions of atrium buildings need to be thought carefully for the integration of interventions to avoid heat gain from outside and remove heat built up from indoors. There is evidence that high and prolonged periods of elevated temperatures in covered atriums could have a larger impact on surrounding occupied spaces in overheating than would be expected from a simple shallow or deep plan form in this research. This is due to the heat stress posed by a covered atrium in addition to the same from surrounding outer façades. It is therefore useful to understand not only the geometric considerations of atriums but also how and to what extent they interact with climate in imposing heat stress on indoors. Providing proper openings on the leeward side and at higher levels of the atrium could immensely benefit for a stack flow of air and removal of hot air.

Decrement Delay

The term Decrement delay in building design refers to the time the building takes for heat to pass through its outer façades and roofs to inside. This is quantified to be in delaying hours between the peak temperatures of the outside ambient climate or exterior surface of the building and the resulting temperature inside the building or of the internal surface of the façade. The decrement delay indicates an idea that thermal behavior of buildings with thermal mass is dynamic.

Decrement delays for the two buildings (A—shallow and B—deep) were calculated. With glazing percentage of the façades close to 50%, and similar material characteristics in façades in both buildings, A and B, a higher decrement delay of 4.5 was observed (Figure 13) from the ambient to the Center zone (C) despite the recorded diversities in maximum air temperatures occurred in ambient and indoor levels. It was noted during previous discussion that higher air temperatures were recorded in Center zone of the shallow form than the deep form meaning that heat dissipation rate inside shallow plan form is greater than in deep form.

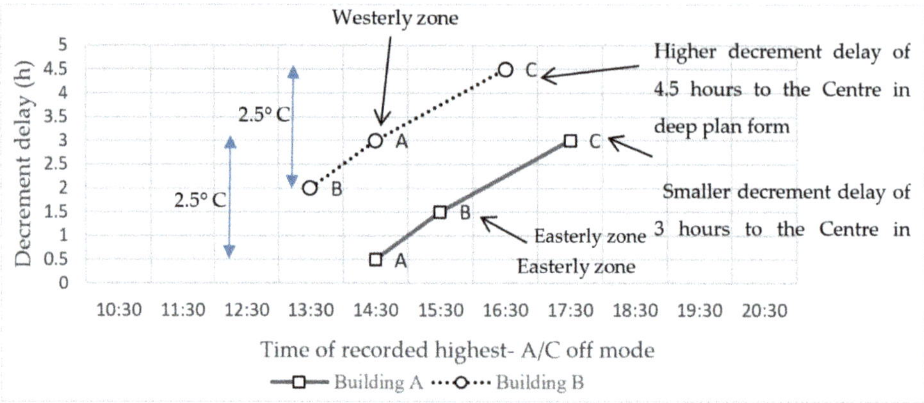

Figure 13. Equal decrement delays of air temperature from perimeter zones (A and B) to the Center (C) in shallow (Building A) and deep plan (Building B) forms in non-AC modes.

Despite higher and longer ambient air recorded with deep plan form Building B, decrement delay from average of internal peripheral zones to the Centre zone in both plan forms is 2.5 h. Highest ambient range recorded with deep plan form was 34 °C–36 °C for 7 h from 9:30 to 16:30 p.m. compared to the lesser values of highest ambient range and its duration in shallow plan form Building A, where highest ambient range was 34 °C–35 °C for just 3 h from 13:30 to 16:30 p.m.—meaning that deep plan form has worked effectively in delaying heat stress to the Center. The outcome suggests a potential of deep plan forms in addressing heat stress from outside.

Decrement delay in the internal surface of the façade in deep plan form close to the peripheral zones A and B is 3.5 h from the highest ambient record whereas the internal surface in shallow form has a shorter decrement delay of 1.5 h. Since façade materials and glazing proportions of both forms are of the same kind, the results indicate that the higher average indoor air temperature due direct heat gain through façades could be a contributing factor for this behavior. This concern however, needs a more detailed investigation. Shorter decrement delay in shallow plan forms indicates a disadvantageous situation in addressing heat stress from outside (Figure 14).

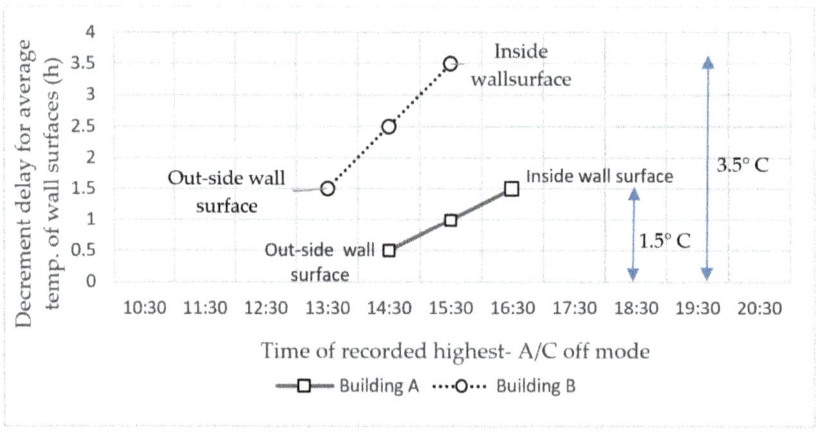

Figure 14. Shorter decrement delay of 1.5 °C with inside wall surface temperature in shallow plan form (Building A) as compared to the larger decrement delay of 3.5 °C with the same in deep plan form (Building B).

The average decrement delay along the height in atrium was just 1.5 h—meaning that covered atriums could be problematic in addressing environmental heat stress. Increase of decrement delays during the daytime along the height shows a thermal stratification due to direct heat gain but any air flow moment cannot be predicted due to the complete closure of the atrium at the top (Figure 15). This stagnation of heat stress and its behavior inside the atrium can be used to show the need to have a heat removal method in atrium form buildings, a major driver in addressing heat stress. More research involving a larger population of atrium buildings with varied plan forms is expected to be carried out.

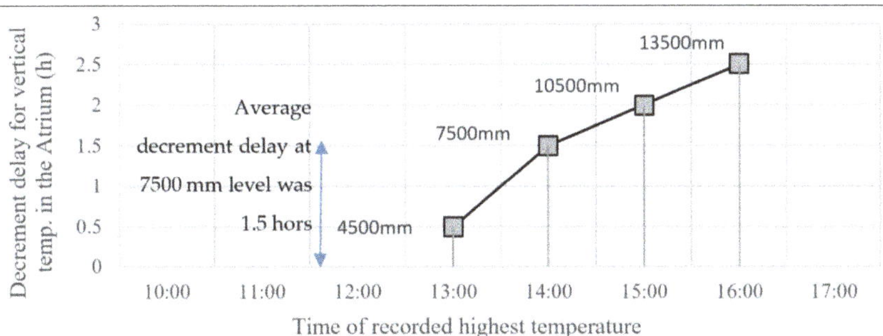

Figure 15. Decrement delay inside the covered atrium in Building C at different vertical levels.

5. Conclusions

Despite growing need to avoid indoor overheating due to environmental heat stress, a lack of evidence exists on the scale of the problem in air-conditioned office buildings in tropics. The research analyzed a high prevalence of heat stress in contemporary office building types in Colombo, indicating the need to understand the plan depth as an important design consideration in dispersing the indoor air temperature due to heat stress through building façades.

The aim of the research was to question the impact of plan depth of multi-level office buildings in tropics in minimizing the elevation of indoor air temperature due to heat stress on the façades. The current body of knowledge which is less supportive for professionals working in urban office building sector and minimum performance standards, particularly in Sri Lanka, do not meaningfully address the issue of heat stress on façades.

From the investigated cases, shallow plan, deep square and deep plan with a closed atrium with a glass roof, it is evident that indoor air temperature distribution in deep plan forms are more favorable in respect to controlling heat stress coming from outside or dispersing the heat across a plan depth. Findings suggested a distribution of relatively lower air temperatures in deep plan forms than in shallow plan forms. This was evident when buildings investigated were in non-air conditioned modes indicating relatively less heat stress situations in deep plan air conditioned buildings in tropics.

Warm climates in tropics with urbanization and dense built mass create more devastating heat gain risks on buildings. An appropriate bio-climatic influence in the building design is required to build resilience to the impacts of heat gain risks even on air-conditioned buildings. The literature that the research evidences are based on is still limited in breath, for tropics in particular. Manipulating plan depth for reducing heat stress in peripheral zones and thus across the entire plan depth could be a promising and supportive initiative in bio-climatic design as well.

Even accounting for fundamental issues like global warming, heat island effects and urbanization, the man parameters causing overheating due to heat stress appear to be those relating to building design itself. Heat gain risk assessment is required for designing adaptive and climate responsive design, and it needs to be emphasized in building regulations in countries with severe warming climates. Based on the results, a method, using the range of indoor air temperature elevation above the corresponding ambient levels during air-conditioning off-mode, to assess heat gain risks can be useful as a simple method for assessing the existence of heat stress problems of existing buildings, including air-conditioned environments. For new building proposals, simulation programs could be developed. For more accurate assessments and predictions multi-zone method is more appropriate to estimate the heat gain risk in real situations, because it can take account of thermal dynamics of the building form, occupancy, lighting and heat dissipation between zones. Use of covered atriums and unprotected façades will create heat stress on indoor spaces due to direct environmental loads from glass atrium roofs and indirect internal gains as well. Considerations of integrating cool atriums with buildings for internal heat removal, night ventilation and heat sink effect of thermal mass is left for future research. These performance evidence are expected to be further validated though an extended field investigation to continue on the same direction of thinking with a long term duration for at least 6 months with larger samples representing a variety of buildings.

Restrictions on land size due to urbanization have led to an increase in compact sallow plan forms with walls made of thermally light-weight materials. Most contemporary buildings have lower ceiling heights and large windows to have a modern glassy appearance. With security concerns, urban air pollution and traffic noise, architects and occupants often tend to render buildings with sealed façades limiting the potential for heat removal from indoors even with atrium buildings. The current monitored experiments of this work clearly support the general level at which current heat stress threshold on building façades and indoors are visible. Although they cannot pass judgement on the veracity of the heat stress threshold for use in any commercial building, results do clearly show the very wide range of conditions with regard to plan form depth over which heat stress on indoor is visible. The work highlights lessons on the ways that plan form can interfere in addressing heat stress from outside.

Limitations and Strengths

Factors other than building envelope, plan form and orientation can influence climate interaction of buildings. The factors include, but not limited to structural components like floor and column system, space lighting systems, occupancy pattern which were not considered due to similarity in physical properties and behavior. The case study approach used in this work depends only on 12 buildings initially and then on three for detailed investigations. However, the methodology of the research could be contended to offer generalizing the findings. The outcome implies that application of fundamentals arising from the observed data could be appropriate and integrated to similar other buildings within the same climatic and urban contexts. This is justifiable because of the similar and related building typologies, environments and occupancy patterns in which the buildings operate.

During the data collection, analysis and presentation names and locations of buildings remained anonymous respecting the owners, occupants and participants of the research as suggested by Creswell [52].

Author Contributions: The author is responsible for the the entire research involving literature review, investigation plan, outcome analysis and conclusions.

Funding: The research was partially supported by a National Research Council (NRC) Grant 13-109 of Sri Lanka by giving funding for purchasing equipment.

Acknowledgments: Author acknowledges the support given by H.W.K. Jayathilake and W.S. Jayasinghe in data collection in the field investigations.

Conflicts of Interest: The author declares no conflicts of interest.

References

1. Santamouris, M.; Balaras, C.A.; Dascalaki, E.; Argiriou, A.; Gaglia, A. Energy conservation and retrofitting potential in Helenic hotels. *Energy Build.* **1996**, *24*, 1–83. [CrossRef]
2. Santamouris, M.; Kolokotsa, D. Passive cooling dissipation techniques for buildings and other structures: The state of the art Review Article. *Energy Build.* **2013**, *57*, 74–94. [CrossRef]

3. Szokolay, S.V. *Introduction to Architectural Science: The Basis of Sustainable Design*; Architectural Press: Hudson, New York, NY, USA; Elsevier: Amsterdam, The Netherlands, 2005.

4. Santamouris, M. *Advances in Passive Cooling*; Earthscan Publishers: London, UK, 2006.

5. Yau, Y.H.; Hasbi, S. A review of climate change impacts on commercial buildings and their technical services in the tropics. *Renew. Sustain. Energy Rev.* **2013**, *18*, 430–441. [CrossRef]

6. IPCC. *Climate Change 2014: Synthesis Report. Contribution of Working Groups I, II and III to the Fifth Assessment Report of the Intergovernmental Panel on Climate Change*; Core Writing Team, Pachauri, R.K., Meyer, L.A., Eds.; IPCC: Geneva, Switzerland, 2014.

7. Coley, D.; Kershaw, T. Changes in internal temperatures within the built environment as a response to a climate change. *Build. Environ.* **2010**, *45*, 89–93. [CrossRef]

8. Howden, S.M. Effect of climate and climate change on electricity demand in Australia. In *Integrating Models for Natural Resources Management Across Discipline, Issue and Scale*; Ghassemi, F., Whetton, P., Little, R., Littleboy, M., Eds.; MSSANZ Inc.: Canberra, Australia, 2001; pp. 655–660.

9. Guan, L. Implication of global warming on Air-Conditioned office buildings in Australia. *Build Res. Inf.* **2009**, *37*, 43–54. [CrossRef]

10. Garnaut, R. *The Garnaut Climate Change Review, Final Report*; The Cambridge University Press: Cambridge, UK, 2008.

11. Radhi, H. Can envelope codes reduce electricity and CO2 emissions in different types of buildings in the hot climate of Bahrain? *Energy* **2009**, *34*, 205–215. [CrossRef]

12. Santamouris, M.; Asimakopoulos, D. *Passive Cooling of Buildings*; Earthscan Publishers: London, UK, 2006.

13. Givoni, B. Effectiveness of mass and night ventilation in lowering the indoor daytime temperatures. *Energy Build.* **1998**, *28*, 25–32.

14. Givoni, B. Indoor temperature reduction by passive cooling systems. *Sol. Energy* **2011**, *85*, 1692–1726. [CrossRef]

15. Rajapaksha, U.; Gunesekara, A.; Rajapaksha, I.; Hyde, R. Developing a conceptual model for passive cooling in buildings in the tropics: a case study of Maritime Museum Galle, Sri Lanka. Proceeding of 48th Conference of Architectural Science Association, Genoa, Italy, 10–13 December 2014; Madeo, F., Novi, E., Eds.; pp. 581–592.

16. Rajapaksha, I.; Nagai, H.; Okumiya, M. A ventilated courtyard as a passive cooling strategy in warm humid tropics. *Renew. Energy* **2003**, *28*, 1755–1778. [CrossRef]

17. Aynsley, R.; Shiel, J.J. Ventilation strategies for a warming world. *Archit. Sci. Rev.* **2017**, *60*, 249–254. [CrossRef]

18. Szokolay, S.V. PLEA Notes. In *Thermal Comfort*; The University of Queensland: Brisbane, Austrilia, 1997.

19. Voss, K.; Herkel, S.; Pfafferott, J.; Lohnert, G.; Wagner, A. Energy efficient office buildings with passive cooling—Results and experiences from a research and demonstration programme. *Sol. Energy* **2007**, *81*, 424–434. [CrossRef]

20. Lam, J. Energy analysis of commercial buildings in subtropical climates. *Build. Environ.* **2000**, *35*, 19–26. [CrossRef]

21. Sanchez, E.; Rolandoa, A.; Sant, R.; Ayuso, L. Influence of natural ventilation due to buoyancy and heat transfer in the energy efficiency of a double skin facade building. *Energy Sustain. Dev.* **2016**, *33*, 139–148. [CrossRef]

22. Barbosa, S.; Kenneth, I. Perspectives of double skin façades for naturally ventilated buildings: A review. *Renew. Sustain. Energy Rev.* **2014**, *40*, 1019–1029. [CrossRef]

23. Valladares-Rendón, L.G.; Schmid, G.; Lo, S. L. Review on energy savings by solar control techniques and optimal building orientation for the strategic placement of façade shading systems. *Energy Build.* **2017**, *140*, 458–479. [CrossRef]

24. Carmen, A.; Fernando, M.-C.; Ignacio, O.; Eloy, A.; Gloria, P.; Isabel, M. Effect of façade surface finish on building energy rehabilitation. *Sol. Energy* **2017**, *146*, 470–483.

25. Prieto, A.; Ulrich, K.; Tillmann, K.; Thomas, A. 25 Years of cooling research in office buildings: Review for the integration of cooling strategies into the building façade (1990–2014). *Renew. Sustain. Energy Rev.* **2017**, *71*, 89–102. [CrossRef]

26. Halawa, E.; Ghaffarianhoseini, A.; Ghaffarianhoseini, A.; Trombley, J.; Hassan, N.; Baig, M.; Yusoff, S.Y.; Ismail, M.A. A review on energy conscious designs of building façades in hot and humid climates: Lessons for (and from) Kuala Lumpur and Darwin. *Renew. Sustain. Energy Rev.* **2018**, *82*, 2147–2161. [CrossRef]

27. Yong, S.; Kim, J.; Gim, Y.; Kim, J.; Cho, J.; Hong, H.; Baik, Y.; Koo, J. Impacts of building envelope design factors upon energy loads and their optimization in US standard climate zones using experimental design. *Energy Build.* **2017**, *141*, 1–15. [CrossRef]

28. Mirrahimi, S.; Mohamed, M.F.; Haw, L.C.; Ibrahim, N.L.N.; Yusoff, W.M.F.; Aflaki, A. The effect of building envelope on the thermal comfort and energy saving for high-rise buildings in hot–humid climate. *Renew. Sustain. Energy Rev.* **2016**, *53*, 1508–1519. [CrossRef]

29. Friess, W.A.; Rakhshan, K. A review of passive envelope measures for improved building energy efficiency in the UAE. *Renew. Sustain. Energy Rev.* **2017**, *72*, 485–496. [CrossRef]

30. Bui, V.P.; Liu, H.Z.; Low, Y.Y.; Tang, T.; Zhu, Q.; Shah, K.W.; Shidoji, E.; Lim, Y.M.; Koh, W.S. Evaluation of building glass performance metrics for the tropical climate. *Energy Build.* **2017**, *157*, 195–203. [CrossRef]

31. Lei, J.; Yang, J.; Yang, E. Energy performance of building envelopes integrated with phase change materials for cooling load reduction in tropical Singapore. *Appl. Energy* **2016**, *162*, 207–217. [CrossRef]

32. Fathoni, A.M.; Chaiwiwatworakul, P.; Mettanant, V. Energy analysis of the day lighting from a double-pane glazed window with enclosed horizontal slats in the tropics. *Energy Build.* **2016**, *128*, 413–430. [CrossRef]

33. Auliciems, A. Global differences in indoor thermal differences requirements. In Proceedings of the 51st Conference of the Australian and New Zealand Association for the Advancement of Science, Brisbane, Australia, 11–15 May 1981.

34. ASHRAE. *ANS/ASHRAE/IESStandards 90.1-2010 Performance Rating Method Reference Manual*; Pacific Northwest National Lab.: Richland, WA, USA, 2016.

35. Nicol, J.F.; Humphreys, M.A. Derivation of the Adaptive Equations for Thermal Comfort in Free-Running Buildings in European Standard EN15251. *Build. Environ.* **2010**, *45*, 11–17. [CrossRef]

36. Brotas, L.; Nicol, F. Estimating overheating in European dwellings. *Archit. Sci. Rev.* **2017**, *60*, 180–191. [CrossRef]

37. Vellei, M.; Alfonso, P.; Ramallo-González; Coley, D.; Lee, J.; Gabe-Thomas, E.; Lovett, T.; Natarajan, S. Overheating in vulnerable and non-vulnerable households. *Build. Res. Inf.* **2017**, *45*, 102–118. [CrossRef]

38. Emmanuel, R. Performance standard for tropical outdoors: A critique of current impasse and a proposal for way forward. *Urban Clim.* **2018**, *23*, 250–259. [CrossRef]

39. Olgyay, V.; Olgyay, A. *Design with Climate: Bio-Climatic Approach to Architectural Regionalism*; Princeton University Press: Princeton, NJ, USA, 1963.

40. Hyde, R. *Climate Responsive Design: A Study of Buildings in Moderate and Hot Humid Climates*; E and FN Spon: London, UK, 2000.

41. Roaf, S. *Ecohouse 2: A Design Guide*; Architectural Press: Hudson, NY, USA, 2003.

42. Hawkes, D. *The Environmental Tradition: Studies in the Architecture of Environment*; E&FN Spon: London, UK, 1996; 212p.

43. Gokhale, M. Between the building and the macroclimate. In Proceedings of the 31st International ANZAScA Conference, Principles and Practice, The University of Queensland, Brisbane, Australia, 29 September–3 October 1997; pp. 27–39.

44. Amos-A, S.; Akuffo, F.O.; Kutin-Sanwu, V. Effects of Thermal Mass, Window Size, and Night-Time Ventilation on Peak Indoor Air Temperature in the Warm-Humid Climate of Ghana. *Sci. World J.* **2013**, *2013*, 1–9. [CrossRef]

45. Md Mohataz, H.; Lau, B.; Wilson, R.; Ford, B. Effect of changing window type and ventilation strategy on indoor thermal environment of existing garment factories in Bangladesh. *Archit. Sci. Rev.* **2017**, *60*, 299–315. [CrossRef]

46. Rajapaksha, U.; Rupasinghe, H.T; and Rajapaksha, I. Resolved duality: external double skin envelopes for energy sustainability of office buildings in the tropics. Proceedings of 31st International Conference of Passive and Low Energy Architecture Conference, Bologna, Italy, 9–11 September 2015.

47. Amarathunga, A.A.N.D.; Rajapaksha, U. A critical review on high rise buildings in the context of bio-climatic design: A case of vertical diversity in Tropical Colombo. In Proceedings of the 9th International Research Conference of FARU Building the Future -sustainable and resilient environments, Colombo, Sri Lanka, 9–10 September 2016; Rajapaksha, U., Ed.; pp. 425–439.

48. ECG19. Energy Consumption Guide: Energy Use in Offices. 2003. Available online: http:www.cibse.org (accessed on 9 September 2019).
49. Rajapaksha, U. Heat stress pattern in conditioned office buildings with shallow plan forms in Metropolitan Colombo. *Buildings* **2019**, *9*, 35. [CrossRef]
50. Simson, R.; Kurnitski, J.; Kuusk, K. Experimental validation of simulation and measurement-based overheating assessment approaches for residential buildings. *Archit. Sci. Rev.* **2017**, *60*, 192–204. [CrossRef]
51. CMC. *Engineers Department Data*; Colombo Municipal Council: Colombo, Sri Lanka, 2017.
52. Creswell, J.W. *Research Design: Qualitative, Quantitative, and Mixed Method Approaches*, 3rd ed.; Sage: Thousand Oaks, CA, USA, 2009.

Article

Impact of Climate Change on the Energy and Comfort Performance of nZEB: A Case Study in Italy

Serena Summa *, **Luca Tarabelli**, **Giulia Ulpiani and Costanzo Di Perna**

Industrial Engineering and Mathematical Sciences Department, Università Politecnica delle Marche, via Brecce Bianche 1, 60131 Ancona, Italy; l.tarabelli@univpm.it (L.T.); giuliaulpiani@gmail.com (G.U.); c.diperna@univpm.it (C.D.P.)

* Correspondence: s.summa@pm.univpm.it

Received: 2 October 2020; Accepted: 26 October 2020; Published: 2 November 2020

Abstract: Climate change is posing a variety of challenges in the built realm. Among them is the change in future energy consumption and the potential decay of current energy efficient paradigms. Indeed, today's near-zero Energy buildings (nZEBs) may lose their virtuosity in the near future. The objective of this study is to propose a methodology to evaluate the change in yearly performance between the present situation and future scenarios. Hourly dynamic simulations are performed on a residential nZEB located in Rome, built in compliance with the Italian legislation. We compare the current energy consumption with that expected in 2050, according to the two future projections described in the Fifth Assessment Report (AR5) by the Intergovernmental Panel on Climate Change (IPCC). Implications for thermal comfort are further investigated by assuming no heating and cooling system, and by tracking the free-floating operative temperature. Compared to the current weather conditions, the results reveal an average temperature increase of 3.4 °C and 3.9 °C under RCP4.5 and RCP8.5 scenarios, estimated through ERA-Interim/UrbClim. This comes at the expense of a 47.8% and 50.3% increase in terms of cooling energy needs, and a 129.5% and 185.8% decrease in terms of heating needs. The annual power consumption experiences an 18% increase under both scenarios due to (i) protracted activation of the air conditioning system and (ii) enhanced peak power requirements. A 6.2% and 5.1% decrease in the hours of adaptive comfort is determined under the RCP4.5 and RCP8.5's 2050 scenarios out of the concerted action of temperature and solar gains. The results for a newly proposed combined index for long-term comfort assessments reveal a milder future penalty, owing to less pronounced excursions and milder daily temperature swings.

Keywords: climate change; near-zero energy buildings; future scenarios; energy efficiency; adaptive comfort; long-term performance

---

1. Introduction

CO_2 emissions are causing a prolonged and clear increase in global temperatures [1]. In 2010, the building sector accounted for about 32% of global energy consumption, 19% of CO_2 emissions and 51% of global electricity consumption [1]. If, on one side, buildings and their related activities are responsible for a significant portion of greenhouse gas emissions, on the other side they represent a great opportunity for mitigation and adaptation to climate change effects [2].

In order to limit the temperature increase to 2 °C compared to pre-industrial levels, the Fifth Assessment Report (AR5) by the Intergovernmental Panel on Climate Change (IPCC) considers four different future scenarios (RCP; Representative Concentration Pathways) which show how the climate will likely change by 2100, depending on different levels of counteraction [3]. When bold mitigation strategies are taken into consideration, the greenhouse gas emissions could be halved by 2050 with a maximum temperature increase of 2 °C, while, with a "business-as-usual" approach, the CO_2 in the atmosphere would increase fourfold compared to pre-industrial levels, with temperature differences exceeding 4 °C. The corresponding change in global and local climatic conditions will impact the energy needs of the existing building stock and, consequently, the primary energy demand [3].

Speaking of buildings and climate change, two main aspects are highlighted in the literature: (i) the assessment of the repercussions in different geographical areas and for different uses of the built space, and (ii) the development of a broad spectrum of techniques to enhance buildings' resilience (no nZEB) and thus mitigate the energy penalty associated with climate change.

Regarding the first point, many studies in the literature report on the major impact of climate change on buildings. For instance, Ciancio et al. [4] compared the current energy needs of a residential building in the context of 19 different European cities, with those expected in 2080. The results show an increase in energy needs for cooling of up to 272% in Mediterranean cities, and a decrease in energy needs for heating up to 45% in Northern European countries. In the same vein, Olonscheck et al. [5] used projections of the regional statistical climate model STAR II and demonstrated that the energy demand for air conditioning in a residential building in Germany will decrease during winter, while remaining almost constant during summer for the next 40 years.

In Chile, Verichev et al. [6] described how temperature increases of 0.68 °C (under RCP2.6) and 1.51 °C (under RCP8.5) will lead to decreases in annual heating degree-days of about 72% and 92% by 2065, respectively. Moreover, Angeles et al. [7] predicted increases in energy demand of 9.6 and 23 kWh/month per person in Southern Greater Antilles and the inland of South America, which will lead to increases in cooling loads of 7.57 GW (under RCP2.6) and 8.15 GW (under RCP8.5) by the end of the 21st century.

Other than residential buildings, those with glass surfaces and predominant internal gains will be the ones which will suffer most from the effects of climate change, i.e., offices and schools, where, according to Frank T. [8], cooling energy demand will be up to 1050% higher than the present one for the RCP 8.5 scenario.

In order to contribute to climate change mitigation and, at the same time, tackle the increase in primary energy demand, new buildings are expected to implement not just appropriate envelope designs [9], but also energy production systems from renewable sources, thermal and/or electrical storage systems [10] or passive solar systems [11]. Differently, several strategies will need to be introduced for existing buildings, such as: (i) the installation of more efficient heating, ventilation and air conditioning (HVAC) systems [12,13], (ii) the installation of adequate solar shading, [8] and/or (iii) proper night ventilation [14].

Indeed, beyond materials [15], technology is another key ingredient in nZEB design. Typically, a hybrid combination of active and passive technologies realizes nZEB-like performances. Among the emerging renewable energy-based solutions are micro cogeneration systems, such as fuel cells, photovoltaic thermal, solar thermal reversible heat pump/organic Rankine cycles and cogeneration solar thermoelectric generators [16]. These hybrid systems may also be empowered with load-sharing concepts [17] and advanced energy storage systems based on integrated phase change materials [18] and optimized schedules [19,20]. On a general note, finding the most appropriate matching between envelope features and HVAC system configurations is pivotal, just like working on demand-driven energy flows, the reduced primary energy uptake and the electricity consumption of auxiliaries, such as pumps and fans. Heat pumps are gaining ground owing to their versatility [21] and technological variety [22]. For instance, the use of polyvalent heat pumps or variable air volume systems [23] possibly mated with grid-tied photovoltaic (PV) systems [24] has proven efficient in reducing the overall energy consumption of buildings. Further, heat recovery systems (from sensible heat exchangers up to run around coils or enthalpy, sensible assisted systems by indirect adiabatic cooling) may be implemented with additional perks [23]. Solar energy systems and passive solar concepts are being increasingly used and refined through optimized control systems based on advanced solar irradiance forecasting models [25] and dynamic occupancy profiles [26]. Solar-based advanced technologies include compact collectors for polygenerative applications, high concentrating PV systems [27], and building integrated photovoltaic systems (BIPV) that not only generate electrical energy but also behave like skin for the buildings [28]. These technologies have the potential to become a source of income for the buildings, even without subsidies, due to the increasing efficiency and decreasing costs of PV systems [29]. Smart management through building control and automation systems is also key, as demonstrated by the introduction of the Smart Readiness Indicator with the latest revision of the EPBD in July 2018 [30]. This applies to any technical systems [31] and to ventilation strategies [32]. Innovative lines of research are further looking into refrigerant-free cooling appliances based on caloric materials [33] and year-round passive daytime radiative cooling [34].

Generally speaking, whatever the specific strategy, the results emphasize that a one-fits-all recipe for nearly-zero energy buildings does not exist [26]. Climate, among other factors, calls for the resolution of diversified optimization problems [35,36]. Climate is a spatial and temporal variable. Here, we focus on the temporal variability by challenging established nZEB paradigms in the context of increased global warming.

Concerning the method, the need to reduce global energy consumption and CO_2 emissions has induced the European Committee for Standardization (CEN) to provide an hourly dynamic calculation method that allows buildings' consumptions to be assessed in a more realistic and detailed evaluation, especially during the summer season [37]. This method, described in EN ISO 52016−1:2017 [38], replaces the one described in ISO 13790:2008 [39] by introducing a new methodology to calculate energy needs for heating and cooling, on both an hourly and a monthly basis.

The need for knowledge on future scenarios emerges, specifically about a better understanding of the most effective energy retrofit strategies for existing nZEB buildings in the Mediterranean climate, to guide future legislative amendments, and the identification of which mitigation policies will be most appropriate for new buildings to limit both CO_2 production and global energy consumption.

Therefore, this study aims at assessing the impact of climate change on (i) the heating and cooling consumption of an nZEB multi-family house, located in Rome and designed according to the most recent Italian regulations [40], and (ii) the level of comfort achieved indoors.

The EURO-CORDEX5 [41] models combined with the ERA-Interim/UrbClim model [42,43], used for predicting future scenarios according to the Fifth Assessment Report (AR5) by the Intergovernmental Panel on Climate Change (IPCC), shows that Rome is the place with the highest temperature increase in Italy; therefore, this city has been chosen as a case study for the present study.

To this end, hourly dynamics simulations were performed in TRNSYS, which is a well-established building dynamic simulation software worldwide, capable of fine assessments of both energy and comfort levels [44], thanks to a vast variety of components that can be implemented in different models in order to simulate a wide range of simple to complex systems [45]. Its visual interface, which implements a component-based approach and the possibility of the addition of new mathematical sub-models, motivates its use for building energy simulation (BES) [46].

The rest of the paper is structured as follows: in Section 2, we present methods; in Section 3, we show the results of the simulations and, in Section 4, we draw conclusions and discuss future work.

2. Methods

In this work, the dynamic simulation software TRNSYS [47] was chosen to assess the thermal behavior of a building located in Rome classified as a nearly-zero energy building (nZEB) according to the Italian regulation enforced on an hourly basis. The focus is on the effects of the expected climate change and urban heat island exacerbation by 2050, both in terms of energy needs and comfort.

To this end, hourly dynamic simulations were performed in TRNSYS according to the following procedure: (i) the production of current and 2050 meteorological input files, (ii) simulations and evaluation of the year-round energy consumption, assuming an infinite power system for heating and cooling, (iii) simulations and evaluation of the free-floating operative temperature assuming no heating/cooling systems, (iv) application of the adaptive comfort [48] theory to rate the quality of the indoor environments and (v) evaluation of the long-term thermal comfort [49].

2.1. Climate and Geographical Data

The hourly climate data of the typical year adopted in this study refer to both the current climate conditions and those in 2050. About future scenarios, two Representative Concentration Pathways (RCPs) were analyzed, which refer to climate change projections developed in the AR5 report by IPCC.

The two future projections (RCP8.5 and RCP4.5) compute the expected temperature rise as a function of different greenhouse gas emissions determined by both human activities and different mitigation activities implemented by local policies:

- RCP8.5 represents a 'business-as-usual' approach, which considers that atmospheric concentrations of CO_2 will triple or quadruple by 2100 compared to pre-industrial levels, thus increasing the air temperature by about 4 °C;
- RCP4.5 contemplates control measure to curtail the greenhouse emissions, assuming a trend reversal (decrease below current levels) by 2070. Consequently, the atmospheric concentrations of CO_2 will be about twice as high as pre-industrial levels by 2100. The increase in air temperature will be capped at 2 °C.

Therefore, three typical annual weather data were retrieved from Meteonorm 7.3 [50]: (i) current climate conditions, (ii) 2050–RCP8.5 climate conditions, and (iii) 2050–RCP4.5 climate conditions.

The current climatic data are developed in a data set of temperatures and solar irradiances measured between 2000 and 2009 and 1991 and 2009, respectively, while future scenarios are simulated by Meteonorm 7.3 through a set of EURO-CORDEX5 [41] models combined with the ERA-Interim/UrbClim model [42,43]. The evaluation of the increase in solar radiation is based on the IPCC AR4 A2 models for CPR 8.5 and A1B for CPR 4.5 [51].

In detail, EURO-CORDEX is the European branch of the CORDEX initiative and produces ensemble climate simulations based on multiple dynamical and empirical-statistical downscaling models forced by multiple global climate models from the Coupled Model Intercomparison Project Phase 5 (CMIP5) [41]. The major aims of the CORDEX initiative are to provide a coordinated model evaluation framework, a climate projection framework, and an interface to the applicants of the climate simulations of climate change impact, adaptation, and mitigation studies [52].

The ERA-Interim/UrbClim model allows you to evaluate the intensity of urban heat of any city in the world with a resolution up to 100 m, providing the necessary input data from international satellite-based land cover, vegetation and soil sealing data and meteorological databases [42,43].

In order to perform simulations through TRNSYS, the following input data were considered: (i) latitude, (ii) dry-bulb external temperature, (iii) relative humidity, and (iv) solar global irradiance on the horizontal plane. Data were provided by the WMO weather station #162390 located in Rome/Ciampino. At 41.8° N (latitude), 12.58° E (longitude) and 131 m above sea level (altitude), the location features a temperate climate, specifically Csa (Mediterranean hot summer climate), in accordance with the Köppen–Geiger climate classification system.

2.2. TRNSYS Model Simulation

Outdoor air temperature, relative humidity and global solar irradiance on the horizontal plane for each of the three different scenarios described in Section 2.1 were provided as inputs through TRNSYS component Type 9 (data reader). Then, these data were transferred to Type 16c, a solar radiation processor which implements the Reindl and Perez 1990 models to obtain beam and diffuse solar irradiances on horizontal and tilted surfaces, respectively. The virtual model of the nZEB case-study was designed through a SketchUp [53] plugin for creating the multi-zone building envelope (TRNSYS3d) [54], and then was provided as input in TRNSYS [47]. Each room of the building was defined by creating a thermal zone for each of them. An interface for the detailed TRNSYS multi-zone building (TRNBuild/Type56) was used for modeling walls, gains and ventilation profiles. In order to comply with the Italian regulation for nZEBs [40], overhangs were integrated to reduce solar gains; therefore, Type 34 was used. An hourly basis zone-by-zone calculation was performed, the results of which in terms of energy needs and indoor operative temperatures were plotted via Type 65. Figure 1 displays the flowchart.

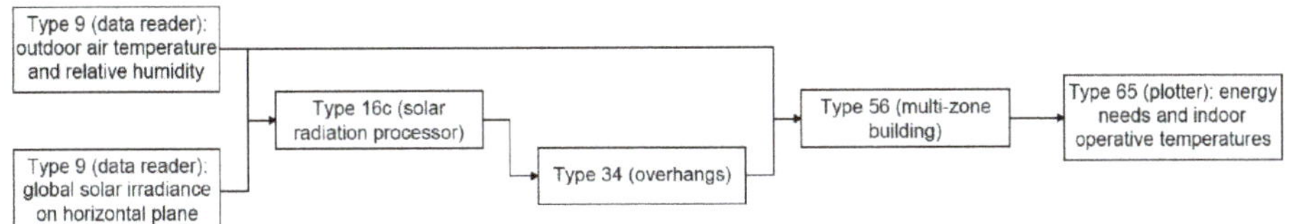

Figure 1. TRaNsient SYstem Simulation (TRNSYS) model flowchart.

2.3. Case-Study

The studied building is a new 3-floor residential nZEB, composed of two apartments per floor. Each apartment has an open space that combines kitchen and living room, a double bedroom, a single bedroom and a bathroom, as shown in Figure 2. Table 1 summarizes the geometrical features of each apartment.

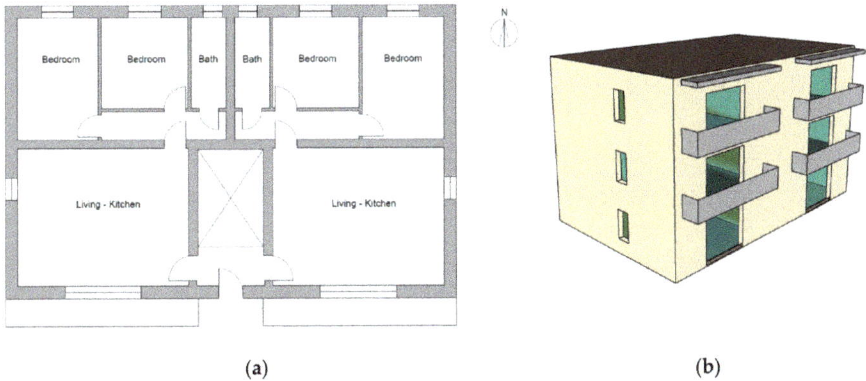

(a)   (b)

Figure 2. (a) Analyzed building: building plan (1:200) and (b) southwest view.

Table 1. Geometrical features of the apartment.

	Unit	Value
Net floor area	m^2	63.73
Net height	m	2.70
Net volume	m^3	172.07
Window-to-wall ratio	%	15

In order to comply with the Italian nZEB requirements the building envelope is made of 2 cm of external plaster, 35 cm of pored bricks, 4 cm of external insulation, and 1 cm of internal render. The 24 cm hollow-core concrete slab of the flat and non-walkable roof is well insulated with 10 cm of expanded polystyrene. The stairwell wall is made of two layers of 12 cm hollow bricks separated by 6 cm of insulation and external plaster, while the partition wall between the two apartments is a cavity wall with two layers of 12 cm of hollow bricks and external plaster. The windows are double glazed with low-emissivity glass. The horizontal overhangs on the south-exposed facades and proper shading devices on all the windows are modeled in order to provide sufficient sunlight control and the reduction of solar gains. The thermo-physical parameters of the main building envelope elements, previously described, are given in Table 2, while in Table 3 the nZEB requirements for the Italian territory are described.

Table 2. Thermo-physical parameters of building envelope elements.

	Thermo-Physical Parameter	Unit	Value
External wall	Thermal transmittance (U)	W m^{-2}K^{-1}	0.30
	Internal areal heat capacity (k1)	kJ m^{-2}K^{-1}	41.70
	External areal heat capacity (k2)	kJ m^{-2}K^{-1}	19.00
	Periodic thermal transmittance (Yie)	W m^{-2}K^{-1}	0.013
	Time shift (Δt)	h	17.72
Stairwell wall	Thermal transmittance (U)	W m^{-2}K^{-1}	0.30
	Internal areal heat capacity (k1)	kJ m^{-2}K^{-1}	44.90
	External areal heat capacity (k2)	kJ m^{-2}K^{-1}	44.90
	Periodic thermal transmittance (Yie)	W m^{-2}K^{-1}	0.048
	Time shift (Δt)	h	13.48
Separating wall between two apartments	Thermal transmittance (U)	W m^{-2}K^{-1}	0.59
	Internal areal heat capacity (k1)	kJ m^{-2}K^{-1}	45.00
	External areal heat capacity (k2)	kJ m^{-2}K^{-1}	45.00
	Periodic thermal transmittance (Yie)	W m^{-2}K^{-1}	0.148
	Time shift (Δt)	h	11.58
Roof	Thermal transmittance (U)	W m^{-2}K^{-1}	0.27
	Internal areal heat capacity (k1)	kJ m^{-2}K^{-1}	65.70
	External areal heat capacity (k2)	kJ m^{-2}K^{-1}	7.40
	Periodic thermal transmittance (Yie)	W m^{-2}K^{-1}	0.037
	Time shift (Δt)	h	11.53
Window	Thermal transmittance (U)	W m^{-2}K^{-1}	1.80
	Solar heat gain coefficient (SHGC)	-	0.71

Table 3. Standard nZEBs energy assessment for the considered case studies.

	Unit	
Energy performance indicator for heating (EP$_{H,nd}$)	[kWh m^{-2}]	5.32
Energy performance indicator for cooling (EP$_{C,nd}$)	[kWh m^{-2}]	28.60
Global average heat transfer coefficent (H'$_t$)	[W m^{-2}K^{-1}]	0.37
Equivalent solar area/Floor area (A$_{sol,est}$/A$_{sup,utile}$)	-	0.013

According to the Italian technical specifications in UNI/TS 11300 [55], the following assumptions were made: usage profiles 24/24 h, internal heat gains rate at 5.72 W/m^2 and ventilation rate at 0.5 h^{-1}. Two different models of the building were created:

- The first one was equipped with an infinite power system for heating and cooling, to evaluate the energy consumption. In this case, the operative temperature, used to control the system, was set to 20 °C in winter and 26 °C in summer, in accordance with the type of building and categories identified in UNI EN 16798–1–Annex A [48], assuming category II as the reference;

- The second one was equipped without any heating and cooling system, to allow free-floating operative temperature for the assessment of thermal comfort. The level of thermal comfort was analyzed through two different approaches. Firstly, we applied the adaptive method, according to which the operative temperatures are compared to the indoor operative temperature ranges defined in UNI EN 16798–1–Annex A [48] for buildings without mechanical cooling systems to identify different comfort levels as a function of the outdoor running mean temperature, calculated as follows:

$$\theta_{rm} = (1 - \alpha)\cdot\left\{\theta_{ed-1} + \alpha\cdot\theta_{ed-2} + \alpha^2\theta_{ed-3}\right\} \tag{1}$$

In Equation (1), α is a constant between 0 and 1 (recommended value is 0.8) and θ_{ed-i} is the daily mean outdoor air temperature on the i-th previous day.

Secondly, we calculated a newly proposed long-term comfort index (Equation (2)). It outperformed the other 6 types of existing indices (23 total) found in the standards and 5 types of new indices (36 total) for comfort assessments in the long run in a recent comparative study [49]. The correlation with the long-term thermal satisfaction of building occupants was based on continuous thermal comfort measurements and post-occupancy evaluation surveys, but in air-conditioned office buildings. Despite the different casuistry, this combined index proved better at taking into account the triggers behind long-term comfort, namely the pronounced excursions beyond some acceptable temperature ranges and the larger variations in daily temperature than the average experience over time [49]. As such, it is worth investigating its expected variation in future scenarios. This combined, normalized percent index is calculated as follows:

$$index = (\frac{\% \text{ To outside specified ranges} + \%\text{To daily range} > \text{a threshold}}{2})$$ (2)

In this study, the hourly operative temperature range was set between 20 °C and 26 °C, in accordance with the type of building and categories identified in UNI EN 16798–1–Annex A [48], assuming category II as reference, and the daily threshold was set to the 80th percentile of the entire simulated database.

3. Results

In this section the modeling results are illustrated and discussed. They refer to the two apartments at the second floor of the building described in Section 2.3. Considering the dispersing surface of the roof, the second floor is the most responsive to external climatic variations. For this reason the results obtained for this floor are evaluated and discussed.

In Section 3.1, the results in terms of kWh obtained from the modeling of the building in the presence of a heating/cooling system are analyzed, while in Section 3.2 the results in terms of comfort obtained with the modeling of the building in free-floating are discussed.

3.1. Simulations to Assess the Energy Consumption

Based on dynamic hourly simulations, the average annual increases in outdoor temperatures (+3.4 °C under RCP4.5 and +3.9 °C under RCP8.5) were found to enhance the global energy consumption from 5079 kWh × year (in 2020) to 6196 kWh × year (+18.0%), and 6248 kWh × year (+18.7%) by 2050, respectively. The breakdown in heating and cooling needs revealed a divergent trend (see Table 4): the heating energy needs showed a decrease in the order of 129.5% (RCP4.5) and 185.8% (RCP8.5), whereas the cooling energy needs increased by 47.8% (RCP4.5) and 50.3% (RCP8.5).

Table 4. Air temperature (T_{air}), relative humidity (UR), horizontal global solar irradiance ($I_{g,H}$), heating energy needs ($\varphi_{h,tot}$), cooling energy needs ($\varphi_{c,tot}$), peak power for heating ($\varphi_{h,max}$) and peak power for cooling ($\varphi_{c,max}$) for the three analyzed scenarios.

Year	T_{air} (°C)	UR (%)	$I_{g,H}$ (kWh/m^2)	$\varphi_{h,tot}$ (kWh)	$\varphi_{c,tot}$ (kWh)	$\varphi_{h,max}$ (kW)	$\varphi_{c,max}$ (kW)
2020	15.8	74.4	160.3	2390.7	2688.4	2.5	3.0
2050 (RCP4.5)	19.2	70.1	188.7	1041.8	5153.8	2.0	3.8
2050 (RCP8.5)	19.7	69.0	188.8	836.4	5411.9	1.9	4.0

This trend was reflected in terms of maximum hourly peak power too—that of the heating system dropped from 2.5 kW (2020) to 1.9 kW (RCP8.5), while that of the cooling system rose from 3.0 kW (2020) to 4.0 kW (RCP8.5).

By performing a frequency distribution analysis of hourly energy demand (see Figure 3), an increase of about 6% was shown for the 1–2 kWh (light blue) and 2–3 kWh (blue) cooling ranges for both future scenarios, while a decrease of about 8% was shown for the 1–2 kWh (orange) heating range.

Moreover, for both future scenarios, the 2–3 kWh (red) heating range disappeared and the 3–4 kWh (black) cooling range was introduced due to an increase in outdoor temperature and solar irradiances. Only the heating range 0–1 kWh (yellow) remained unchanged among the different scenarios.

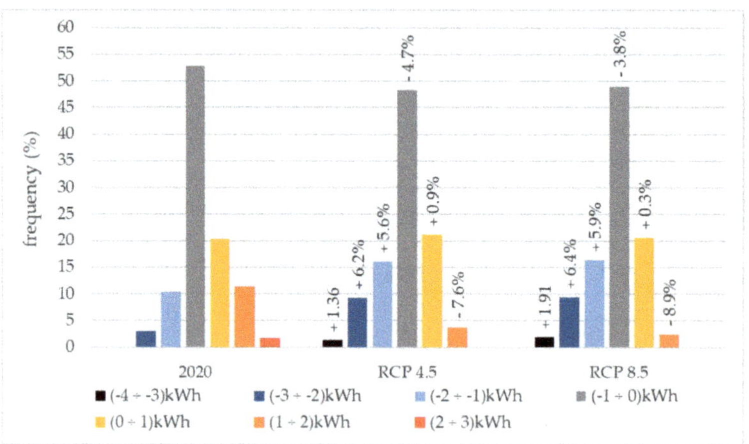

Figure 3. Frequency distribution of energy needs for heating and cooling. Negative ranges refer to cooling, positive to heating.

In addition to the increases in peak power and cooling energy demand, the number of hours over which the cooling system was active increased too—from 3033 h in 2020 to 3983 h (+23.8%) and 4022 h (+24.6%) under RCP4.5 and RCP8.5 scenarios, respectively. Indeed, the length of the cooling season (calculated from the first to the last day the system was on) varied for the different scenarios—174 days in 2020, 224 in 2050 (RCP4.5) and 208 in 2050 (RCP8.5). At the same time, the percentage of hours over which the conditioning system stayed off ($T_{air} \leq 26\ °C$) was found to be higher for RCP4.5 than RCP8.5 (25.9% and 19.5%, respectively). Therefore, the number of hours the cooling system was switched on for in the two future scenarios was similar—3983 h for RCP4.5 and 4023 h for RCP8.5.

In combination with the increase in temperature, the model ERA-Interim/UrbClim [42] provides also an increase in global solar irradiance on the horizontal plane ($I_{g,H}$), and thus on tilted and oriented surfaces. As shown in Figure 4, an evident increase in global solar irradiance was observed for both future scenarios compared to 2020, resulting in an increase in solar gains, especially during the months of January, October, November and December. The increase in solar irradiance in the summer months, although greater than in the winter months, did not lead to an evident increase in internal gains. This is because, in order to design an nZEB building, Italian requirements [40] force the use of mobile shading throughout the summer period.

The average annual percentage of global solar irradiance on the horizontal plane was found to increase (see Table 5) by about 19.7% and 15.5% for the cooling period (May–October) and the heating period (November–April), respectively, for both future scenarios. Figure 4 and Table 5 show that the average monthly solar gains do not increase in summer between 2020 and 2050, due to movable shading devices operating in summertime only. Assuming equal internal gains and ventilation rates, the increase in average summer energy consumption, accounting for 249.3% (+2463.4 kWh) and 293.5% (+2723.5 kWh) under RCP4.5 and RCP8.5, respectively, was ascribable to the corresponding average temperature increase (2.7 °C and 3.2 °C). Differently, the decrease in average winter consumption, accounting for 70.3% (−1350.9 kWh) and 77.0% (−1554.3 kWh) under RCP4.5 and RCP8.5, respectively, was associated with both the average increase in temperature (2.1 °C and 2.5 °C) and the increase in solar radiation.

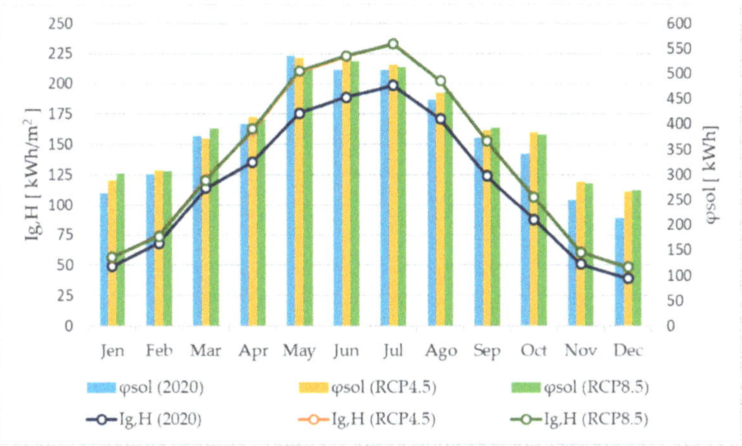

Figure 4. Monthly sum of solar gains (φsol) and solar global irradiance on the horizontal plane ($I_{g,H}$) for 2020, 2050 (RCP4.5) and 2050 (RCP8.5).

Table 5. Monthly percentage variation of the main input and output values between 2020 and the RCP4.5 and RCP8.5 (2050) scenarios.

	2020 vs. 2050 (RCP4.5)						2020 vs. 2050 (RCP8.5)				
Mtly	T_{air}	$I_{g,H}$	φ_{sol}	φ_h	φ_c	**Mtly**	T_{air}	$I_{g,H}$	φ_{sol}	φ_h	φ_c
Jen	54.2%	14.3%	9.4%	−54.2%	-	Jen	58.8%	15.5%	14.8%	−59.2%	-
Feb	38.2%	9.3%	2.6%	−48.2%	-	Feb	46.0%	8.2%	1.8%	−58.3%	-
Mar	24.4%	6.9%	−1.3%	−68.4%	-	Mar	27.8%	5.7%	4.0%	−78.1%	-
Apr	26.1%	18.9%	3.6%	−100.0%	-	Apr	26.6%	20.0%	2.3%	−100.0%	-
May	15.3%	19.6%	−0.7%	-	160.8%	May	16.3%	20.1%	−4.7%	-	157.2%
Jun	15.3%	18.0%	4.8%	-	69.6%	Jun	16.9%	18.3%	3.6%	-	74.9%
Jul	19.7%	17.5%	1.9%	-	64.1%	Jul	22.4%	17.3%	1.2%	-	71.9%
Ago	21.2%	18.2%	2.8%	-	75.4%	Ago	23.7%	18.2%	3.3%	-	84.9%
Sep	19.8%	23.4%	3.8%	-	160.9%	Sep	22.5%	23.5%	4.9%	-	184.1%
Oct	15.2%	21.3%	12.2%	-	964.6%	Oct	18.6%	20.9%	10.8%	-	1187.8%
Nov	20.9%	19.7%	14.2%	−95.8%	-	Nov	25.3%	19.5%	12.8%	−100.0%	-
Dec	33.0%	24.0%	24.1%	−55.0%	-	Dec	38.8%	24.2%	25.9%	−66.7%	-

By performing a non-linear regression analysis between daily heating/cooling needs and outdoor air temperatures (see Figure 5), we observed how future distributions gradually shift towards higher temperatures and consumptions—from a maximum cooling demand of 40.4 kWh/day in 2020 to 64.3 kWh/day under RCP4.5 and 68.0 kWh/day under RCP8.5. Precisely for this reason, the distribution in graphs b and c of Figure 5 loses symmetry, and the density of the points in the left side (heating needs) decreases. Figure 5 also shows that the minima (φh/c = 0) of the trendlines for future scenarios (graph b and c) narrow down and get slightly shifted towards higher temperatures (to 16.0 °C and 16.5 °C, respectively), compared to the current scenario (graph a) where 0 φh/c conditions occur between 14.5 °C and 16.6 °C.

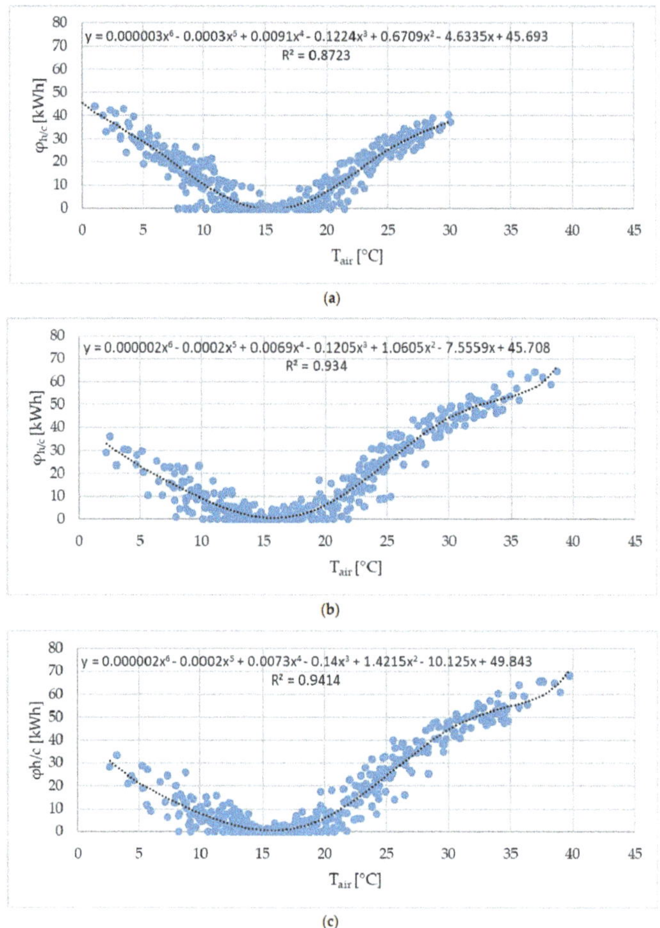

Figure 5. Non-linear regression between daily energy demand of the building and outdoor air temperature: (**a**) 2020 current scenario, (**b**) future scenario 2050 RCP4.5 and (**c**) future scenario 2050 RCP8.5. Trendlines, respective polynomial equations and R-squares are reported on each graph.

3.2. Simulations to Assess Thermal Comfort

In this section, the two approaches employed for the assessment of indoor thermal comfort are sequentially presented: (i) the adaptive method and (ii) the combined long-term index proposed by Li et al. [49].

The assessment of adaptive comfort can be performed only on buildings used mainly for human occupancy engaged in sedentary activities (e.g., residential buildings and offices), where thermal conditions can be regulated by occupants through accessing operable windows and adapting their behavior. On the other hand, the new combined index considers the long-term thermal satisfaction in response to pronounced excursions and daily variability.

The analysis was performed for two rooms (see Figure 6), one due southeast (living room) and one due northwest (double bedroom), to account for different exposures and thus different solar gains. The evaluation focused on the cooling periods as identified in Section 3.1 (174 days in 2020, 224 in 2050 (RCP4.5) and 208 in 2050 (RCP8.5)); therefore, the assessment of thermal comfort is performed for each climate scenario over different periods, assuming the absence of cooling systems.

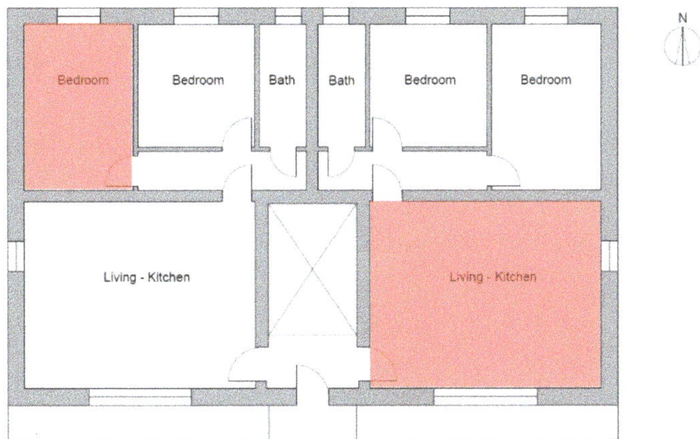

Figure 6. Representative rooms for adaptive comfort assessment: southeast living room and northwest double bedroom.

Regarding the adaptive method, as shown in Table 6, the living room provided comfort conditions (operative temperature cooling set point of 26 °C) for shorter time periods than the double bedroom, regardless of the scenario. Assuming category II as a reference, the percent decrease between 2020 and 2050 RCP4.5 reached 4.1% and 3.7% for the two considered rooms, against 9.0% for both rooms under RCP8.5.

Table 6. Adaptive comfort assessment for the two analyzed rooms under different climate scenarios: comfort hours and percentages over the relevant cooling periods.

Year	Cooling Period			Room	Cat. I		Cat. II		Cat. III	
	Data	Days	h		h	%	h	%	h	%
2020	6 May–26 October	174	4176	Living room	681	16.3%	1091	26.1%	1566	37.5%
				Bedroom	1266	30.3%	1883	45.1%	2169	51.9%
2050 (RCP4.5)	6 April–15 November	224	5376	Living room	733	13.6%	1186	22.1%	1788	33.3%
				Bedroom	1840	34.2%	2223	41.4%	2422	45.1%
2050 (RCP8.5)	6 April–30 October	208	4992	Living room	513	10.3%	855	17.1%	1299	26.0%
				Bedroom	1184	23.7%	1800	36.1%	2035	40.8%

Results for the combined index are shown in Table 7. The double bedroom experienced percent increases in discomfort events of 0.9% and 3.6% between 2020 and 2050 for RCP4.5 and RCP8.5, respectively. In contrast, for the living room the variation stayed below 1% under both the considered scenarios.

Table 7. Combined index assessment for the two analyzed rooms under different climatic scenarios: percentages over the relevant cooling periods.

Year	Cooling Period			Room	Combined Discomfort Index
	Data	Days	h		%
2020	26 May–26 October	174	4176	Living room	56.1%
				Bedroom	47.9%
2050 (RCP4.5)	6 April–15 November	224	5376	Living room	55.0%
				Bedroom	48.8%
2050 (RCP8.5)	6 April–30 October	208	4992	Living room	55.9%
				Bedroom	51.5%

4. Discussion and Conclusions

The global climate is undergoing major upheavals, posing a serious risk of premature obsolescence for the current nZEB paradigms.

In accordance with the experimental study by A. Martinelli et al. [56], wherein it is shown that the Urban Heat Island phenomenon is more accentuated in the areas closer to the city center, the ERA-Interim/CliUrbm predictive model [42,43] used in this study identified Rome as the Italian city that will undergo the greatest temperature increase by 2050. For this reason, the impact of climate change on the energy needs and indoor comfort of an nZEB located in this location was assessed.

Based on the simulations made and detailed in Sections 2 and 3, the following considerations can be made:

The respective rises of 3.4 °C and 3.9 °C by 2050 for the RCP4.5 and RCP8.5 scenarios do not lead to decreases in heating needs great enough to meet the sharp increase in cooling needs. Specifically, in fact, compared to the current energy needs, there will be an average annual increase of 1143 kWh (+22.4%). This increase is consistent with the increase obtained by Ciancio et al. [4] (+34.4%) for the building located in Rome, which is not an nZEB, but has similar transmittances to our case study.

a. Peak electricity demand is especially worrisome since it is usually covered by low-efficiency power plants, yet it is strongly associated with typical nZEB paradigms. In fact, while air conditioning is only a fraction of all building energy uses, it is the primary driver of peak electricity demand [57]. Efficiently curbing the air conditioning needs by targeting a resilient nZEB design will be key in the future.

b. According to the ERA-Interim/UrbClim model, by 2050, not only will temperatures rise, but a 19.7% increase in global solar irradiance on the horizontal plane is also to be expected during the summer months, thus triggering higher solar gains. The increase in the solar irradiance, often underestimated by the models [58,59], is consistent with the study conducted by M. Wild et al. [60], which predicts a decrease in solar radiation (clear-sky condition) for many regions of the world, except for parts of China and Europe. In our case study, the increase in the solar irradiance does not imply an increase in the solar contributions inside the building because the national nZEB regulation foresees the use of mobile shading devices for the whole summer period. These systems appear rather effective, since the increase in solar gains is negligible in summer, but becomes evident in winter (particularly in January, October, November and December).

c. By performing simulations in the absence of cooling systems, 6.2% and 5.1% reductions in the hours of adaptive comfort are determined under the RCP4.5 (2050) and RCP8.5 (2050) scenarios, respectively, out of the concerted actions of temperature and solar gains. The results of the newly proposed combined index for long-term comfort assessments revealed a milder future penalty. The index estimates how the level of occupant adaptation and sensitivity to variation would be affected in the future. It was demonstrated that the comfort implications of the pronounced excursions and large variations in daily temperature will be marginal under both scenarios, with greater influence on northwest rather than southeast oriented thermal zones, likely owing to the effect of the combination of higher temperatures and higher solar irradiation in levelling out the daily swings.

In conclusion, this study adds to the current body of knowledge on the preservation of nZEB performance in the future by performing hourly dynamic simulations on a reference building in Rome, modeled in accordance with the latest legislations. It quantifies potential changes in terms of energy and comfort levels and provides useful recommendations to legislators on building standards, both for the design of new nZEBs, e.g., the presence of solar shading devices, and for the renovation of the existing building stock. Further analysis may target climate dependencies and may include technological variants.

Author Contributions: Conceptualization, S.S. and L.T.; methodology, S.S., L.T., G.U. and C.D.P.; software, L.T.; validation, S.S.; formal analysis, S.S.; investigation, S.S.; resources, S.S.; data curation S.S.; writing—original draft preparation, S.S.; writing—review and editing, S.S., L.T. and G.U.; visualization, S.S., L.T. and C.D.P.; supervision, C.D.P. All authors have read and agreed to the published version of the manuscript.

Funding: This research received no external funding.

Conflicts of Interest: The authors declare no conflict of interest.

References

1. Lucon, O.; Urge-Vorsatz, D.; Ahmed, A.Z.; Akbari, H.; Bertoldi, P.; Cabeza, L.F.; Liphoto, E. Gadgil Chapter 9—Buildings. In *Climate Change 2014: Mitigation of Climate Change. IPCC Working Group III Contribution to AR5*; Cambridge University: Cambridge, UK, 2014.
2. Ürge-Vorsatz, D.; Lucon, O.; Zain Ahmed, A.; Akbari, H.; Bertoldi, P.; Cabeza, L.F.; Eyre, N.; Gadgil, A.; Harvey, L.D.D.; Jiang, Y.; et al. Buildings. In *Mitigation. Working Group III contribution to the Fifth Assessment Report of the Intergovernmental Panel of Climate Change*; Cambridge University: Cambridge, UK, 2014; pp. 671–738.
3. IPCC. *Climate Change—The Synthesis Report*; IPCC: Geneva, Switzerland, 2014.
4. Ciancio, V.; Salata, F.; Falasca, S.; Curci, G.; Golasi, I.; de Wilde, P. Energy demands of buildings in the framework of climate change: An investigation across Europe. *Sustain. Cities Soc.* **2020**, *60*, 102213. [CrossRef]
5. Olonscheck, M.; Holsten, A.; Kropp, J.P. Heating and cooling energy demand and related emissions of the German residential building stock under climate change. *Energy Policy* **2011**, *39*, 4795–4806. [CrossRef]
6. Verichev, K.; Zamorano, M.; Carpio, M. Effects of climate change on variations in climatic zones and heating energy consumption of residential buildings in the southern Chile. *Energy Build.* **2020**, *215*, 109874. [CrossRef]
7. Angeles, M.E.; González, J.E.; Ramírez, N. Impacts of climate change on building energy demands in the intra-Americas region. *Theor. Appl. Climatol.* **2018**, *133*, 59–72. [CrossRef]
8. Frank, T. Climate change impacts on building heating and cooling energy demand in Switzerland. *Energy Build.* **2005**, *37*, 1175–1185. [CrossRef]
9. Tribuiani, C.; Tarabelli, L.; Summa, S.; Di Perna, C. Thermal performance of a massive wall in the Mediterranean climate: Experimental and analytical research. *Appl. Sci.* **2020**, *10*, 4611. [CrossRef]
10. Cabeza, L.F.; Chàfer, M. Technological options and strategies towards zero energy buildings contributing to climate change mitigation: A systematic review. *Energy Build.* **2020**, *219*, 110009. [CrossRef]
11. Ulpiani, G.; Summa, S.; di Perna, C. Sunspace coupling with hyper-insulated buildings: Investigation of the benefits of heat recovery via controlled mechanical ventilation. *Sol. Energy* **2019**, *181*, 17–26. [CrossRef]
12. Troup, L.; Eckelman, M.J.; Fannon, D. Simulating future energy consumption in office buildings using an ensemble of morphed climate data. *Appl. Energy* **2019**, *255*, 113821. [CrossRef]
13. Wang, L.; Liu, X.; Brown, H. Prediction of the impacts of climate change on energy consumption for a medium-size office building with two climate models. *Energy Build.* **2017**, *157*, 218–226. [CrossRef]
14. Favarolo, P.A.; Manz, H. Temperature-driven single-sided ventilation through a large rectangular opening. *Build. Environ.* **2005**, *40*, 689–699. [CrossRef]
15. Petcu, C.; Petran, H.-A.; Vasile, V.; Toderasc, M.-C. Materials from Renewable Sources as Thermal Insulation for Nearly Zero Energy Buildings (nZEB). In *Conference on Sustainable Energy*; Springer: Cham, Switzerland, 2017; pp. 159–167.
16. Carmo, C.; Dumont, O.; Nielsen, M.P.; Elmegaard, B. Assessment of Emerging Renewable Energy-based Cogeneration Systemsfor nZEB Residential Buildings Assessment of Emerging Renewable Energy-based Cogeneration Systems for nZEB Residential Buildings. *Environ. Sci. Eng.* **2016**.
17. De Santoli, L.; Basso, G.L.; Spiridigliozzi, G.; Garcia, D.A. Innovative Hybrid Energy Systems for Heading Towards NZEB Qualification for Existing Buildings. In Proceedings of the 2018 IEEE International Conference on Environment and Electrical Engineering and 2018 IEEE Industrial and Commercial Power Systems Europe (EEEIC/I&CPS Europe); Institute of Electrical and Electronics Engineers (IEEE): New York, NY, USA, 2018; pp. 1–6. [CrossRef]
18. Stritih, U.; Tyagi, V.V.; Stropnik, R.; Paksoy, H.; Haghighat, F.; Joybari, M.M. Integration of passive PCM technologies for net-zero energy buildings. *Sustain. Cities Soc.* **2018**, *41*, 286–295. [CrossRef]

19. Georgiou, G.S.; Christodoulides, P.; Kalogirou, S.A. Optimizing the energy storage schedule of a battery in a PV grid-connected nZEB using linear programming. *Energy* **2020**, *208*, 118177. [CrossRef]

20. Kazmi, H.; D'Oca, S.; Delmastro, C.; Lodeweyckx, S.; Corgnati, S.P. Generalizable occupant-driven optimization model for domestic hot water production in NZEB. *Appl. Energy* **2016**, *175*, 1–15. [CrossRef]

21. Wemhoener, C.; Schwarz, R.; Rominger, L. IEA HPT Annex 49-Design and integration of heat pumps in nZEB. *Energy Procedia* **2017**, *122*, 661–666. [CrossRef]

22. Gondal, I.A. Prospects of Shallow geothermal systems in HVAC for NZEB. *Energy Built Environ.* **2020**. [CrossRef]

23. Becchio, C.; Corgnati, S.P.; Vio, M.; Crespi, G.; Prendin, L.; Ranieri, M.; Vidotto, D. Toward NZEB by optimizing HVAC system configuration in different climates. *Energy Procedia* **2017**, *140*, 115–126. [CrossRef]

24. Kim, D.; Cho, H.; Koh, J.; Im, P. Net-zero energy building design and life-cycle cost analysis with air-source variable refrigerant flow and distributed photovoltaic systems. *Renew. Sustain. Energy Rev.* **2020**, *118*, 109508. [CrossRef]

25. Naveen Chakkaravarthy, A.; Subathra, M.S.P.; Jerin Pradeep, P.; Manoj Kumar, N. Solar irradiance forecasting and energy optimization for achieving nearly net zero energy building. *J. Renew. Sustain. Energy* **2018**, *10*, 035103. [CrossRef]

26. Reda, F.; Fatima, Z. Northern European nearly zero energy building concepts for apartment buildings using integrated solar technologies and dynamic occupancy profile: Focus on Finland and other Northern European countries. *Appl. Energy* **2019**, *237*, 598–617. [CrossRef]

27. Synnefa, A.; Laskari, M.; Gupta, R.; Pisello, A.L.; Santamouris, M. Development of Net Zero Energy Settlements Using Advanced Energy Technologies. *Procedia Eng.* **2017**, *180*, 1388–1401. [CrossRef]

28. Gholami, H.; Nils Røstvik, H.; Manoj Kumar, N.; Chopra, S.S. Lifecycle cost analysis (LCCA) of tailor-made building integrated photovoltaics (BIPV) façade: Solsmaragden case study in Norway. *Sol. Energy* **2020**, *211*, 488–502. [CrossRef]

29. Pikas, E.; Kurnitski, J.; Thalfeldt, M.; Koskela, L. Cost-benefit analysis of nZEB energy efficiency strategies with on-site photovoltaic generation. *Energy* **2017**, *128*, 291–301. [CrossRef]

30. European Community (EC). Directive (EU) 2018/844 of the European Parliament and of the Council of 30 May 2018 Amending Directive 2010/31/EU on the Energy Performance of Buildings Directive 2012/27/EU on Energy Efficiency; L156/75. In *Official Journal of the European Union*; European Union: Brussels, Belgium, 2018.

31. Karlessi, T.; Kampelis, N.; Kolokotsa, D.; Santamouris, M.; Standardi, L.; Isidori, D.; Cristalli, C. The Concept of Smart and NZEB Buildings and the Integrated Design Approach. *Procedia Eng.* **2017**, *180*, 1316–1325. [CrossRef]

32. Rey-Hernández, J.M.; González, S.L.; San José-Alonso, J.F.; Tejero-González, A.; Velasco-Gómez, E.; Rey-Martínez, F.J. Smart energy management of combined ventilation systems in a nZEB. *E3S Web Conf.* **2019**, *111*. [CrossRef]

33. Ulpiani, G.; Ranzi, G.; Bruederlin, F.; Paolini, R.; Fiorito, F.; Haddad, S.; Kohl, M.; Santamouris, M. Elastocaloric cooling: Roadmap towards successful implementation in the built environment. *AIMS Mater. Sci.* **2019**, *6*, 1135–1152. [CrossRef]

34. Ulpiani, G.; Ranzi, G.; Shah, K.W.; Feng, J.; Santamouris, M. On the energy modulation of daytime radiative coolers: A review on infrared emissivity dynamic switch against overcooling. *Sol. Energy* **2020**, *209*, 278–301. [CrossRef]

35. Harkouss, F.; Fardoun, F.; Biwole, P.H. Optimization approaches and climates investigations in NZEB—A review. *Build. Simul.* **2018**, *11*, 923–952. [CrossRef]

36. Feng, W.; Zhang, Q.; Ji, H.; Wang, R.; Zhou, N.; Ye, Q.; Hao, B.; Li, Y.; Luo, D.; Lau, S.S.Y. A review of net zero energy buildings in hot and humid climates: Experience learned from 34 case study buildings. *Renew. Sustain. Energy Rev.* **2019**, *114*, 109303. [CrossRef]

37. Di Giuseppe, E.; Ulpiani, G.; Summa, S.; Tarabelli, L.; Di Perna, C.; D'Orazio, M. Hourly dynamic and monthly semi-stationary calculation methods applied to nZEBs: Impacts on energy and comfort. *IOP Conf. Ser. Mater. Sci. Eng.* **2019**, *609*, 072008. [CrossRef]

38. International Organization for Standardization. *Energy Performance of Buildings—Energy Needs for Heating and Cooling, Internal Temperatures and Sensible and Latent Heat Loads—Part 1: Calculation Procedures*; International Organization for Standardization: Geneva, Switzerland, 2017; p. 204.
39. CEN European Commitee for Standardization. *EN ISO 13790: 2008 Energy Performance of Buildings—Calculation of Energy Use for Space Heating and Cooling*; European Commitee for Standardization: Brussels, Belgium, 2008.
40. Italian Republic. *Italian Iterministerial Decree 26th June 2015: Application of Calculation Methods for Energy Performance and Definition of Minimum Building Requirements*; Ministry of Economic Development: Roma, Italy, 2015; pp. 1–8.
41. World Climate Research Program. EURO-CORDEX—Coordinated Downscaling Experiment—European Domain. Available online: https://euro-cordex.net/ (accessed on 28 October 2020).
42. Flemish Research & Technology Organisation VITO. Urban Climate Service Centre. Available online: www.urban-climate.be (accessed on 28 October 2020).
43. Remund, J.; Grossenbacher, U. Urban Climate-Impact on Energy Consumption and Thermal Comfort of Buildings Urban Climate—Impact on Energy Consumption and Thermal Comfort of Buildings. In Proceedings of the International Conference on Solar Energy for Buildings and Industry EuroSun2018, Rapperswil, Switzerland, 10–13 September 2018. [CrossRef]
44. Duffy, M.J.; Hiller, M.; Bradley, D.E.; Keilholz, W.; Thornton, J.W. TRNSYS-features and functionalitity for building simulation 2009 conference. In Proceedings of the 11th International IBPSA Conference-Building Simulation, Glasgow, UK, 27–30 July 2009; pp. 1950–1954.
45. Crawley, D.B.; Hand, J.W.; Kummert, M.; Griffith, B.T. Contrasting the capabilities of building energy performance simulation programs. *Build. Environ.* **2008**, *43*, 661–673. [CrossRef]
46. Vadiee, A.; Dodoo, A.; Gustavsson, L. A Comparison Between Four Dynamic Energy Modeling Tools for Simulation of Space Heating Demand of Buildings. In *Cold Climate HVAC Conference*; Springer: Cham, Switzerland, 2019; pp. 701–711. [CrossRef]
47. Duffy, M.J.; Hiller, M.; Bradley, D.E.; Keilholz, J.W.W. *TRNSYS 17: A Transient System Simulation Program*; Solar Energy Laboratory, University of Wisconsin: Madison, WI, USA, 2010.
48. CEN European Commitee for Standardization. *EN 16798-1:2019 Energy Performance of Buildings—Ventilation for Buildings—Part 1: Indoor Environmental Input Parameters for Design and Assessment of Energy Performance of Buildings Addressing Indoor Air Quality, Thermal Environment, Lighting and Acous*; European Commitee for Standardization: Brussels, Belgium, 2019.
49. Li, P.; Parkinson, T.; Schiavon, S.; Froese, T.M.; de Dear, R.; Rysanek, A.; Staub-French, S. Improved long-term thermal comfort indices for continuous monitoring. *Energy Build.* **2020**, *224*, 110270. [CrossRef]
50. Meteotest. *Meteonorm 7 V7.3.3*; Meteotest: Bern, Switzerland, 2018.
51. Remund, J.; Müller, S.; Studer, C.; Cattin, R. *Handbook Part II: Theory Global Meteorological Database Version 7 Software and Data for Engineers, Planers and Education. Handbook Part II: Theory*; Meteotest AG: Bern, Switzerland, 2019.
52. Giorgi, F.; Jones, C.; Asrar, G.R. Addressing climate information needs at the regional level: The CORDEX framework. *WMO Bull.* **2009**, *58*, 175–183.
53. Trimble, I.N.C. SketchUp. 2020. Available online: https://www.sketchup.com/node/4481 (accessed on 28 October 2020).
54. Ellis, P. OpenStudio Plugin for Google SketchUp3D. 2009. Available online: https://www.openstudio.net/ (accessed on 28 October 2020).
55. *UNI-Ente Italiano di Normazione, UNI/TS 11300-1:2014-Energy Performance of Buildings-Part 1: Evaluation of Energy Need for Space Heating and Cooling*; Comitato Termotecnico Italiano (CTI): Milan, Italy, 2014.
56. Martinelli, A.; Kolokotsa, D.D.; Fiorito, F. Urban heat island in Mediterranean coastal cities: The case of Bari (Italy). *Climate* **2020**, *8*, 79. [CrossRef]
57. Ortiz, L.; González, J.E.; Lin, W. Climate change impacts on peak building cooling energy demand in a coastal megacity. *Environ. Res. Lett.* **2018**, *13*, 094008. [CrossRef]
58. Allen, R.J.; Norris, J.R.; Wild, M. Evaluation of multidecadal variability in CMIP5 surface solar radiation and inferred underestimation of aerosol direct effects over Europe, China, Japan, and India. *J. Geophys. Res. Atmos.* **2013**, *118*, 6311–6336. [CrossRef]

59. Wild, M.; Schmucki, E. Assessment of global dimming and brightening in IPCC-AR4/CMIP3 models and ERA40. *Clim. Dyn.* **2011**, *37*, 1671–1688. [CrossRef]
60. Wild, M.; Folini, D.; Henschel, F.; Fischer, N.; Müller, B. Projections of long-term changes in solar radiation based on CMIP5 climate models and their influence on energy yields of photovoltaic systems. *Sol. Energy* **2015**, *116*, 12–24. [CrossRef]

Publisher's Note: MDPI stays neutral with regard to jurisdictional claims in published maps and institutional affiliations.

Article

Thermal Environment Design of Outdoor Spaces by Examining Redevelopment Buildings Opposite Central Osaka Station

Hideki Takebayashi

Department of Architecture, Kobe University, Nada Ward 657-8501, Japan; thideki@kobe-u.ac.jp;
Tel.: +81-78-803-6062

Received: 24 October 2019; Accepted: 13 December 2019; Published: 14 December 2019

Abstract: Thermal environmental design in an outdoor space is discussed by focusing on the proper selection and arrangement of buildings, trees, and covering materials via the examination of redevelopment buildings in front of Central Osaka Station, where several heat island countermeasure technologies have been introduced. Surface temperatures on the ground and wall were calculated based on the surface heat budget equation in each 2 m size mesh of the ground and building wall surface. Incident solar radiation was calculated using ArcGIS and building shape data. Mean radiant temperature (MRT) of the human body was calculated using these results. Distribution of wind velocity was calculated by computational fluid dynamics (CFD) reproducing buildings, obstacles, trees, and the surroundings. The effect of MRT on SET* was greater than that of wind velocity at 13:00 and 17:00 on a typical summer day. SET* reduction was the highest by solar radiation shading, followed by surface material change and ventilation. The largest ratio of the area considered for the thermal environment was 83% on Green Garden, which consists of 44% of building shade, 21% of tree shade, 7% of water surface, and 11% of green cover. It is appropriate to consider the thermal environment design of outdoor space in the order of shade by buildings, shading by trees, and improvement of surface materials.

Keywords: outdoor space; thermal environment; radiation environment; wind environment

1. Introduction

In our previous study [1], the effects of solar radiation shading by trees in open spaces was evaluated through a case study. Outdoor open spaces are used for various purposes such as walking, resting, talking, meeting, studying, exercising, playing, performing, eating, and drinking. Therefore, providing various thermal environments according to the various purposes above-mentioned is desirable. The results from one of our previous studies [2] are reprinted as follows: "By investigating the redevelopment building in front of Central Station in Osaka, the radiation environment was evaluated with a focus on ground cover materials and solar radiation shielding. ArcGIS and building shape data were used to calculate the spatial distribution of solar shading. A surface heat balance equation was calculated to determine the surface temperature of the ground and walls. Assuming the human body is a sphere, the mean radiation temperature (MRT) of the human body was calculated. Solar radiation shielding and improvements in surface coverage were the most dominant factors in the radiation environment. On a typical summer day (August) when air temperature is high, improvements in solar shading and surface coverage did not provide a comfortable standard new effective temperature (SET*) in the afternoon. However, there were several places where people did not feel uncomfortable, especially in the rooftop garden and green gardens, which have large areas of shaded grass and water."

This study is the continuation of our previous studies [1,2]. The study site and the calculation method of solar radiation, surface temperature, MRT, and wind velocity distribution were the same as those of our previous studies [1,2]. Results from another of our previous studies [1] are reprinted as follows. "At 10:00, 13:00, and 17:00 on a typical summer sunny day, we analyzed building and tree awnings at 25 and 32 measurement points in Station Plaza and Green Garden. Assuming various heights of buildings, the need for sunshade by trees was pointed out at 10 m or more from the south building and 6 m or more from the west or east building." The subject of this study was the thermal environmental design in an outdoor space, focusing on the proper selection and arrangement of buildings, trees, and covering materials through the examination of redevelopment buildings in front of Central Osaka Station where several heat island countermeasure technologies such as terrace gardens on medium height rooftops (Rooftop Gardens), mist and waterscape in Station Plaza, ground garden with trees, water, and green cover between buildings (Green Garden), rows of trees, and water streams around the buildings were introduced. In particular, the results of the case study were analyzed from the perspective of how effective it is to proceed with the thermal environmentally-friendly design of outdoor spaces to increase the generic understanding for better outdoor thermal environment design.

2. Calculation of Thermal Element Distribution

The study site layout is shown in Figure 1 [3], which is the same as those in our previous studies [1,2]. The layouts of Station Plaza, Rooftop Gardens, and Green Garden are shown in Figure 2. The ratio of each ground cover type is shown in Table 1. An outline of the study sites is shown in Table 2. The trees were reproduced based on the actual situation, and the average height of the trees was about 6 m because this study was conducted just after completion of the site. The calculation methods for surface temperature, solar radiation, MRT, and wind velocity distribution were also the same as those used previously [1,2], and an outline of the calculation methods is shown in Table 3. Daytime air temperatures obtained from the Osaka Meteorological Observatory on a typical summer day (11 August 2013), the day of the autumnal equinox (23 September 2013), and the day of the summer solstice (21 June 2013) are shown in Figure 3. The air temperature was over 30 °C in the morning and over 35 °C in the afternoon on a typical summer day, and it was around 30 °C in the afternoon on the day of the autumnal equinox and day of the summer solstice. MRT was calculated by integrating the amount of solar radiation and infrared radiation incident on the human body. The incident solar radiation was calculated by the method described above, and the incident infrared radiation was calculated using the surface temperature and the view factor of the surrounding objects. The objective area was divided into meshes according to the form of the buildings. The calculation conditions for computational fluid dynamics (CFD) are shown in Table 4, referring to Tominaga et al. [4]. The applicability of this software for an urban area such as Osaka City was verified using a verification database provided by Tominaga et al. [4].

Table 1. Ratio of each ground cover type.

Site	Concrete	Wood Deck	Grass	Water Surface	Asphalt
Station Plaza	79%	0%	0%	10%	11%
Rooftop Gardens	37%	25%	38%	0%	0%
Green Garden	40%	0%	26%	15%	19%
Total	52%	5%	12%	6%	25%

Figure 1. Layout of the study site [1,2].

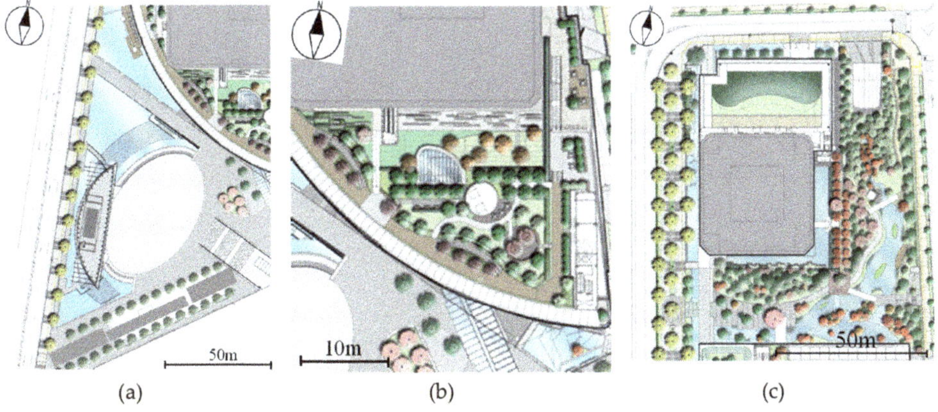

Figure 2. Layout of (**a**) Station Plaza; (**b**) Rooftop Gardens; and (**c**) Green Garden [1,2].

Table 2. Outline of the study sites.

Study Sites	Location	Land Cover Characteristics
Station Plaza	The site is beside the north-eastern (180 m high) and the southern (150 m high) high-rise buildings.	There is little vegetation cover, and open spaces (concrete surfaces) and water surfaces dominate.
Rooftop Gardens	The site is on the southern (45 m high) and the central (43 m high) middle-rise buildings.	The ratios of concrete, wood deck, and grass are similar, ~30%.
Green Garden	The site is between the northern (174 m high) and the central (154 m high) high-rise buildings.	The site features green grassy areas, water surfaces, medium-height trees, and concrete walkways.

Table 3. Outline of calculation methods, which is a reprint of our previous study [1,2].

Element	Method
Surface temperature	It is calculated based on the surface heat budget equation in each 2 m size mesh of the ground and building wall. Air temperature, air absolute humidity, underground temperature, convection heat, and moisture transfer coefficients of the function of wind velocity are set by the observation values as boundary conditions.
Incident solar radiation	It is calculated using ArcGIS and building shape data, as per the method described by Takebayashi et al. [5]. The visible area of the upper hemisphere is calculated by ArcGIS tool considering the influence of the adjacent buildings. The visible area is then overlain with the sun-map and sky-map raster to calculate the diffuse and direct solar radiation received from each direction.
Mean Radiant Temperature	MRT of the human body is calculated using surface temperature and incident solar radiation. The human body is assumed to be a sphere, and solar radiation absorption ratio of the human body is assumed to be 0.5, considering the clothing conditions in summer.
Wind velocity	It is calculated by computational fluid dynamics (CFD) reproducing buildings, obstacles, trees, and the surroundings. The standard k-ε turbulence model (one of the Reynolds–Averaged Navier–Stokes equation (RANS) models) is selected for use in the simulation. A general purpose CFD software (STREAM, version 9, Software Cradle Co. Ltd., Osaka, Japan) is used for calculation. The Navier–Stokes equations are discretized using a finite volume method, and the SIMPLE algorithm is used to handle pressure-velocity coupling. Inflow boundary conditions are given based on weather conditions.

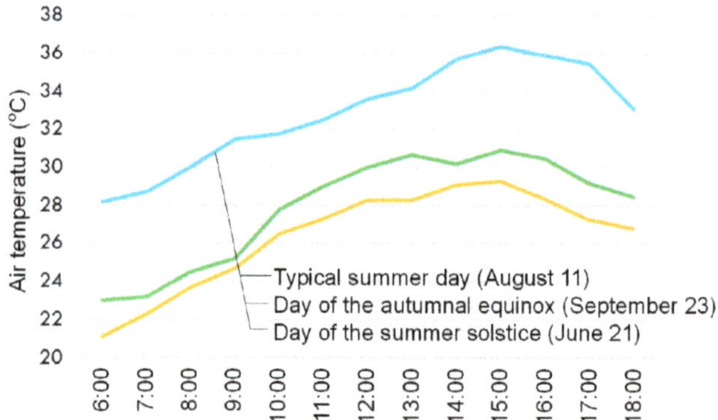

Figure 3. Daytime air temperature at Osaka Meteorological Observatory on a typical summer day (11 August 2013), the day of the autumnal equinox (23 September 2013), and the day of the summer solstice (21 June 2013).

Table 4. The calculation conditions for computational fluid dynamics (CFD).

Software	STREAM ver. 9
Turbulence model	Standard k-ε model
Advection term	Up-wind difference scheme
Inflow boundary	Power low, 3.9 m/s, WSW at 50.9 m high, power: 0.27
Outflow boundary	Zero gradient condition
Up, side boundary	Free-slip condition
Wall, ground surface	Generalized log-low
Convergence criterion	10^{-5}

The calculation results of surface temperature, MRT, wind velocity, and SET* distribution at 13:00 on a typical summer day (11 August 2013) are shown in Figure 4. Surface temperature and MRT distributions are reprints of our previous study [2]. The explanation concerning SET* is the same as per a previous study [2] and is reprinted as follows: "The equivalent dry bulb temperature in an isothermal environment with a relative humidity of 50% is the definition of SET* [6]. The subject has the same thermal stress and temperature regulation strain as the actual test environment while wearing standardized clothing for the relevant activity. It is used for thermal environmental evaluation. Gagge et al. proposed SET* by improving the new effective temperature (ET*) [6]. Gagge et al. also proposed ET*, an index based on human energy balance and a two-node model [7]. SET* is frequently used as an indoor and outdoor comfort indicator. The metabolic rate was assumed to be 2.0 met and the clothes was assumed to be 0.6 clo. The thermal equilibrium calculation program for the thermos-physiological model of the human body, which has already been verified in previous studies [6,7], calculates SET*." Air temperature and relative humidity were set by the observation values at the Osaka Observatory. We compared the temporal and spatial distributions of MRT in our previous study [2] from 5:00 to 18:00 on the day of the summer solstice, a typical summer day in August, and the day of the autumnal equinox. The result is reprinted as follows; "Incident solar radiation dominates the characteristics of MRT spatial distribution. Sunny and shaded points cause large differences in MRT time changes. In the afternoon of the summer solstice and autumnal equinox, a comfortable thermal environment was realized by sun shade. However, on a typical summer day (August), since air temperature is too high, it is difficult to make SET* comfortable in the afternoon, both with sun shade and with improved surface cover." In this study, 13:00 on a typical summer day (11 August 2013) was chosen as a representative time, together with 17:00 on a typical summer day (11 August 2013) and 13:00 on a summer solstice day (21 June 2013). Weak wind regions affected by buildings, obstacles, and planting were confirmed in the wind velocity distribution. The influence of MRT was dominant in the SET* distribution, despite the high wind velocity in Rooftop Gardens, which is shown in the upper right in Figure 4c as GL + 46.5 m.

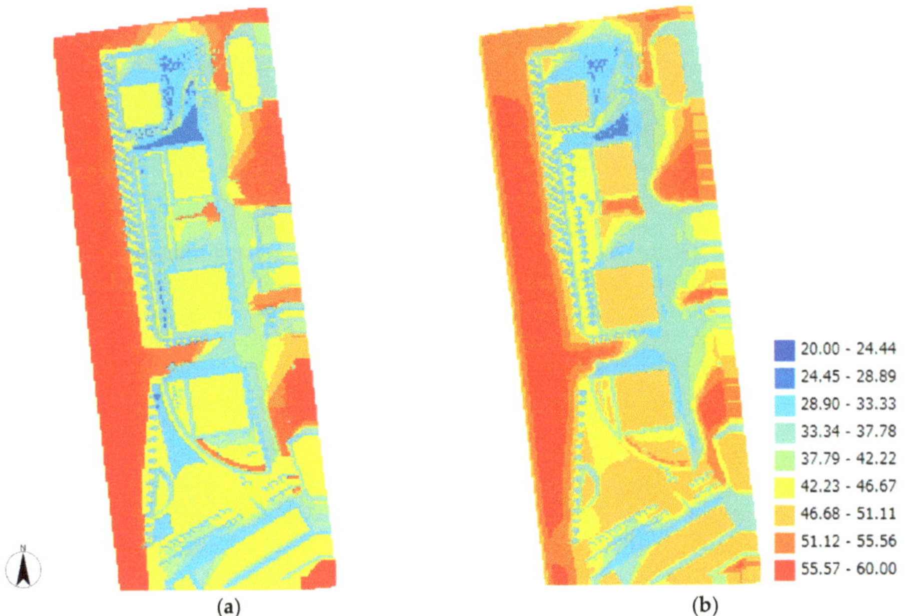

Figure 4. *Cont.*

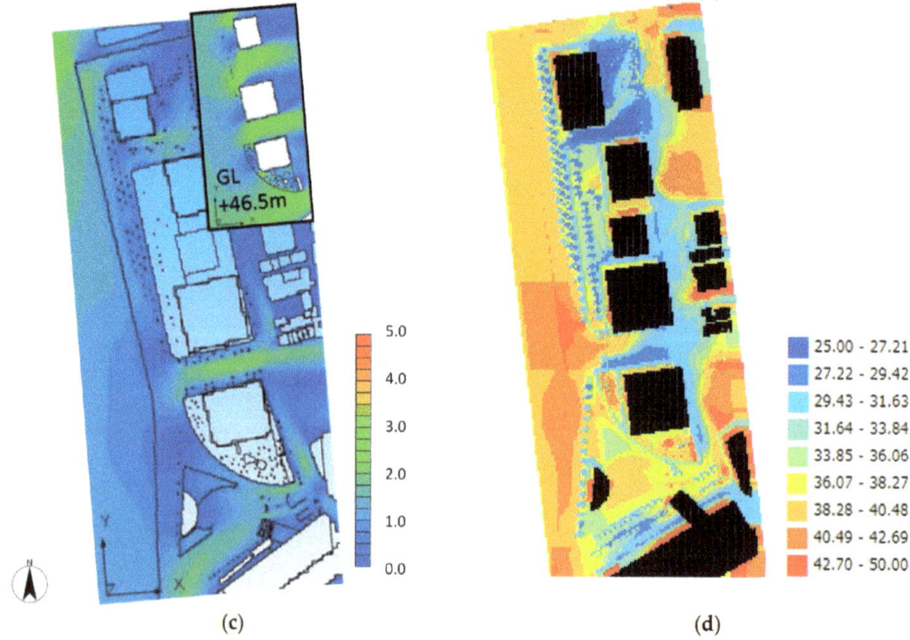

(c) (d)

Figure 4. Calculation results of (**a**) surface temperature (°C), (**b**) mean radiant temperature (MRT) (°C), (**c**) wind velocity (m/s), and (**d**) standard new effective temperature (SET*) (°C) distribution at 13:00 on the typical summer day (11 August 2013).

3. Relationship between MRT, Wind Velocity, and SET*

3.1. Evaluation under Various Conditions

The relationships between MRT, wind velocity, and SET* at 13:00 and 17:00 on a typical summer day (11 August 2013) and at 13:00 on the summer solstice day (21 June 2013) are shown in Figure 5. Air temperature is high (34.0 °C) at 13:00 and is still high (33.6 °C) at 17:00 on the typical summer day. Furthermore, it was a little low (28.3 °C) at 13:00 on the summer solstice day. The average wind velocity and MRT values in sunny and shaded locations on Station Plaza, Rooftop Gardens, and Green Garden are presented by the vertical and horizontal axes. The standard deviation is expressed by the length of the bar, and the number of corresponding points is expressed by the size of bubbles. The numbers inside and beside the bubbles indicate the number of corresponding points. SET*, which is the center of the bubble, is recognized from the background contour lines. Air temperature and relative humidity given uniformly, are also shown in the figure. The translucent bubbles denote sunny points, and opaque bubbles denote shaded points. The relationship between SET* and thermal comfort evaluation reported by Ishii et al. [8] is as follows: comfortable < 26.5 °C < slightly comfortable < 27.5 °C < neither comfortable nor uncomfortable < 29.5 °C < slightly uncomfortable < 31.5 °C < uncomfortable < 32.5 °C < very uncomfortable. The comfortable range is shown in the blue colored background with 29.5 °C as the boundary in Figure 5, which is the boundary between neither comfortable nor uncomfortable and slightly uncomfortable by Ishii et al. [8].

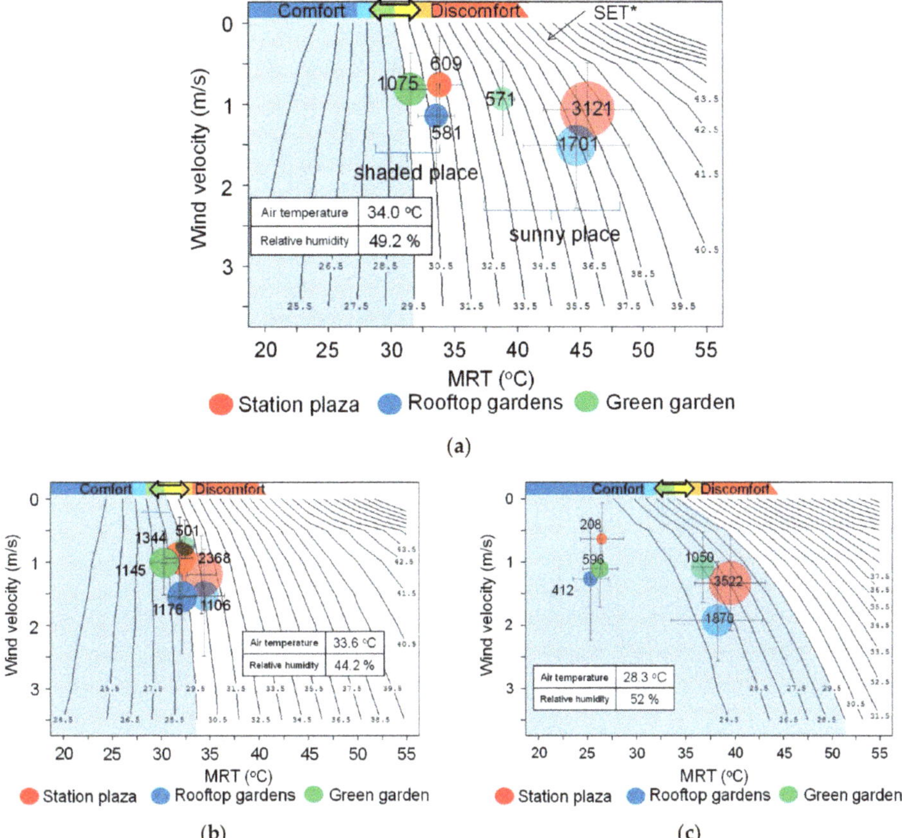

Figure 5. Relationship between MRT, wind velocity, and SET*. (**a**) 13:00 on the typical summer day (11 August 2013); (**b**) 17:00 on the typical summer day (11 August 2013); (**c**) 13:00 on the summer solstice day (21 June 2013).

The effect of MRT on SET* was greater than that of the wind velocity at 13:00 and 17:00 on the typical summer day. SET* values on both sunny and shaded points are in a relatively comfortable range at 13:00 on the summer solstice day because of the relatively lower air temperature. The difference in MRT values between the sunny and shaded points was about 8 to 12 °C, so the difference in SET* values was also large, which was about 5 to 8 °C at 13:00 on the typical summer day. While the ratio of shaded points was small at Station Plaza and Rooftop Gardens, it was slightly higher in Green Garden. As a result, the number of points with low surface temperature was slightly high, so the averaged MRT and SET* were low, even in sunny points at Green Garden. The difference in wind velocity between Rooftop Gardens, Station Plaza, and Green Garden was approximately less than 1.0 m/s, so the difference in SET* values was slightly lower at sunny points. SET* approached a comfortable range at all sites, especially in shaded points, at 17:00 even on the typical summer day. Although the difference in SET* values between sunny and shaded points was large, the SET* values at any site were in a comfortable range at 13:00 on the summer solstice day.

3.2. Influence of Surface Materials

The relationship between MRT, wind velocity, and SET* on Station Plaza, Rooftop Gardens, and Green Garden at 13:00 on the typical summer day (11 August 2013) is shown in Figure 6. While the difference in SET* values between sunny and shaded places was about 8 °C at Station Plaza, it was only 1 to 2 °C between the concrete and water surfaces. The SET* values on sunny wooden decks in Rooftop Gardens were high because of the high MRT. The differences between sunny and shaded places on wooden decks, concrete, and green cover were 6 to 9 °C. While SET* on the shaded green cover was a little lower than that on shaded concrete, it was almost the same on the sunny green cover and sunny concrete because the difference in MRT was small due to their mixed presence on the slightly narrow Rooftop Gardens. While the difference in SET* values between sunny and shaded places was about 4.5 to 6 °C in Green Garden, it was only 1 to 2.5 °C between the water surface, green cover, and concrete. Summaries of SET* reduction by solar radiation shading, surface material change, and ventilation are shown in Table 5. SET* reduction was the highest by solar radiation shading, followed by surface material change and ventilation.

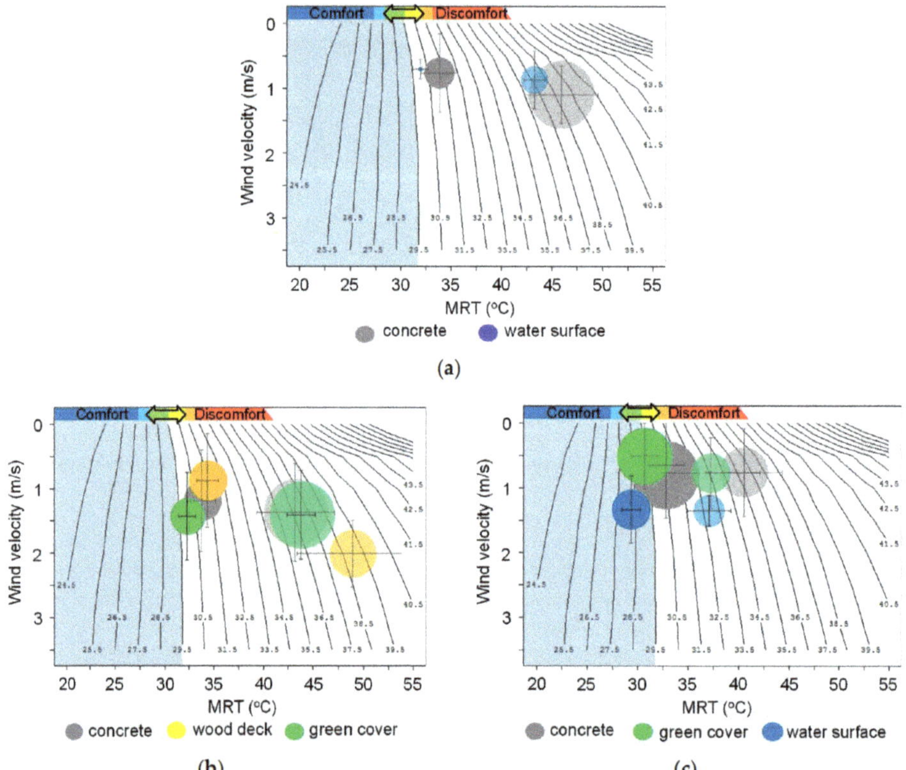

Figure 6. Relationship between MRT, wind velocity, and SET* at 13:00 on the typical summer day (11 August 2013). (**a**) Station Plaza; (**b**) Rooftop Gardens; and (**c**) Green Garden.

Table 5. Summaries of SET* reduction by solar radiation shading, surface material change, and ventilation.

SET* Reduction	Station Plaza	Rooftop Gardens	Green Garden
Solar radiation shading	8.0 °C	6.0 °C	6.0 °C
Surface materials change from concrete	to water surface 2.0 °C (sunny) 1.0 °C (shade)	to green cover 0 °C (sunny) 1.5 °C (shade)	to green cover 2.5 °C (sunny) 1.0 °C (shade) to water surface 2.5 °C (sunny) 1.5 °C (shade)
Ventilation	0.3 °C	0.5–1.5 °C	0 °C

4. Discussion

Changes in shaded areas by buildings, trees, and surface cover to water surface and green cover on Station Plaza, Rooftop Gardens, and Green Garden are shown in Figure 7. The outer circle indicates the shaded area by buildings in blue, the middle circle indicates the shaded area by trees in green, and the inner circle indicates water surface in light blue and green cover in light green. Finally, areas where these were not considered (sunny places) are indicated in red in the inner circle.

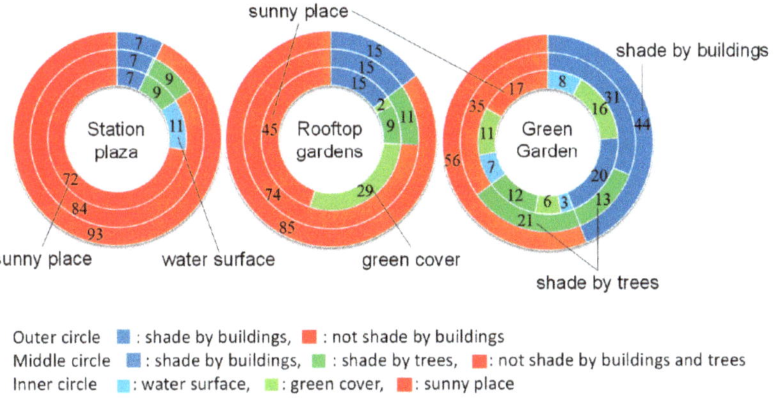

Figure 7. Shaded area by buildings and trees and surface cover change to water surface and green cover on Station Plaza, Rooftop Gardens, and Green Garden.

Shaded areas by trees are required on Station Plaza and Rooftop Gardens because shaded areas by buildings (7% and 15%, respectively) were much smaller than those in Green Garden (44%). As a result, as shown in Figure 4, SET* values at Green Garden were a little mitigated than those in other sites. This is in accordance with the results of previous studies by Ali-Toudert and Mayer [9,10], who showed that shading is the key strategy to mitigating outdoor heat stress under hot summer conditions. However, the shading effect by trees does not completely contribute to the shaded area at Green Garden because shaded areas by trees are located behind buildings (13%) and are included in shaded areas by buildings (44%). As a result, shaded areas by trees were 34% in all, but trees contributed to only 21% appearance of shade in Green Garden. This is consistent with the results of Algeciras et al. [11], who showed that the spatial distribution of thermal conditions at the street level depends strongly on the aspect ratio and street direction. Nevertheless, shaded areas by trees on Station Plaza and Rooftop Gardens (9%, 11%) were smaller than those at Green Garden (21%). Tree growth and tree arrangement have important effects on the thermal environment of the outdoor space, as pointed out by Liang et al. [12] and Zhao et al. [13], respectively. Although previous researchers such as Ali-Toudert and Mayer [10],

Lee et al. [14], and Chen et al. [15] have pointed out that building geometry and vegetation play the most significant role in affecting the thermal comfort index, unfortunately there is an insufficient number of trees at Station Plaza and Rooftop Gardens. Therefore, surface material changes such as water surface and green cover is also required at Station Plaza and Rooftop Gardens. Finally, water surface (11%) was 27% of the area considered for the thermal environment at Station Plaza (7% of building shade, 9% of tree shade, and 11% of water surface), and green cover (29%) was 55% of the area considered for the thermal environment at Rooftop Gardens (15% of building shade, 11% of tree shade, and 29% of green cover). As a result, as shown in Figure 6 and Table 5, the SET* values at Station Plaza and Rooftop Gardens were mitigated by surface material changes. In other words, the effect of the surface material countermeasures shown in Figure 2 was added in this process. Therefore, it is appropriate to consider the thermal environment design of outdoor spaces in the order of shade by buildings, shade by trees, and improvement by surface materials. The largest ratio of the area considered for the thermal environment was 83% at Green Garden, which consists of 44% of building shade, 21% of tree shade, 7% of water surface, and 11% of green cover.

5. Conclusions

Thermal environmental design in outdoor space is discussed by focusing on proper selection and arrangement of buildings, trees, and covering materials via the examination of redevelopment buildings in front of Central Osaka Station, where several heat island countermeasure technologies have been introduced. The effect of MRT on SET* was greater than that of wind velocity at 13:00 and 17:00 on a typical summer day. SET* values on both sunny and shaded points were in a relatively comfortable range at 13:00 on the summer solstice day because of the relatively lower air temperature. SET* reduction was the highest by solar radiation shading (about 6 to 8 °C at 13:00 on the typical summer day), followed by surface material change (about 0 to 2.5 °C at 13:00 on the typical summer day) and ventilation (about 0 to 1.5 °C at 13:00 on the typical summer day). From the analysis of shaded area by buildings and trees and surface cover change to water surface and green cover at Station Plaza, Rooftop Gardens, and Green Garden, the largest ratio of the area considered for the thermal environment was 83 % at Green Garden, which consists of 44% of building shade, 21% of tree shade, 7% of water surface, and 11% of green cover. Areas shaded by trees are required at Station Plaza and Rooftop Gardens because the shaded area by buildings (7% and 15%, respectively) was much smaller than that at Green Garden (44%). Furthermore, because the shaded area by trees at Station Plaza and Rooftop Gardens (9% and 11%, respectively) was smaller than that at Green Garden (21%), surface material changes such as water surface and green cover are also required at Station Plaza and Rooftop Gardens. It is appropriate to consider the thermal environment design of outdoor space in the order of shade by buildings, shading by trees, and improvement of surface materials.

Author Contributions: Conceptualization, H.T.; investigation, H.T.

Funding: This research received no external funding.

Acknowledgments: The author thanks Makiko Kasahara, Shingo Tanabe, and Makoto Kouyama for their cooperation with respect to our analysis.

Conflicts of Interest: The authors declare no conflicts of interest.

References

1. Takebayashi, H.; Kasahara, M.; Tanabe, S.; Kouyama, M. Analysis of Solar Radiation Shading Effects by Trees in the Open Space around Buildings. *Sustainability* **2017**, *9*, 1398. [CrossRef]
2. Takebayashi, H.; Kyogoku, S. Thermal Environmental Design in Outdoor Space Focusing on Radiation Environment Influenced by Ground Cover Material and Solar Shading, through the Examination on the Redevelopment Buildings in Front of Central Osaka Station. *Sustainability* **2018**, *10*, 337. [CrossRef]
3. Grand Front Osaka. Available online: http://umeda-connect.jp/special/vol10/part01.html (accessed on 9 October 2019).

4. Tominaga, Y.; Mochida, A.; Yoshie, R.; Kataoka, H.; Nozu, T.; Yoshikawa, M.; Shirasawa, T. AIJ guidelines for practical applications of CFD to pedestrian wind environment around buildings. *J. Wind Eng. Ind. Aerodyn.* **2008**, *96*, 1749–1761. [CrossRef]

5. Takebayashi, H.; Ishii, E.; Moriyama, M.; Sakaki, A.; Nakajima, S.; Ueda, H. Study to examine the potential for solar energy utilization based on the relationship between urban morphology and solar radiation gain on building rooftops and wall surfaces. *Sol. Energy* **2015**, *119*, 362–369. [CrossRef]

6. Gagge, A.; Fobelets, P.; Bergland, L. A standard predictive index of human response to thermal environment. *Ashrae Trans.* **1986**, *92*, 709–731.

7. Gagge, A.P.; Stolwik, J.A.J.; Nishi, Y. An effective temperature scale based on a simple model of human physiological regulatory response. *Ashrae Trans.* **1971**, *77*, 247–257.

8. Ishii, A.; Katayama, T.; Shiotsuki, Y.; Yoshimizu, H.; Abe, Y. Experimental study on comfort sensation of people in the outdoor environment. *J. Archit. Plan. Environ. Eng.* **1988**, *386*, 28–37.

9. Ali-Toudert, F.; Mayer, H. Numerical study on the effects of aspect ratio and orientation of an urban street canyon on outdoor thermal comfort in hot and dry climate. *Build. Environ.* **2006**, *41*, 94–108. [CrossRef]

10. Ali-Toudert, F.; Mayer, H. Effects of asymmetry, galleries, overhanging façades and vegetation on thermal comfort in urban street canyons. *Sol. Energy* **2007**, *81*, 742–754. [CrossRef]

11. Algeciras, J.A.R.; Consuegra, L.G.; Matzarakis, A. Spatial-temporal study on the effects of urban street configurations on human thermal comfort in the world heritage city of Camagüey-Cuba. *Build. Environ.* **2016**, *101*, 85–101. [CrossRef]

12. Liang, X.; Tian, W.; Li, R.; Niu, Z.; Yang, X.; Meng, X.; Jin, L.; Yan, J. Numerical investigations on outdoor thermal comfort for built environment: Case study of a Northwest campus in China. *Energy Procedia* **2019**, *158*, 6557–6563. [CrossRef]

13. Zhao, Q.; Sailor, D.; Wentz, E. Impact of tree locations and arrangements on outdoor microclimates and human thermal comfort in an urban residential environment. *Urban For. Urban Green.* **2018**, *32*, 81–91. [CrossRef]

14. Lee, H.; Mayer, H.; Chen, L. Contribution of trees and grasslands to the mitigation of human heat stress in a residential district of Freiburg, Southwest Germany. *Landsc. Urban Plan.* **2016**, *148*, 37–50. [CrossRef]

15. Chen, L.; Yu, B.; Yang, F.; Mayer, H. Intra-urban differences of mean radiant temperature in different urban settings in Shanghai and implications for heat stress under heat waves: AGIS-based approach. *Energy Build.* **2016**, *130*, 829–842. [CrossRef]

 *climate*

Article

Modeling and Analysis of Barriers to Climate Change Adaptation in Tehran

Behnam Ghasemzadeh [1] and Ayyoob Sharifi [2,3,4,*

[1]   PhD in Urban Planning, Science and Research Branch, Islamic Azad University, Tehran 19395/1495, Iran;
     behnam.ghasemzadeh@yahoo.com
[2]   Graduate School of Humanities and Social Sciences, Hiroshima University, Higashi-Hiroshima 739–8530, Japan
[3]   Graduate School of Advanced Science and Engineering, Hiroshima University,
     Higashi-Hiroshima 739–8530, Japan
[4]   Network for Education and Research on Peace and Sustainability (NERPS), Hiroshima University,
     Higashi-Hiroshima 739–8530, Japan
*   Correspondence: sharifi@hiroshima-u.ac.jp; Tel.: +81-82-424-6826

Received: 18 August 2020; Accepted: 21 September 2020; Published: 24 September 2020

Abstract: Since the impacts of climate change will last for many years, adaptation to this phenomenon should be prioritized in urban management plans. Although Tehran, the capital of Iran, has been subject to a variety of climate change impacts in recent years, appropriate adaptation measures to address them are yet to be taken. This study primarily aims to categorize the barriers to climate change adaptation in Tehran and analyze the way they interact with each other. The study was done in three steps: first, the focus group discussion (FGD) method was used to identify the barriers; next, the survey and the structural equation modeling (SEM) were used to validate the barriers, identify their importance, and examine their possible inter-relationships; and finally, the interpretive structural modeling (ISM) was applied to categorize and visualize the relationships between the barriers. Results show that barriers related to the 'structure and culture of research', 'laws and regulations', and 'planning' belong to the cluster of independent barriers and are of greater significance. The 'social' barrier and barriers related to 'resources and resource management' are identified as dependent barriers and are of lesser importance. Barriers related to 'governance', 'awareness', 'education and knowledge', 'communication and interaction', and 'economy' are identified at the intermediate cluster. The findings of this study can provide planners and decision makers with invaluable insights as to how to develop strategies for climate change adaptation in Tehran. Despite the scope of the study being confined to Tehran, its implications go far beyond this metropolis.

Keywords: climate change adaptation; barriers; focus group discussion; Tehran; structural equation modeling; urban management

1. Introduction

Hosting more than half of the world population, cities are exposed to the threats and consequences of climate change. While many countries and cities are increasingly developing and implementing mitigation plans, it is argued that, due to historical greenhouse gas emissions, climate change impacts cannot be avoided [1,2]. There is also strong evidence suggesting that climate change will increase intensity and frequency of adverse events that may threaten urban lives and livelihoods [3]. Accordingly, it would not be a stretch to claim that climate change is the most outstanding environmental challenge of this era [4]. Among others, flash floods, storms, hails, tropical storms, rising sea levels, shrinking arctic sea ice, and heat and cold waves are induced and/or intensified by climate change [5]. Accordingly, recognizing the potentially dire consequences of these climatic events, cities around the globe are increasingly taking a more active role in developing plans and policies for climate change adaptation [6,7].

Planning for climate change adaptation is among the most complex challenges confronting many cities around the world [8]. There is a growing consensus that adaptation to climate change impacts is now a necessity for cities [9,10]. Although, understanding this need, many cities have made significant progress in this regard and have developed adaptation strategies, climate change adaptation is yet to be at the top of city governors and planners' agenda, particularly in developing countries. Therefore, there is a need for more research to understand the underlying complexities and barriers to adaptation, provide effective responses to challenges, and identify opportunities for enhanced adaptation [9]. Overall, it can be said that although some progress has been made in finding a common language about climate change adaptation, the path is still fraught with many challenges when it comes to providing frameworks to identify and evaluate the adaptation actions and developing criteria to assess the progress towards adaptation [7,11].

Recognizing the need to better understand the challenges and barriers to adaptation, the literature on this topic has been expanding over the past few years. Existing literature provides useful information related to various barriers linked with governance, institutions, resources, finance, policies, communication, and culture [12]. Barrier in this context refers to any undesirable factor that may undermine efforts aimed at adaptation to climate change impacts. However, the existing literature is mainly focused on identifying barriers without analyzing their interrelations. Due to the space limit, only a brief summary is provided here. Biesbroek, et al. [13] conducted semi-structured interviews with policy makers and scientists in the field of climate change adaptation in different sectors and administrative levels to identify and categorize barriers. The main questions were: "(1) What barriers to adaptation can be identified from the literature? and (2) what do actors in the governance of adaptation experience as the most important barriers to adaptation?" Their study proposed seven clusters of adaptation barriers, of which the most important one was the "conflicting timescales". Other clusters included: "conflicting interests, lack of financial resources, unclear division of tasks and responsibilities, uncertain societal costs and future benefits; and fragmentation within and between scales of governance". The analysis indicated that the respondents' position in the hierarchy affects the understanding of the barriers so that actors from the lower levels of government consider the barriers more serious than those in the higher levels.

Measham, et al. [14] tried to identify the barriers to adaptation in Sydney, Australia, in 2008. The data were gathered through in-depth interviews with 33 participants from three municipal councils of Mosman, Leichhardt, and Sutherland. The identified barriers were related to leadership, competing priorities, planning processes, information constraints, and institutional constraints. Among these, lack of information, lack of resources, and institutional constraints stood out. Lack of resources, in terms of skilled personnel and economic capacities, is also highlighted in a recent study of England's coastal urban areas [15].

Similarly, in the San Francisco Bay Area, California, USA, Ekstrom and Moser [16] performed 43 interviews with key informants, observed public meetings relevant to climate change adaptation, and performed document analysis to identify barriers to climate change adaptation. The identified barriers were related to: "(1) institutions and governance, (2) attitudes, values, and motivations, (3) resources and funding, (4) politics, (5) leadership, (6) adaptation options/processes, (7) understanding, (8) science, (9) expertise, (10) communication, (11) personality issues, and (12) technology".

Examining activities and processes in Norwegian municipalities, Amundsen, et al. [17] identified four key barriers to climate change adaptation at the municipal level, namely: "unfamiliarity with existing data on climate change, lack of concrete data, lack of local expertise for dealing with effects of climate change, and an unclear role for local governments when working with adaptation policies and measures."

Some studies have only highlighted barriers related to the planning process. For instance, Rivera, et al. [18] stated that fragmentation in disaster risk management systems is a barrier to achieving integrated planning for climate change adaptation in Nicaragua. Taking a similar approach, Aylett [19] analyzed the challenges of climate change planning and implementation. The top 10 challenges identified in his study were: "lack of funding for implementation, competing priorities, lack of funding to hire sufficient staff, lack of staff or staff time, difficulty factoring climate change into infrastructure budgeting procedures, political focus on short-term goals, lack of understanding of local government responses, lack of understanding among staff, local government's lack of jurisdiction over key policies areas, and difficulty mainstreaming climate change into existing departmental functions".

In the same vein, Anguelovski, Chu and Carmin [8] analyzed three approaches to climate change adaptation in three cities of Quito, Surat, and Durban and found similar results. They also found that continued and strong leadership commitment and sustained stakeholder involvement are critical for effective climate change adaptation. Finally, analyzing a large geographical scale, Whitney and Ban [20] assessed the social-ecological barriers in coastal British Columbia and identified "Policy action, management understanding, management action, scientific research, policy understanding, and community uptake" as the major barriers.

What can be understood from this overview of the literature is that the barriers and their importance vary across cities and areas. This variation may arise from differences in socio-economic structures, as well as, differences in the nature of climate change threats that they are faced with. This indicates the need for more research to identify locally-specific barriers. Yet, despite such variations, lack of funding and financial resources, inadequate awareness and understanding, and institutional and governance barriers can serve as common themes linking all these studies together. This brief literature review also reveals that most studies have only identified the barriers; however, how all these barriers might interact with one another is far from clear and settled [7]. To fill this gap, by focusing on Tehran, the capital city of Iran, this study aims to contribute to the context-specific knowledge of adaptation barriers and to provide insights into potential interactions between them.

Iran has been subject to several impacts and consequences of climate change in recent years [21]. The capital city, Tehran, has been singled out as one of the most high-risk cities in Iran in terms of environmental challenges [22]. Environmental issues such as climate change and the necessity to scale up effective measures towards it are some of the challenges facing the urban management of Tehran more than ever. By some estimates, the worst is yet to come with Tehran's average temperature projected to rise even further [21]. Tehran is located in a basin area surrounded by mountains, exposing it to acute air pollution and local climate change. Studies also suggest that Tehran is prone to climate change-induced events such as floods, sandstorms, unprecedented heat, and water scarcity [23,24]. Despite this, urban managers are yet to provide appropriate responses to these threats. The fact that there is a lack of research exploring barriers to climate change adaptation and analyzing the interrelationships between the barriers may have contributed to this issue. To our best understanding, there is only one study identifying barriers to climate change adaptation in Tehran [7]. To build on that study, here the barriers are further explored and possible interrelationships between them are also examined. The main objectives of this study are as follows: 1) to discuss the barriers to climate change adaptation in Tehran; 2) to examine the significance of different barriers; 3) to confirm the factor structure of the barriers using confirmatory factor analysis (CFA); 4) to analyze the interrelationships between the barriers using interpretive structural modeling (ISM); 5) to classify the barriers into various categories; and 6) to experimentally test the model using the ISM.

2. The Case Study Area

Tehran is located between 51°5′ and 51°37′ Eastern longitude and 35°32′ and 35°49′ Northern latitude [24]. It is the most populous city in Iran with a population of over 13 million people. Geographically, it touches Alborz mountains in the north and the plains of Shahriar and Varamin in the south [25]. Figure 1 shows the geographical location of Tehran [7].

Tehran also serves as the economic and political hub of the country. Additionally, characteristics such as diverse ethnic and cultural makeup due to massive migrations from all around the country;

the existence of industrial zones, valuable historical buildings, and natural tourism attractions; and problems such as the inequitable access to utilities and opportunities in the city [26], make Tehran an ideal case for studying urban issues.

Tehran was selected as the subject of our study for the following reasons:

- Due to the top-down nature of planning in Iran, Tehran is considered as a model city and therefore its policies are copied [27]. Therefore, the identification and analysis of the barriers in Tehran can also serve as a model for other cities.
- Tehran is subject to both direct and indirect climate threats such as flash floods and extreme heat events. Such threats have caused significant damages over the past few years [28].
- Low resilience of many districts of Tehran has been demonstrated in previous studies [29].

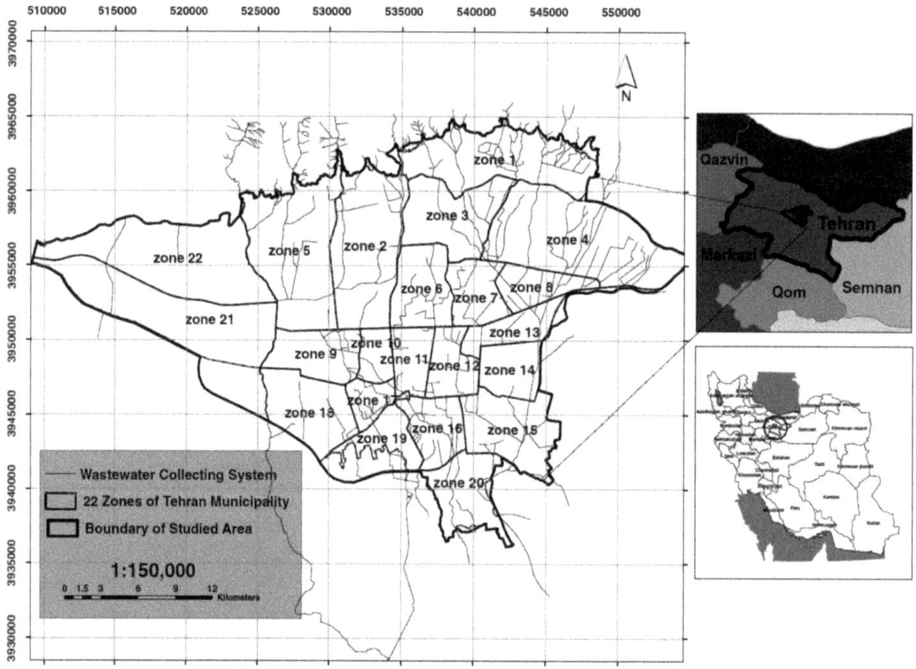

Figure 1. Tehran's geographical position [30].

3. Materials and Methods

This study adopts a mixed-methods approach, including a focus group discussion (FGD) approach [7], a questionnaire-survey, interpretive structural modeling (ISM) approach, and CFA and path analysis approach to accomplish its objectives. An overview of the adopted methods and their correspondence to the research objectives is shown in Figure 2. These methods are separately discussed in the following sections.

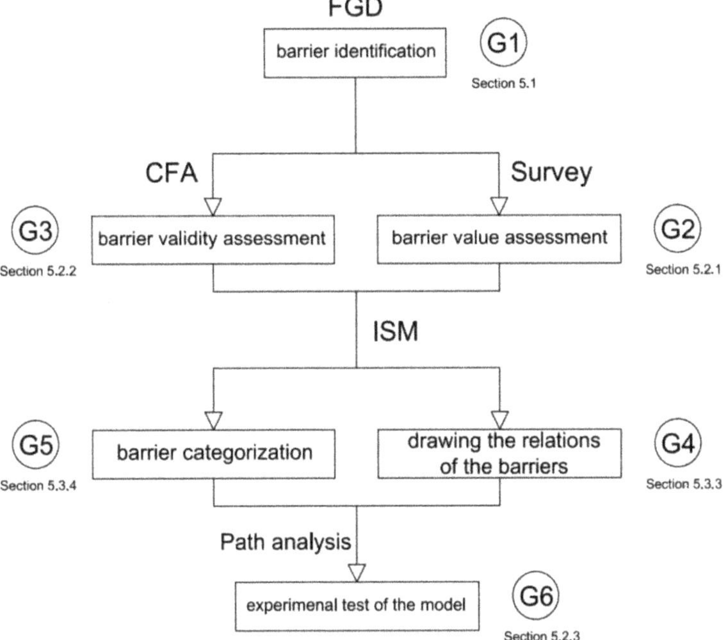

Figure 2. Flowchart of the study (G refers to research goals).

The FGD was used to identify the barriers [7,31]. The survey method was used to assess the importance and validity of the identified variables. To this end, the importance of each barrier was first determined using the one-sample t-test. Their validity and reliability were then assessed using the CFA and Cronbach's alpha [32,33]. The path analysis was also used to examine the causal relationships between the variables [34]. The ISM was applied to determine the hierarchical levels of the variables, to draw the structural model of the relations, and to categorize them into four groups of autonomous, independent, dependent, intermediate [35,36]. The final categorization of the barriers was based on their driving power and their level of dependence.

3.1. Step1: Identification of Barriers (FGD)

Barriers were identified in a qualitative research method that relied on FGDs [7]. Nine FGDs of 4 to 8 people were held where, overall, 59 experts participated in the process (specifically, one four-member FGD, three seven-member FGDs, three six-member FGDs, and two eight-member FGDs). According to the literature, this size range allows distilling diverse opinions from the participants in an effective and efficient manner [31]. The experts represented diverse fields ranging from architecture, urban design, urban economy, urban development, urban sociology, urban environment, and urban planning. They were selected using the purposeful sampling method [37], and on the basis of their age, education and technical, scientific, and executive backgrounds. Participants were selected based on the authors' knowledge about their expertise and the aims of the study. As will be explained in the remainder of this section, sample size was determined based on the state of progress in achieving the aims of the activity. In other words, the activity continued until reaching data saturation. Table S1 of the Supplementary material shows the characteristics of the FGD participants [7].

The FGDs were held at the Urban Development and Architecture Research Center of the Science and Research Branch of Islamic Azad University. The time and place of the FGDs were chosen in consultation with the participants. All FGDs were held in the morning and conducted in Persian as the official language of the country. Each session lasted for about 90 to 120 minutes. At the beginning of each session, after introducing the researchers to the participants, the process of the FGD and the objectives of the study were explained to them. The interview questions were semi-structured and developed based on expert opinions and literature review. The questions were assessed in terms of content validity and approved by a panel of experts with academic degrees in related fields. The discussions began with the general and open question of "What do you know about climate change?" Then, the opinions expressed determined the course of the interview. When discussions seemed to veer off track, some more direct and relevant questions were asked by the facilitator to put them back on track. Some explorative questions, such as "please provide some examples," were also asked to deepen the discussions. Further, some other techniques were also used to elicit the highest possible amount of information from the participants; some examples of such techniques were providing feedback, asking for further clarification and elaboration, and the use of non-verbal language [7].

The questions were asked by the first author as the interviewer and all discussions and non-verbal communications were noted by the two trained facilitators acting as transcribers and observers. Discussions were recorded by a digital voice recorder. The interviewer did his best to ensure that all participants contributed to the discussions and the facilitators noted the discussion process and the non-verbal behaviors such as movements, postures, gestures, silence, tone of speech, emphasis, eye contacts, and changes in facial expression. To further consolidate and substantiate the accuracy of the data, at the end of each FGD session, the above-mentioned key issues were reviewed so that the participants could confirm what they meant and modify the statements, if needed. This respondent validation was performed with the aim of increasing the internal validity of the research. After each FGD session, the recordings were transcribed in the MS Word 2010 program and checked with the field notes made during the FGDs [7].

Data collection continued until reaching data saturation, which is an indicator of the adequacy of sample size in qualitative research. Data saturation is reached when: a) it seems that no new ideas will be identified in subsequent interviews; b) the discussions have been rich enough; c) the relations between the ideas are in place [38,39]. In this research, data saturation was reached after 7 sessions of FGDs, but for good measure, two more sessions were held, which generated no new data.

The steps involved in the data analysis were as follows: debriefing the facilitators and the field notes, listening to the recorded voices and transcribing them, comparing the transcriptions with the notes of the FGD, and considering the non-verbal observations [40].

Data analysis and coding were performed manually, using the inductive content analysis method [41]. This method is based on not separating the data collection from the analysis so that the analysis is performed simultaneously with the data collection. In other words, after conducting the first interview, the transcribed text of the interview was read carefully in order to formulate categories of barriers to climate change adaptation. While reading the first text passage, whenever material that can be considered as a barrier was found, a barrier category was constructed (using a short sentence as a label to represent the barrier category). As the next passage was examined, it was checked whether the barrier that is discussed falls under the previously defined barrier category or should be categorized as a distinct one. This way, a set of barrier categories was developed upon finishing the coding process for the first round of the survey. Following this, the next survey was performed, and similar procedures were taken for coding and barrier categorization. In other words, new barrier categories were only constructed when the identified barriers could not be subsumed to one of the previously formulated barrier categories. This process was continued for each interview until data saturation was reached. In other words, the categorization was finalized when additional interviews did not result in the formulation of new barrier categories.

In the Supplementary material, measures taken to comply with ethical issues, when conducting the FGDs, have been explained.

3.2. Step2: Validation of the Barriers (Survey)

After the identification of the barriers through FGDs, they were validated by a questionnaire-based survey. The purposes of the survey were: (1) identifying the less important barriers, (2) assessing the construct validity of the items through the confirmatory factor analysis, and (3) studying the effects of the barriers on each other. To do so, a self-administered questionnaire was developed which consisted of 31 items related to the 9 main barriers identified by the FGDs. In the first part of the questionnaire, respondents' personal information was recorded, namely their field of study, age, gender, education level, work experience. Descriptive statistics related to the 200 respondents are presented in Table S2 of the Supplementary material. The procedures related to the sample size selection and also the way the survey was administered are explained later in this section.

In the second part, they were asked to make their own judgments on the importance of the barriers on the basis of their experience in climate change adaptation in Tehran's urban management activities. The questions were multiple-choice and had been designed based on a five-point Likert scale, ranging from very high to very low [42]. The initial questionnaire was sent to three municipal experts and three academics of the related domain for their likely revision. The experts' suggestions were incorporated in the modified/final questionnaire to improve the understanding of the question statements.

Once verbal consent was obtained from the respondents, the questionnaires were delivered or sent to them by email. Respondents were selected from the pool of experts who had sufficient experience in research and administrative fields related to climate change adaptation. To the best of the authors' knowledge, there is no consensus about the required sample size for confirmatory factor analysis and structural models. However, when calculating the sample size for confirmatory factor analysis, the number of variables does not matter so much as the number of factors. Therefore, when using the structural equation modeling (SEM), 20 respondents are needed for each factor (latent variable) [33]. Since there were 9 latent variables in this study, the minimum sample size was 180. Overall, 280 questionnaires were distributed, only 200 of which were analyzed because there were missing items in the other 80. The studies that have applied interpretive sequential modeling (ISM) often suggest the use of the SEM for validating the variables and for better understanding the relations between them [32]. In addition, the SEM was used as a complementary method of validation in this study. Reliability of the constructs was evaluated by Cronbach's alpha, where values greater than 0.7 were acceptable [43].

3.3. Step3: Analysis of the Barriers (ISM)

3.3.1. Introducing the ISM

The barriers to climate change adaptation were analyzed by the ISM. This approach enables one to understand complicated issues by transforming them into a multi-level structural model. It also enables decision makers to level the complex relations between the elements of an issue in terms of importance and size. The steps involved in the ISM are [35,36,44]:

1. Listing the barriers;
2. Establishing a contextual relationship between the barriers;
3. Developing a structural self-interaction matrix (SSIM) for the barriers that shows pairwise relation of the barriers;
4. "Framing the reachability matrix from the SSIM and verifying the matrix for transitivity. The transitivity of the contextual relations is a basic assumption for the ISM and means that if variable A is related to B and B to C, variable A is, then, related to variable C" [44];
5. "Partitioning the reachability matrix into various levels" [44];
6. Drawing a directed graph based on the relations achieved in the reachability matrix and removing the transitive links;

7. Transforming the graph drawn in step 7 into an ISM by replacing the barrier nodes with a statement; and finally

8. Examining the ISM framework to check any likely conceptual inconsistencies and to make any necessary modifications [45,46]. Figure 3 shows the steps involved in the ISM.

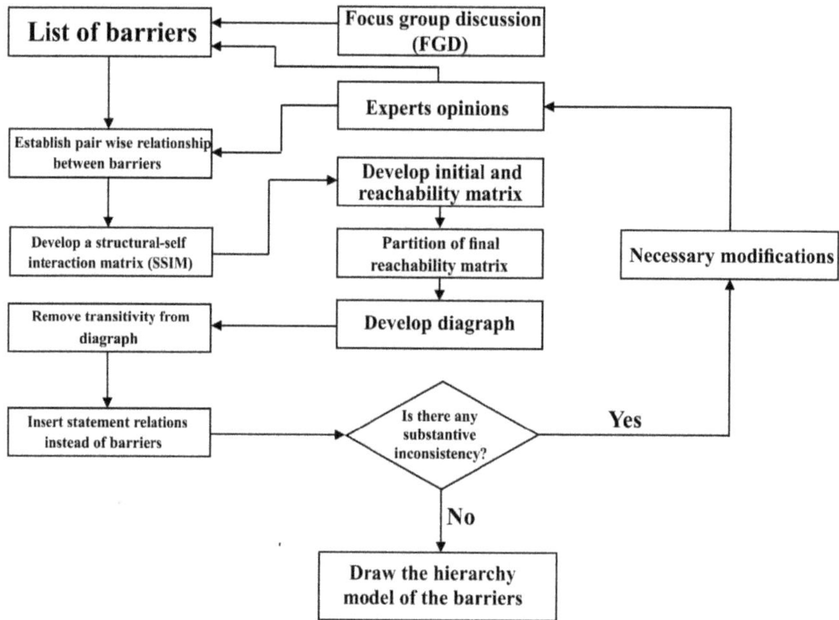

Figure 3. Flowchart for preparing interpretive structural modeling (ISM).

3.3.2. Data Collection in the ISM

Previous studies using ISM have relied on the opinions of 5 to 10 experts [46,47]. In this study, the group of experts included 7 participants and they were asked to state their opinions about the pair matrix comparisons of the impact of variables on each other. In case of disagreement between the experts, the issue was discussed till consensus was reached. Inclusion criteria for selecting the experts were: theoretical mastery, practical experience, ability and willingness to participate, and availability. Characteristics of the experts who participated in the ISM are shown in Table S3 of the Supplementary material.

4. Results

4.1. Findings of FGDs

Nine themes and 31 sub-themes emerged from the FGDs, which are presented in Table 1. The theme 'structure and culture of research' highlights the lack of research institutes that specifically work on climate change adaptation and inform planners and decision makers of the adaptation needs and priorities. This issue is further exacerbated by the limited academic capacity related to developing/implementing climate adaptation plans and/or limited opportunities for climate–policy interactions. As the focus of the second theme indicates, these issues are probably partly rooted in the lack of awareness of climate change and its impacts among planners, policy makers, and citizens, and limited education and communication efforts to enhance their awareness. Overcoming these awareness-related barriers is hampered by the lack of society-wide efforts that can increase citizens' interest in climate action. In particular, limited capacity of non-governmental organizations (NGOs) is a key social barrier. Issues related to limited availability of resources for developing and implementing climate adaptation plans are also critical and constitute a major theme. These include issues such as lack of comprehensive local databases, and difficulties in accessing and processing data.

Table 1. Themes and sub-themes identified through focus group discussions (FGDs) as barriers to climate change adaptation [7].

Row	Themes	Sub-Themes
1	Structure and culture of research (A)	A1: The absence of a center or research institute to support decision making and policy making. A2: The absence of a centralized and specialized mechanism for defining, assessing, and applying studies on climate change and adaptation to it. A3: Failed attempts, unsavory experiences, and the inability of academics in conducting research on climate change adaptation.
2	Awareness, education, and knowledge (B)	B1: Low awareness of climate change and its related strategies. B2: The absence of appropriate programs for continuous advancement of knowledge and awareness. B3: The underperformance of specialized and public media.
3	Social (C)	C1: Citizens' disinterest. C2: The limited number of NGOs. C3: Low responsibility and commitment.
4	Resources and resource management (D)	D1: Grudging governmental institutes in granting access to data. D2: Insufficient national and local data. D3: Logistical challenges both in terms of software and hardware.
5	Laws and regulations (E)	E1: Drastic underrepresentation of climate change and adaptation to it in laws and regulations. E2: Incongruity of local and national plans and regulations. E3: Legal loopholes in regulations on climate change. E4: The lack of supervision of the performance of municipalities.
6	Communication and interaction (F)	F1: The lack of communication with other countries with successful experience. F2: Poor interaction of the related bodies with domestic and foreign experts.
7	Economy (G)	G1: The lack of research funds for climate change adaptation. G2: The low and unstable incomes of municipalities.
8	Governance (H)	H1: Structural characteristics and old-fashioned management procedures. Contradictions in engineering standards and instructions. H2: The lack of integrated urban management. H3: Previous failed attempts. H4: The poor participation and utilization of municipalities in related programs.
9	Planning (I)	I1: The absence of a local independent body within municipalities for policy making. I2: Fragmented approaches toward land use planning. I3: The absence of a domestic model of adaptation. I4: The poor performance of authorities in charge of tackling climate change. I5: The absence of proper mechanisms for the evaluation of urban programs. I6: The absence of an integrated local plan for climate change adaptation.

In terms of laws and regulations, the key barriers are the lack of mechanisms to update laws and regulations and align them with climatic concerns and priorities, lack of compatibility and consistency between local and national regulations, and the weaknesses in terms of enforcement of plans and regulations. The sixth theme highlights the lack of efforts and platforms for communication and experience/knowledge sharing with other countries and organizations with expertise related to climate action planning. Such interactions would be essential for awareness raising and may also provide opportunities for updating planning laws and regulations.

The last three themes are concerned with barriers related to governance, management, and planning. Limited economic capacity of municipalities and their lack of access to financial resources makes it difficult for them to allocate a budget for research and practice related to climate change adaptation. Furthermore, urban planning and governance are suffering from major barriers such as outdated management approaches, dominance of engineering-based approaches that fail to consider socio-economic factors, limited local authority and independence, absence of integrated management approaches that consider interactions between various planning and design activities that may influence the capacity to adapt to climate change impacts, lack of appropriate skills for climate adaptation planning, absence of evaluation and assessment programs, and limited local capacity to develop and implement context-specific adaptation plans.

4.2. Findings of the Survey

As stated above, this step of the study aimed to assess the importance of the variables, their validity, and their effects on each other. The findings are separately presented here.

4.2.1. Barrier Value Assessment

The first goal of the survey was to evaluate the importance of the variables. In doing so, the one-sample t-test was applied, the results of which are presented in Table 2 and Figure S1 of the Supplementary material. Results show that all identified barriers have an importance score of above 3 and the average score of the barriers is significantly higher than the threshold 3 (all significant at $P < 0.01$). Table 2 shows that all variables had a Cronbach's alpha above 0.7, indicating good reliability. Based on the t coefficient and the average score of the variables, the most important variable themes were social ($t = 11.096$, mean = 3.71), communication and interaction ($t = 8.683$, mean = 3.71), and resources and resource management ($t = 7.570$, mean = 3.51). The less important ones, on the other hand, were laws and regulations ($t = 3.179$, mean = 3.24), and the economy ($t = 5.423$, mean = 3.47).

Table 2. Statistics of barriers.

Barriers	Notation	Mean	Questions	Median	Mode	Standard Deviation	Range	Cronbach's Alpha	One Sample t-test Test Value = 3
Structure and culture of research	A	3.3983	1–3	3.33	3	0.91314	4	0.725	6.169
Awareness, education, and knowledge	B	3.4483	4–6	3.67	4	0.91857	4	0.714	6.902
Social	C	3.7117	7–9	4	4	0.907	4	0.72	11.096
Resources and resource management	D	3.51	10–12	3.67	4	0.95277	4	0.73	7.57
Laws and regulations	E	3.245	13–16	3.25	3	1.08986	4	0.785	3.179
Communication and interaction	F	3.7175	17–18	4	5	1.16866	4	0.754	8.683
Economy	G	3.475	19–20	3.5	5	1.23877	4	0.714	5.423
Governance	H	3.509	21–25	3.6	4	1.01463	4	0.814	7.095
Planning	I	3.495	26–31	3.67	5	1.01523	4	0.866	6.895

4.2.2. Validity of the Variables

The second goal of the survey was to assess the validity of the variables. In doing so, the CFA, which is a major component of the SEM, was applied. The SEM is a technique to determine, estimate, and evaluate the linear correlation models among a set of observed variables that are fewer than unobserved ones [34]. The SEM includes endogenous or dependent, and exogenous or independent variables with an observable or non-observable categorization. Since the SEM enables us to make assessments at the structural level, it is more flexible than other statistical methods. On the other hand, it is also referred to as causal modeling because it can assess causal relations [43]. Path analysis and CFA are two common forms of the SEM. Confirmatory factor analysis is used to confirm the structure of a set of observed variables. The SEM is similar to path analysis and makes parameter estimates based on direct and indirect relations between observed variables [34].

In this study, the validity of the variables was assessed in terms of face validity, convergent validity, and discriminant validity as parts of construct validity. Construct validity is the extent to which a set of measured variables reflects the latent construct for which it is designed [48]. Since the identified barriers and their measures are the results of the unanimous consensus of a group of experts with good face validity, it can be said that the measures reflect their latent construct. Convergent validity was calculated through factor loadings and average variance extracted (AVE) (Table 3). All standardized factor loadings were statistically significant ($p < 0.01$). Moreover, the AVE for each construct is above 0.5, which points to the convergent validity of that construct. Discriminant or divergent validity is calculated through comparing the AVE and the correlation coefficients of the corresponding inter-construct squared correlation estimates [34,49].

Table 4 shows the discriminant validity, where the diameter refers to the AVE and the other cell values represent inter-construct squared correlation. To satisfy the discriminant validity criterion, inter-construct squared correlation values should not exceed the AVEs of either the constructs [49]. As seen in Table 4, all inter-construct combinations, except one (i.e., combination of F and G) are in line with this criterion. This points to the discriminant validity of the factors. Figure 4 illustrates the results of the CFA.

Table 3. Results of confirmatory factor analysis.

No.	Barriers	Variables/Items	Standardized Estimate	AVE
1	A	A1	0.773	
		A2	0.643	0.77
		A3	0.711	
2	B	B1	0.876	
		B2	0.678	0.68
		B3	0.455	
3	C	C1	0.526	
		C2	0.854	0.7
		C3	0.67	
4	D	D1	0.671	
		D2	0.733	0.69
		D3	0.677	
5	E	E1	0.596	
		E2	0.684	
		E3	0.727	1.12
		E4	0.766	
6	F	F1	0.771	
		F2	0.744	0.65
7	G	G1	0.707	
		G2	0.764	0.59

Table 3. *Cont.*

No.	Barriers	Variables/Items	Standardized Estimate	AVE
8	H	H1	0.649	
		H2	0.659	
		H3	0.735	1.13
		H4	0.784	
		H5	0.95	
9	I	I1	1	
		I2	1.176	
		I3	1.109	
		I4	1.114	1.69
		I5	0.971	
		I6	1.22	

Table 4. Discriminant validity.

	A	B	C	D	E	F	G	H	I
A	0.77								
B	0.4624	0.68							
C	0.046656	0.042436	0.7						
D	0.710649	0.390625	0.037636	0.69					
E	0.106276	0.061009	0.000144	0.101761	1.12				
F	0.180625	0.111556	0.000169	0.181476	0.872356	0.65			
G	0.178929	0.070225	0.000016	0.190096	0.670761	0.946729	0.59		
H	0.075625	0.030976	0.002916	0.050176	0.725904	0.674041	0.555025	1.13	
I	0.139876	0.058081	0.003844	0.126736	0.719104	0.786769	0.695556	0.923521	1.61

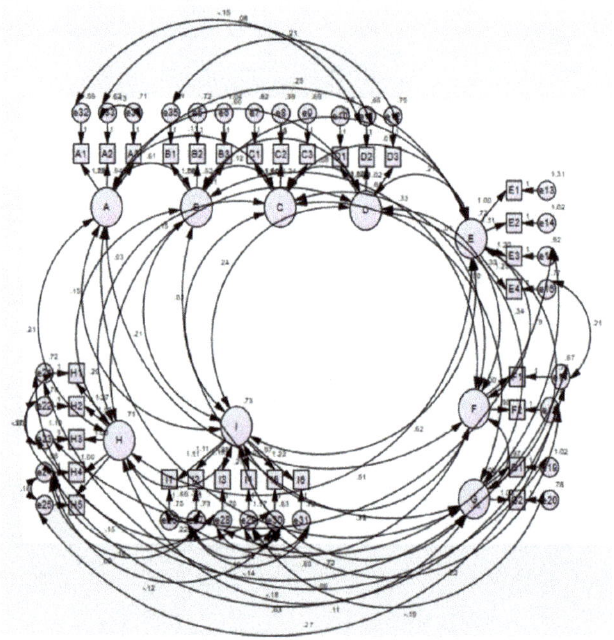

Figure 4. Results of the confirmatory factor analysis (see Table 1 for the codes).

4.2.3. Impact of the Variables on Each Other

The third goal of the survey was to investigate the impacts of the variables on each other. In doing so, the path analysis was used. Results of the path analysis are shown in Table 5. The table shows that the effect of variable A (structure and culture of research) on variable E (laws and regulations) was 0.051, which means as A goes up by 1 standard deviation, E increases by 0.051 standard deviation.

Table 5. Discriminant validity of the barriers to climate change adaptation.

Relations			Estimate	S.E.	C.R.	Standardized Regression Weights
E	<—	A	0.674	0.055	1.1	0.051
H	<—	E	0.961	0.046	14.76	0.723
G	<—	H	1.01	0.092	10.437	0.785
F	<—	H	0.051	0.08	12.635	0.875
D	<—	G	0.071	0.058	0.888	0.068
D	<—	F	0.457	0.062	1.143	0.089
B	<—	D	0.148	0.062	7.437	0.466
C	<—	B	0.059	0.07	2.118	0.149
I	<—	B	0.674	0.048	1.247	0.053

<— Means the causal path from variable A to E.

Other paths, where the barriers affect each other are presented in Table 5 and Figure S2 of the Supplementary material. The effect of variable E (laws and regulations) on H (governance) was 0.723, H on G (economy) was 0.785, H on F (communication and interaction) was 0.875, G on D (resources and its management) was 0.068, F on D was 0.089, D on B (awareness, education, and knowledge) was 0.466, B on C (social) was 0.149, and B on I (planning) was 0.053. The highest values observed was for the effects of governance (H) on communication and interaction (F) and economy (G). This could be explained by the power structure of the country and the top-down nature of the urban governance system that play significant roles in determining the nature of communication and interaction, as well as, the economic capacity of municipalities. The lowest values were obtained for the effect of the structure and culture of research (A) on laws and regulations (E) and the awareness, education, and knowledge (B) on Planning (I). The former is likely an indication of the traditional disconnect between regulations and research activities (lack of research-informed regulations). Therefore, the effect is likely to be limited. Similarly, planning follows a top-down manner and is not informed by knowledge. While raising awareness would be essential for achieving improved planning, these two are currently not directly linked due to the top-down structure of urban governance. Overall, the high effects of urban governance may indicate the importance of making improvements in the governance structure. However, it should be mentioned that further research is needed to better explain the reasons behind these relationships.

The path analysis model indicated an acceptable model fit of chi square ($\chi 2$) = 34.905; degree of freedom (DF) = 17; probability of an exact fit (p) = 0.000; $\chi 2$/DF = 2.053 (<5); comparative fit index (CFI) = 0.98; Tucker–Lewis index (TLI) = 0.957; incremental fit index (IFI) = 0.98; normed-fit index (NFI) = 0.962; relative fit index (RFI) = 0.920; goodness-of-fit index (GFI) = 0.964; root mean square residual (RMR) = 0.106; and root mean square error of approximation (RMSEA) = 0.073. The values of the fit indices indicate a reasonable fit of the path analysis model with the data [50].

Kendall's tau-b two-tailed correlation was applied to the barriers to climate change adaptation in Tehran's urban management to check the possible multi-collinearity. Results show no multi-collinearity between the barriers (see Table S4 of the Supplementary material). Multi-collinearity occurs when two or more independent variables show a high correlation in a multivariate regression. Correlation here means the existence of a linear relation between independent variables. Based on the extent of the correlation, the collinearity will differ. In fact, when there is no multi-collinearity between the variables, we are sure that the observed impacts of the variables on each other are not due to their interactions and correlations.

4.3. Findings of the ISM

4.3.1. Structural Self-Interaction Matrix

To make the self-interaction matrix, the experts were asked to state the relationships among the variables in the form of pairwise comparisons. The agreed relationships among the barriers are presented in Table 6. Four symbols are usually used to express the relationships among the factors in the ISM method as follows:

- "V indicates that factor i directly affects factor j;
- A indicates that factor j directly affects factor i;
- X indicates that factor i and factor j interact with each other; and
- O indicates that factor i has nothing to do with factor j" [51].

Table 6. Structural self-interaction matrix (SSIM).

Planning	Governance	Economy	Communication and Interaction	Laws and Regulations	Resources and Their Management	Social	Awareness, Education, and Knowledge	Structure and Culture of Research
X	V	X	V	O	A	V	X	
X	V	A	X	X	A	A	X	
X		O	X	O	A	A	X	
V			A	A	A	A	V	
V				V	V	V	V	
V					X	A	V	
V						A	V	
X							X	

4.3.2. Final Reachability Matrix

The SSIM was converted into a binary matrix (known as initial reachability matrix) through replacing V, A, X, and O symbols by 1 and 0 digits. The rules are as follows [47]:

- "Put 1 in (i, j) entry and 0 in (j, i) entry of the reachability matrix, if (i, j) entry in SSIM is V;
- Put 0 in (i, j) entry and 1 in (j, i) entry of the reachability matrix, if (i, j) entry in SSIM is A;
- Put 1 in (i, j) entry and 1 in (j, i) entry of the reachability matrix, if (i, j) entry in SSIM is X; and
- Put 0 in (i, j) entry and 0 in (j, i) entry of the reachability matrix, if (i, j) entry in SSIM is O".

Then, the final reachability matrix was developed by introducing the concept of transitivity. The final reachability matrix is shown in Table S5 of the Supplementary material. Driving power in this table refers to the number of other factors (i.e., barriers) that one particular factor affects, whereas dependency refers to the number of other factors affecting one particular factor. According to this table, the highest driving powers belong to "structure and culture of research" and "laws and regulations" with a score of 8 and the lowest belongs to "social" with a score of 3. While more in-depth analysis is needed to explain these results, a likely explanation could be the significance of research for facilitating science-based policy making and the essential role that updated regulations can play in streamlining adaptation in urban planning. The low value for "social" may indicate that, within the context of Tehran, its driving power is comparatively lower and it is mainly dependent and influenced by other factors.

4.3.3. Level Partitions

The reachability set and antecedent set for each factor were obtained from the reachability matrix. The reachability set includes the element and other elements that are all affected by one particular element. The antecedent set includes the element and all elements that affect that particular element. Then, the common elements of these sets are identified. The element that is common in reachability and antecedent sets is located on the first level of the ISM hierarchy. Leveling helps us identify the higher- and lower-importance barriers. Results of level partitioning are shown in Table S6 of the supplementary material. Since the first element in the ISM hierarchy does not affect any other elements above it, it was removed and the leveling continued with the other elements. Finally, the determined levels were used to draw a causal diagram, which is presented in Figures 5 and 6.

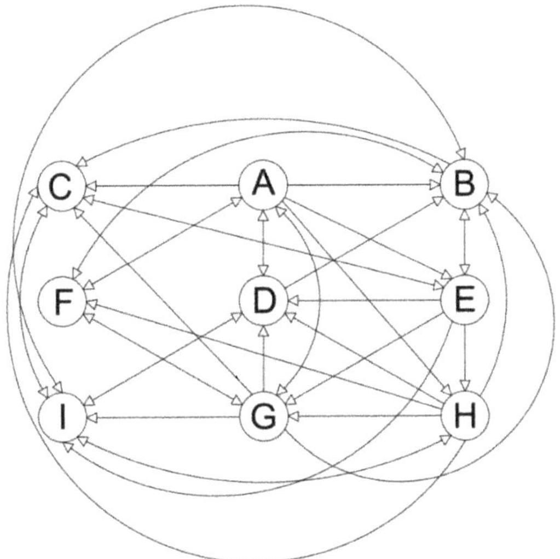

Figure 5. Interrelationships among nine barriers to climate change adaptation in Tehran.

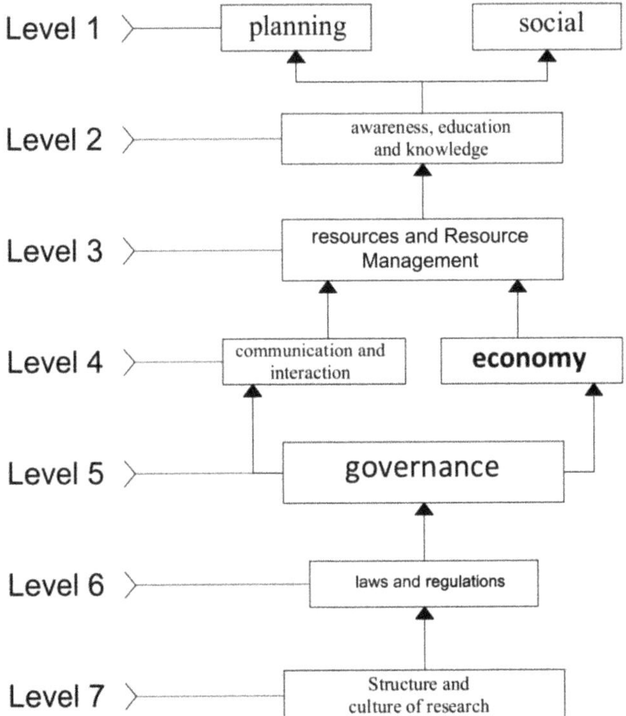

Figure 6. ISM based framework for barriers to climate change adaptation in Tehran's urban management.

The "social" and "planning" barriers are located on the first level. The "awareness, education, and knowledge" barrier is located on the second level, "resources and resource management" on the third level, "communication and interaction" and "economy" on the fourth level, "governance" on the fifth level, "laws and regulations" on the sixth level, and the "structure and culture of research" on the seventh level.

4.3.4. Cross-Impact Matrix Multiplication Applied to Classification (Matrice d'impacts croisés multiplication appliquée á un classment (MICMAC)) Analysis

As mentioned in Section 3, one of the outputs of the ISM is the classification of the barriers into four categories, namely, autonomous, dependent, linkage, and independent. Figure 7 shows this categorization of the barriers. The first category is the autonomous barriers with a poor driving power and dependency. Barriers in this category are almost apart from the system and their relationship with the system is poor and insignificant. The barriers in the dependent category have a poor driving power and strong dependency. The third category includes the linkage barriers with both a strong driving power and strong dependency. These barriers were not steady and any changes in them could affect the other barriers; and in return, the feedback of the effect is seen in the linkage barrier itself. Category four is the independent barriers with a strong driving power but poor dependency. This categorization shows that the "structure and culture of research", "laws and regulations", and "planning" have a high driving power and are located in the category of independent barriers. The "social" and "resources and resource management" with a poor driving power and strong dependency are located in the category of dependent barriers. Further, the barriers of "governance", "awareness, education, and knowledge", "communications and interactions", and "economy" are located in the category of linkage barriers, with both a high driving power and high dependency. No barrier is located in the category of autonomous, which means that all the barriers can be inserted in the causal relations.

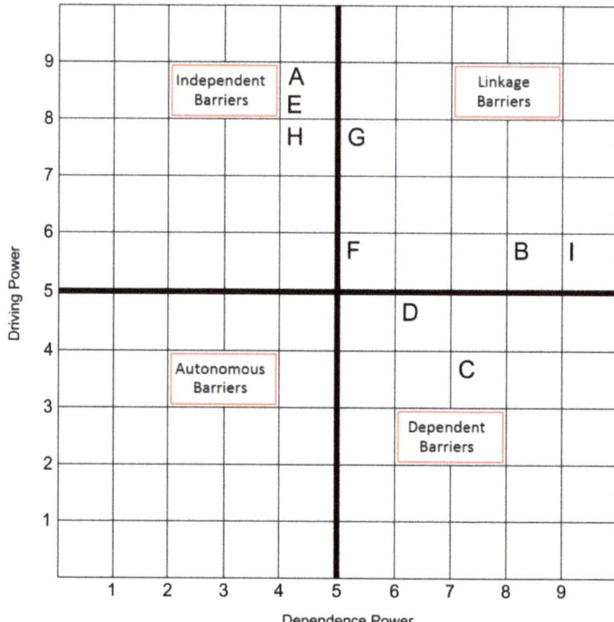

Figure 7. Driving power-dependence diagram for the barriers to climate change adaptation.

5. Discussion

This study primarily aimed to explore different barriers to climate change adaptation in Tehran and analyze their inter-relationships. Cities in general, and metropolitan areas specifically, are complex systems that need to grapple with multiple challenges, only one of which is climate change [52]. To gain improved insights into the significance and magnitude of the problem, one needs to identify and analyze the barriers prior to any decision making or policy considerations as to how to adapt to it.

Among others, the current study had hypothesized that Tehran's urban management is facing some barriers that prevents it from appropriate adaptation to climate change impacts. On the other hand, the absence of a comprehensive study in Iran to investigate the barriers to climate change adaptation and explore their relative importance and inter-relationships is another reason for doing this study. Although some studies have already identified the barriers to climate change adaptation, none has assessed the interactions of the barriers with each other [7,13,14,53]. Sketching out the relationships among the different barriers schematically and classifying them into different categories can enable planners and policy makers to gain a hierarchical insight into the way these diverse barriers might interact. This can be helpful as the intricate and multifarious relationships across the barriers could be hard to manage without a schematic representation.

As mentioned in Table 1, nine themes of 'structure and culture of research'; 'awareness, education, and knowledge'; 'social'; 'resources and resource management'; 'laws and regulations'; 'communication and interaction'; 'economy'; 'governance'; and 'planning' have been identified in a related research as major barriers. Then, drawing on a mixed interpretative-experimental method, the relationships across these barriers and their impact on each other were schematically represented and tested. The barriers of "structure and culture of research", "laws and regulations", and "governance" were identified as key barriers because they had a high driving power to affect the other barriers. These barriers are important in terms of the structures of urban management. In other words, these barriers make up the basis for structural relationships across all barriers and should receive more weight in a causal hierarchy of the barriers. Tehran's authorities should pay greater attention to these three barriers in all policies and plans aimed to adapt to climate change. These barriers are deeply rooted in fundamental weaknesses and in the macro-structure of planning and policy making in the whole country. Therefore, given the universality of the causes of such barriers, the issue should by no means be confined to Tehran. To mitigate the problem, all players in Iran's urban management sector, including municipalities, city councils, and the citizens should take some measures and play a crucial role.

Tehran's city council, as an observer and local legislator, can demand actions from the municipality as well as revise the regulations and develop new ones with incentives and necessary measures to promote adaptation. Creating a working group in the city council for continuous monitoring of the actions can also go a long way. The municipality can further create a multi-disciplinary team to develop policies and executive plans, communicate with educational and research organizations involved in climate change adaptation, benefit from the economic, institutional, and legal capacities of the three institutional powers (i.e., executive, legislative, and judiciary) and the ministries, and identify the conflicting and incongruous regulations to request their amendment.

The barriers of "economy", "communication and interaction", "planning", and "awareness, education, and knowledge" were classified into the 'linkage' category, which means that they have a moderate driving power and moderate dependency. The importance of these barriers lies in the role they play in an interface loop between the independent and dependent barriers. Although the dependent barriers were found to have a poor driving power, the correlation coefficient matrix showed that they had significant positive correlation with some independent and linkage barriers. None of the barriers were classified as autonomous, which means that all the identified barriers have an impact and should not be excluded from the causal relationship structure.

Goodness of fit index of the path analysis model showed that the model had a good fitness according to the experimental data. This means that the interpretive structure of the relationships drawn by the ISM is compatible with the experimental data. The path analysis model also indicated that some barriers have a causal effect on others. For example, the model showed that the "laws and regulation", "governance", "economy", and "resources and resource management" had a significant impact on other barriers.

Finally, the low priority assigned to the problem of climate change is a fundamental weakness of Tehran's urban management sector that can have irreversible physical, social, political, geographical, and ecological consequences. Measures such as integrated urban-regional planning, applying locally-relevant ideas, mainstreaming the policies and actions, and citizen participation can go a long way toward mitigating the problem [6,54].

6. Conclusions

Stressing the importance of looking at the issue of climate change adaptation as a multi-dimensional problem, this study provided new insights into barriers to climate change adaptation. In view of Tehran's social, economic, physical, political, and geographic status, it has always served as a model for other cities in Iran. Therefore, the findings of this study may also be useful for other Iranian cities. The findings may also be applicable to other metropolises outside Iran with similar problems. At least, the methods used in this study can be adopted by researchers and planners in other contexts to better understand the relative importance of climate change adaptation barriers and realize how they may influence each other.

A limitation of this study, however, was the restricted evidence, especially experimental evidence, which caused some difficulties in identifying the barriers. To overcome this problem, the barriers identified using FGDs were used. Another restriction was our limited access to the key informant participants to take part in the survey part of the study, as they held highly-ranked governmental positions with little time to spare. To ensure this limitation would not prove a formidable challenge, the authors first contacted them and obtained their consent and commitment to the study before sending them the questionnaires. Despite this, the researchers sometimes had to lean over backwards to secure a time with them.

In light of the undeniable significance of studying the barriers to climate change adaptation, one line of research that future studies can pursue is to pick where this study has left out by studying this problem in other metropolises around the world, especially those in the Global South. The simultaneous application of quantitative and qualitative methods and drawing on residents' living experiences might also help us better understand this problem and better interpret the findings. Using other methods and approaches such as fuzzy Delphi, Decision Making Trial and Evaluation Laboratory (DEMATEL), neural networks, Shannon entropy, and graph theory along with the ISM and the SEM may enable the researchers to better understand and compare the results. Finally, in view of the numerous stakeholders and actors involved in Tehran's urban management, future studies might want to focus on the stakeholder analysis of adaptation to climate change.

Supplementary Materials: The following are available online at http://www.mdpi.com/2225-1154/8/10/104/s1, Figure S1: Average importance scores of the barriers (see Table 1 for the codes), Figure S2: Path diagram of the SEM for the barriers to climate change adaptation, Table S1: Characteristics of the participants in the FGDs to identify the barriers to climate change adaptation, Table S2: Demographics of the survey respondents, Table S3: Demographics of the participants, Table S4: Correlation coefficients for barriers, Table S5: Final reachability matrix, Table S6: Results of level partitions.

Author Contributions: Conceptualization, B.G. and A.S.; methodology, B.G.; software, B.G.; validation, B.G.; formal analysis, B.G.; investigation, B.G.; resources, B.G.; data curation, B.G.; writing—original draft preparation, B.G. and A.S.; writing—review and editing, B.G. and A.S.; visualization, B.G. All authors have read and agreed to the published version of the manuscript.

Funding: This research received no external funding.

References

1. Bulkeley, H. Navigating climate's human geographies: Exploring the whereabouts of climate politics. *Dialogues Hum. Geogr.* **2019**, *9*, 3–17. [CrossRef]
2. Sharifi, A. Co-benefits and synergies between urban climate change mitigation and adaptation measures: A literature review. *Sci. Total Environ.* **2020**, *750*, 141642. [CrossRef] [PubMed]
3. Paterson, B.; Charles, A. Community-based responses to climate hazards: Typology and global analysis. *Clim. Chang.* **2019**, *152*, 327–343. [CrossRef]
4. Makondo, C.C.; Thomas, D.S. Climate change adaptation: Linking indigenous knowledge with western science for effective adaptation. *Environ. Sci. Policy* **2018**, *88*, 83–91. [CrossRef]
5. Haghtalab, N.; Goodarzi, M.; Habib, N.M.; Yavari, A.R.; Jafari, H.R. Climate modeling in Tehran & Mazandaran provinces by LARSWG and comparing changes in northern and southern central Alborz hillside. *J. Environ. Sci. Technol.* **2013**, *15*, 38–49.
6. Araos, M.; Berrang-Ford, L.; Ford, J.D.; Austin, S.E.; Biesbroek, R.; Lesnikowski, A. Climate change adaptation planning in large cities: A systematic global assessment. *Environ. Sci. Policy* **2016**, *66*, 375–382. [CrossRef]
7. Ghasemzadeh, B.; SaeidehZarabadi, Z.S.; Majedi, H.; Behzadfar, M.; Sharifi, A. Assessment of the impacts of barriers to climate change adaptation in the urban management of tehran: A midex-methods approach. *J. Geogr. Space* **2020**, *81*.
8. Anguelovski, I.; Chu, E.; Carmin, J. Variations in approaches to urban climate adaptation: Experiences and experimentation from the global South. *Glob. Environ. Chang.* **2014**, *27*, 156–167. [CrossRef]
9. Carter, J.G. Climate change adaptation in European cities. *Curr. Opin. Environ. Sustain.* **2011**, *3*, 193–198. [CrossRef]
10. Sharifi, A. Trade-offs and conflicts between urban climate change mitigation and adaptation measures: A literature review. *J. Clean. Prod.* **2020**, *276*, 122813. [CrossRef]
11. Ramesh, A.; Banwet, D.; Shankar, R. Modeling the barriers of supply chain collaboration. *J. Model. Manag.* **2010**, *5*, 176–193.
12. Valente, S.; Veloso-Gomes, F. Coastal climate adaptation in port-cities: Adaptation deficits, barriers, and challenges ahead. *J. Environ. Plan. Manag.* **2020**, *63*, 389–414. [CrossRef]
13. Biesbroek, R.; Klostermann, J.; Termeer, C.; Kabat, P. Barriers to climate change adaptation in the Netherlands. *Clim. Law* **2011**, *2*, 181–199.
14. Measham, T.G.; Preston, B.L.; Smith, T.F.; Brooke, C.; Gorddard, R.; Withycombe, G.; Morrison, C. Adapting to climate change through local municipal planning: Barriers and challenges. *Mitig. Adapt. Strateg. Glob. Chang.* **2011**, *16*, 889–909.
15. Young, D.; Essex, S. Climate change adaptation in the planning of England's coastal urban areas: Priorities, barriers and future prospects. *J. Environ. Plan. Manag.* **2020**, *63*, 912–934. [CrossRef]
16. Ekstrom, J.A.; Moser, S.C. Identifying and overcoming barriers in urban climate adaptation: Case study findings from the San Francisco Bay Area, California, USA. *Urban Clim.* **2014**, *9*, 54–74.
17. Amundsen, H.; Berglund, F.; Westskog, H. Overcoming barriers to climate change adaptation—A question of multilevel governance? *Environ. Plan. C Gov. Policy* **2010**, *28*, 276–289.
18. Rivera, C.; Tehler, H.; Wamsler, C. Fragmentation in disaster risk management systems: A barrier for integrated planning. *Int. J. Disaster Risk Reduct.* **2015**, *14*, 445–456.
19. Aylett, A. Institutionalizing the urban governance of climate change adaptation: Results of an international survey. *Urban Clim.* **2015**, *14*, 4–16. [CrossRef]
20. Whitney, C.K.; Ban, N.C. Barriers and opportunities for social-ecological adaptation to climate change in coastal British Columbia. *Ocean Coast. Manag.* **2019**, *179*, 104808.
21. Daneshvar, M.R.M.; Ebrahimi, M.; Nejadsoleymani, H. An overview of climate change in Iran: Facts and statistics. *Environ. Syst. Res.* **2019**, *8*, 7.
22. Navazi, A.; Navazi, S. Prioritizing climate change risks with fuzzy-AHP method and providing prevention, reduction, and adaption strategies in Tehran Metropolis. *Socio Cult. Strategy J.* **2017**, *5*, 123–142.
23. Ghazal, R.; Ardeshir, A.; Rad, I.Z. Climate change and stormwater management strategies in Tehran. *Procedia Eng.* **2014**, *89*, 780–787. [CrossRef]
24. Delfani, S.; Karami, M.; Pasdarshahri, H. The effects of climate change on energy consumption of cooling systems in Tehran. *Energy Build.* **2010**, *42*, 1952–1957. [CrossRef]

25. Keikhosravi, Q. The effect of heat waves on the intensification of the heat island of Iran's metropolises (Tehran, Mashhad, Tabriz, Ahvaz). *Urban Clim.* **2019**, *28*, 100453. [CrossRef]

26. Aliakbari, E.; Akbari, M. Interpretive-structural modeling of the factors that affect the viability of Tehran Metropolis. *Spat. Plan. (Modares Hum. Sci.)* **2017**, *21*, 1–31.

27. Ziari, K.; Mehrabani, B.F.; Farhadikhah, H. Investigating of reorganization and decentralization strategies of Tehran and offering the optimal pattern. *Town Ctry. Plan.* **2016**, *8*, 1–34.

28. Ahmadi, M.; Motamedvaziri, B.; Ahmadi, H.; Moeini, A.; Zehtabiyan, G.R. Assessment of climate change impact on surface runoff, statistical downscaling and hydrological modeling. *Phys. Chem. Earth Parts A/B/C* **2019**, *114*, 102800.

29. Moghadas, M.; Asadzadeh, A.; Vafeidis, A.; Fekete, A.; Kötter, T. A multi-criteria approach for assessing urban flood resilience in Tehran, Iran. *Int. J. Disaster Risk Reduct.* **2019**, *35*, 101069. [CrossRef]

30. Radmehr, A.; Araghinejad, S. Developing strategies for urban flood management of Tehran city using SMCDM and ANN. *J. Comput. Civ. Eng.* **2014**, *28*, 05014006. [CrossRef]

31. Hennink, M.M. *Focus Group Discussions*; Oxford University Press: Oxford, UK, 2014; 221p.

32. Gupta, A.; Suri, P.K.; Singh, R.K. Analyzing the interaction of barriers in e-governance implementation for effective service quality: Interpretive structural modeling approach. *Bus. Perspect. Res.* **2019**, *7*, 59–75. [CrossRef]

33. Jackson, D.L. Revisiting sample size and number of parameter estimates: Some support for the N: Q hypothesis. *Struct. Equ. Model.* **2003**, *10*, 128–141. [CrossRef]

34. Jain, V.; Raj, T. Modelling and analysis of FMS productivity variables by ISM, SEM and GTMA approach. *Front. Mech. Eng.* **2014**, *9*, 218–232. [CrossRef]

35. Watson, R.H. Interpretive structural modeling—A useful tool for technology assessment? *Technol. Forecast. Soc. Chang.* **1978**, *11*, 165–185. [CrossRef]

36. Janes, F. Interpretive structural modelling: A methodology for structuring complex issues. *Trans. Inst. Meas. Control* **1988**, *10*, 145–154. [CrossRef]

37. Creswell, J.W.; Poth, C.N. *Qualitative Inquiry and Research Design: Choosing Among Five Approaches*; Sage Publications: Singapore, 2016.

38. Palinkas, L.A.; Horwitz, S.M.; Green, C.A.; Wisdom, J.P.; Duan, N.; Hoagwood, K. Purposeful sampling for qualitative data collection and analysis in mixed method implementation research. *Adm. Policy Ment. Health Ment. Health Serv. Res.* **2015**, *42*, 533–544. [CrossRef]

39. Saunders, B.; Sim, J.; Kingstone, T.; Baker, S.; Waterfield, J.; Bartlam, B.; Burroughs, H.; Jinks, C. Saturation in qualitative research: Exploring its conceptualization and operationalization. *Qual. Quant.* **2018**, *52*, 1893–1907. [CrossRef]

40. Sharif, F.; Masoumi, S. A qualitative study of nursing student experiences of clinical practice. *Bmc Nurs.* **2005**, *4*, 6. [CrossRef]

41. Mayring, P. Qualitative Content Analysis: Theoretical Foundation, Basic Procedures and Software Solution, GESIS– Leibniz Institute for the Social Sciences. Available online: https://www.ssoar.info/ssoar/bitstream/handle/document/39517/ssoar-2014-mayring-Qualitative_content_analysis_theoretical_foundation.pdf (accessed on 11 September 2020).

42. Joshi, A.; Kale, S.; Chandel, S.; Pal, D.K. Likert scale: Explored and explained. *Curr. J. Appl. Sci. Technol.* **2015**, *7*, 396–403. [CrossRef]

43. Thirupathi, R.; Vinodh, S. Application of interpretive structural modelling and structural equation modelling for analysis of sustainable manufacturing factors in Indian automotive component sector. *Int. J. Prod. Res.* **2016**, *54*, 6661–6682. [CrossRef]

44. Joshi, K.; Kant, R. Structuring the underlying relations among the enablers of supply chain collaboration. *Int. J. Collab. Enterp.* **2012**, *3*, 38–59. [CrossRef]

45. Sivaprakasam, R.; Selladurai, V.; Sasikumar, P. Implementation of interpretive structural modelling methodology as a strategic decision making tool in a Green Supply Chain Context. *Ann. Oper. Res.* **2015**, *233*, 423–448. [CrossRef]

46. Avinash, A.; Sasikumar, P.; Murugesan, A. Understanding the interaction among the barriers of biodiesel production from waste cooking oil in India-an interpretive structural modeling approach. *Renew. Energy* **2018**, *127*, 678–684. [CrossRef]

47. Attri, R.; Singh, B.; Mehra, S. Analysis of interaction among the barriers to 5S implementation using interpretive structural modeling approach. *Benchmarking Int. J.* **2017**, *24*, 1834–1853. [CrossRef]

48. Morrison, D.F. *Multivariate Statistical Methods*; McGraw-Hill: New York, NY, USA, 1976.

49. Fornell, C.; Larcker, D.F. Evaluating structural equation models with unobservable variables and measurement error. *J. Mark. Res.* **1981**, *18*, 39–50. [CrossRef]

50. Byrne, B.M. Structural equation modeling with AMOS, EQS, and LISREL: Comparative approaches to testing for the factorial validity of a measuring instrument. *Int. J. Test.* **2001**, *1*, 55–86. [CrossRef]

51. Ma, G.; Jia, J.; Ding, J.; Shang, S.; Jiang, S. Interpretive structural model based factor analysis of BIM adoption in Chinese construction organizations. *Sustainability* **2019**, *11*, 1982. [CrossRef]

52. Ruth, M.; Coelho, D. Understanding and managing the complexity of urban systems under climate change. *Clim. Policy* **2007**, *7*, 317–336. [CrossRef]

53. Chanza, N.; Chigona, A.; Nyahuye, A.; Mataera-Chanza, L.; Mundoga, T.; Nondo, N. Diagnosing barriers to climate change adaptation at community level: Reflections from Silobela, Zimbabwe. *GeoJournal* **2019**, *84*, 771–783. [CrossRef]

54. Calthorpe, P. *Urbanism in the Age of Climate Change*; Island Press: Washington, DC, USA, 2015.

 *climate*

Article

Urban Heat Island in Mediterranean Coastal Cities: The Case of Bari (Italy)

Alessandra Martinelli [1], Dionysia-Denia Kolokotsa [2] and Francesco Fiorito [1,3,*] 

[1] Department of Civil, Environmental, Land, Building Engineering and Chemistry, Polytechnic University of Bari, 70125 Bari, Italy; ag.martinelli@libero.it
[2] School of Environmental Engineering, Technical University of Crete, 73100 Chania, Greece; dkolokotsa@enveng.tuc.gr
[3] Faculty of Built Environment, University of New South Wales, Sydney 2052, Australia
* Correspondence: francesco.fiorito@poliba.it; Tel.: +39-080-5963401

Received: 30 April 2020; Accepted: 18 June 2020; Published: 19 June 2020

Abstract: In being aware that some factors (i.e. increasing pollution levels, Urban Heat Island (UHI), extreme climate events) threaten the quality of life in cities, this paper intends to study the Atmospheric UHI phenomenon in Bari, a Mediterranean coastal city in Southern Italy. An experimental investigation at the micro-scale was conducted to study and quantify the UHI effect by considering several spots in the city to understand how the urban and physical characteristics of these areas modify air temperatures and lead to different UHI configurations. Air temperature data provided by fixed weather stations were first compared to assess the UHI distribution and its daily, monthly, seasonal and annual intensity in five years (from 2014 to 2018) to draw local climate information, and then compared with the relevant national standard. The study has shown that urban characteristics are crucial to the way the UHI phenomenon manifests itself. UHI reaches its maximum intensity in summer and during night-time. The areas with higher density (station 2—Local Climate Zone (LCZ) 2) record high values of UHI intensity both during daytime (4.0 °C) and night-time (4.2 °C). Areas with lower density (station 3—LCZ 5) show high values of UHI during daytime (up to 4.8 °C) and lower values of UHI intensity during night-time (up to 2.8 °C). It has also been confirmed that sea breezes—particularly noticeable in the coastal area—can mitigate temperatures and change the configuration of the UHI. Finally, by analysing the frequency distribution of current and future weather scenarios, up to additional 4 °C of increase of urban air temperature is expected, further increasing the current treats to urban liveability.

Keywords: urban heat island; coastal cities; Mediterranean climate; urban heat island intensity; sample year

---

1. Introduction

The climate inside cities is heavily influenced by urbanization that radically changes the landscape by replacing open spaces and vegetation with buildings, roads and other infrastructures.

Materials and characteristics of urban surfaces play a fundamental role in the definition of the local climate, together with anthropogenic heat emissions [1]. Therefore, the climate inside the cities is different from that of rural areas. The very well-known phenomenon of Urban Heat Island (UHI) causes an increase in temperature in dense portions of the cities (downtown, in particular) [2]. UHI phenomenon has a direct consequence on the increase in buildings' energy demand [3–5], on energy poverty [6,7], and on people's health, and on liveability of cities [8–10].

Several factors influence UHI. They include the geographical position of the site and weather conditions such as wind speed and cloudiness [11,12]. Moreover, other factors more directly related to human action, such as the alteration of building fabric and pavement properties [13], urban geometry and density, energy production processes are determinant for the UHI phenomenon [14,15]. Among the uncontrollable variables, wind and cloud cover are two primary weather characteristics that affect UHI development, which generally forms during periods of clear skies and calm winds. Conversely, strong winds and cloud cover hinder the formation of the UHI [16].

The climate and topography of cities, primarily determined by their geographical location, also influence the formation of the UHI. For example, the presence of large bodies of water mitigates temperatures due to the sea breeze phenomenon [17]. In coastal cities this phenomenon overlaps with the UHI one, with consequences on local climate [18–21]. As an example, in the coastal city of Chania (Greece) the UHI reaches its maximum intensity during summer. A strong relation has been demonstrated between UHI and local climate condition. In particular the formation of UHI is strongly influenced by wind speed and direction [22]. Another study conducted in the city of Barcelona (Spain) has shown that in summer, due to the sea breeze effect, the UHI intensity is weaker. As a consequence, the maximum UHI intensity is reached in December [23].

Trees and vegetation also contribute to mitigating temperature increases. The loss of vegetation, together with the presence of impermeable surfaces reduces the evapotranspiration of water, with the consequent increase in surface and air temperatures. Geletič et al. [24] performed a study to analyse the seasonal variability of the Surface Urban Heat Island (SUHI) in three European cities using remote sensing data. The study showed that SUHI differences were more pronounced in summer and that the greatest impact on seasonal variability of SUHI occurred in areas with taller plants.

Urban geometry is another factor that influences UHI development. Urban geometry affects wind flow, energy absorption, and the ability of a given surface to emit long-wave radiation back to space. In urban canyons (i.e., relatively narrow streets lined by tall buildings) [17], during the day, the presence of tall buildings creates shade and thus reduces surface and air temperatures. When the sunlight reaches the surfaces in the canyon, the solar energy is reflected and absorbed by the walls of buildings, creating an increase in temperatures [25]. During the night, buildings and structures can obstruct the heat that is released from urban infrastructures, and urban canyons usually impede cooling. Salvati et al. performed a comparative study of the effects of urban textures on the microclimate in two Mediterranean cities [26]. Five variables (urban morphology, vegetative cover, anthropogenic heat deriving from buildings and traffic and albedo) were considered and related to the variability of the Urban Heat Island Intensity (UHII). The study showed that urban morphology is the variable with the highest impact on UHI phenomenon. A compact urban texture leads to higher UHII because it favours the accumulation of anthropogenic heat in the canopy layer. Moreover, a morphology with higher density causes an increase in temperatures at night, because of the reduction in long-wave radiation losses, due to the lower sky view factor. The findings of Salvati et al. are also confirmed by the results of the study carried out by Paramita et al. in tropical regions [27]. In this case, the additional effect of urban canyons consists of the limitation of wind flows with the consequence increase in both air temperature and relative humidity.

Finally, the microclimate of cities is affected by anthropogenic heat. The intensity of anthropogenic heat emissions varies depending on urban activity and infrastructure [28]. In contrast to what happens in rural areas and in summer, anthropogenic heat can contribute significantly to the development of UHI in winter [29,30].

The present study aims at analysing the urban climate of a coastal Mediterranean city, affected at the same time by the UHI phenomenon and by diurnal sea breeze and nocturnal land-breeze. From this perspective, we have focused the study on the city of Bari (Italy). The analyses presented in this paper are conducted within the Urban Canopy Layer [31], approximatively equivalent to that of the mean height of the main roughness elements (buildings and trees) [32]. Air temperature values recorded in the period between January 2014 and December 2018 at four meteorological stations positioned at ground level (with instruments placed between 2 and 5 m above the ground) are considered. The objective of the study is the assessment of the influence of both phenomena on the modification of the local climate. From this perspective, together with a detailed assessment of extent of the UHI phenomenon in different locations within the city of Bari, a comparison of microclimatic data with the ones included in relevant national standard for the Italian territory [33] is presented. Moreover, an assessment of the impact of UHI in future weather scenarios is presented.

2. Materials and Methods

The study was carried out in the city of Bari (Puglia, Italy 41.12 N, 16.87 E), located on the south-western coast of the Adriatic Sea. The city has a medium density and hosts about 320.000 inhabitants [34]. Bari is a typical Mediterranean city with mild winters and hot summers. The highest temperatures are recorded in August (daily T_{avg} 24.3 °C) while the lowest ones are in January (daily T_{avg} 8.7 °C). On average, there are four days of frost per year and 31 days per year with a maximum temperature equal to or greater than 30 °C. During the period when heat stress is most likely to occur, the highest temperatures are concentrated between 12:00 and 14:00. Average annual rainfall is 563 mm, distributed over 70 days on average, with minimum value in summer, maximum peak in autumn and secondary maximum in winter. November is the wettest month and July the driest. The recorded annual average relative humidity is 71.3% with a minimum of 65% in July and a maximum of 77% in November and December [35].

The analyses presented in this paper have been performed within the Urban Canopy Layer and based on the air temperature data recorded by four fixed meteorological stations located in the metropolitan area. Starting from the hourly temperature data, daily, monthly, seasonal and annual variability of the UHI phenomenon has been studied.

The first step to analyse this phenomenon thus consists of identifying two reference points: an urban and a non-urban (or rural) one. In principle, it would be simple to identify the urban point by making it coincide with the core of a city, for instance. On the contrary, the non-urban point is harder to identify. This occurs because, in the same city, the possible factors that contribute to the increase in temperature are many and differently distributed across the urban area. Thus, although considering the same city, it is possible to identify intermediate areas, more or less urbanized than others or more or less rural than others, which make it possible to better understand how geometric, climatic, environmental variations in the urban structure, in the urban landscape are decisive in affecting temperature variations. Consequently, for this study, the concept of the non-urban area was further extended. A preliminary analysis of the area of Bari and its province was carried out by considering several areas (each one monitored with a standard weather station) to understand how the characteristics of the areas might affect temperatures and, therefore, UHI phenomenon. This analysis is essential to identify the reference weather station, used for calculating the Urban Heat Island Intensity (UHII).

Identification of Locations for Weather Stations and Meteorological Data Collection

In order to assess the UHI phenomenon in the city of Bari, a first screening of all available weather stations was performed. While selecting the weather stations in the area, it was found that many of them were located on the roofs of buildings. This specific position of the sensors makes them—and their recorded data—more sensitive to the sea breeze phenomenon, due to the absence of obstacles [23]. Moreover, since the sensors may be too close to outdoor air conditioning equipment, data recording reliability might be affected. Therefore, we considered only the weather stations positioned at the street level. Moreover, to allow for a more precise and complete analysis, all the meteorological stations that missed more than a month of recorded data during the investigated years were excluded. Therefore, a total of four weather stations located in the metropolitan city of Bari have been considered. Three weather stations are located in the city of Bari (Figure 1a). They are all managed by the regional environmental protection agency (Agenzia Regionale per la Prevenzione e la Protezione dell'Ambiente - ARPA Puglia) [36]. One weather station is located in the city of Valenzano

(Figure 1b), and is managed by the Mediterranean Agronomic Institute (Centre International de Hautes Études Agronomiques Méditerranéennes - CIHEAM) of Bari [37].

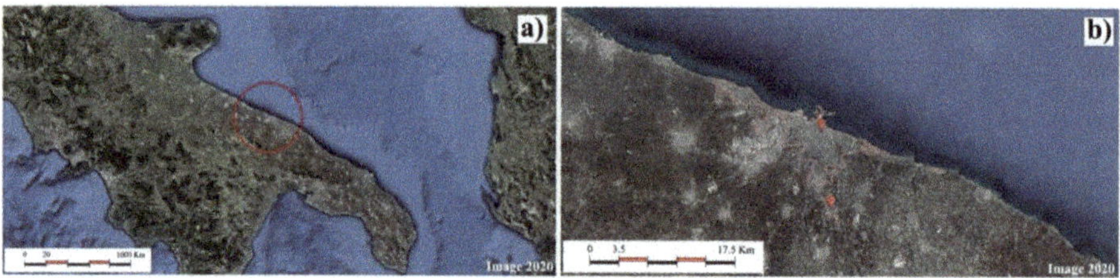

Figure 1. Google Earth image of (**a**) geographic location of Bari and (**b**) Bari and Valenzano areas.

Figure 2 includes the exact location and an aerial image of each site, while Figure 3 shows a detailed image of the weather stations.

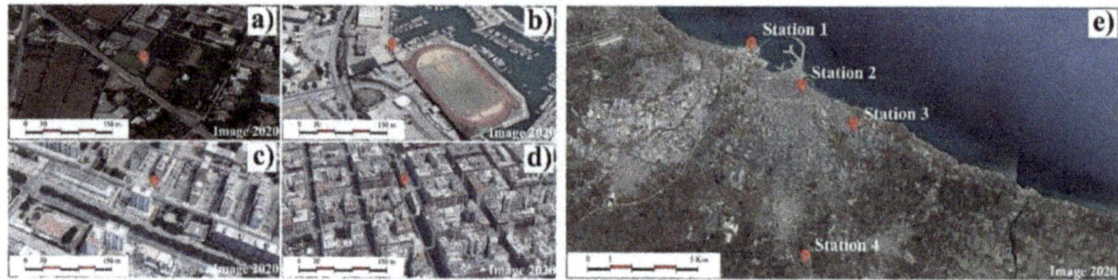

Figure 2. (**a**) Station 4 area, (**b**) station 1 area; (**c**) station 3 area; (**d**) station 2 area; (**e**) locations of the weather stations.

Figure 3. View from the ground of the four weather stations: (**a**) weather station 1 positioned near the University Sport Centre, (**b**) weather station 2 at corso Cavour, (**c**) weather station 3 at via Caldarola, (**d**) weather station 4 at the Mediterranean Agronomic Institute in Valenzano.

Stations 1, 2 and 3 were installed between 2004 and 2012 to monitor the air quality in the area. They detect the presence of pollutants such as CO, C_6H_6, PM10, NO_2, PM2.5 and record climatic data such as air temperature, relative humidity, wind speed and direction, solar radiation, atmospheric pressure and precipitation. Station 4 has been active since 1983 and has been installed for scientific and educational purposes. It records only climatic data, the same as for previous stations.

Weather station 1 is positioned along the promenade Lungomare Starita, in an area next to the sea where the university sport centre is located. It is a non-residential area, not restricted by obstacles such as relatively tall or low buildings and not on direct urban traffic. The values of geometric and surface cover properties (see Table 1) calculated for a squared area (side of 600m) centred on the weather station allow us to classify the zone as Local Climate Zone (LCZ) 9 (sparsely built), according to the local climate zone classification by Stewart and Oke (2012) [38]. Given that the station is positioned near the sea, subclass G—water—is added. Weather station 2 is placed in the city centre, in an area with high population density, and mid-rise residential buildings. The building blocks, very close one to each other, are interspersed with narrow streets, not exceeding 15 m in width. Together they create a very dense square mesh, interrupted by the main street—Corso Cavour—where the weather station is located. Apart from the presence of trees along the main street, the area has no special and significant green spaces. According to the local climate zone classification, the area can be defined as LCZ 2—compact mid-rise. Weather station 3 is located in a predominantly residential area, close to the city centre. Various green spaces are also present. According to the local climate zone classification, the area can be defined as LCZ 5—open mid-rise. Finally, weather station 4 is located 10 km far from the city centre, on agricultural land owned by the CIHEAM. The station is, thus, located in a traffic-free area, free from obstacles and surrounded by sparse buildings not higher than three storeys above ground. According to the local climate zone classification, station 4 belongs to the same climate zone as station 1 (LCZ 9—sparsely built), but with a different subclass (B—scattered trees).

Table 1. Classification of the Local Climatic Zone (LCZ) for each station. In parentheses are limits for each parameter according to Stewart and Oke (2012) [38].

Station nr.	Local Climate Zone	Built Type	Sky-View Factor	Aspect Ratio	Building Surface Fraction	Height of Roughness Elements
1	LCZ 9_G	Sparsely built with water	0.85 (>0.8)	0.28 (0.1–0.25)	12 (10–20)	8 (3–10)
2	LCZ 2	Compact mid-rise	0.45 (0.3–0.6)	1.25 (0.75–2)	50 (40–70)	25 (10–25)
3	LCZ 5	Open mid-rise	0.65 (0.5–0.8)	0.75 (0.3–0.75)	32 (20–40)	15 (10–25)
4	LCZ 9_B	Sparsely built with scattered trees	0.9 (>0.8)	0.16 (0.1–0.25)	19 (10–20)	6 (3–8)

According to the objectives of our study, among all the provided data, it was decided to consider a timeframe of five years for the analyses. From this perspective, five available consecutive years, from January 2014 to December 2018, were considered. During this timeframe, not all data were provided by the weather stations. In these cases, it was decided not to fill the data gaps since any assumption or approximation of data would have meant attributing arbitrary values. Table 2 includes the percentages of missing hourly air temperature data for each control unit and for each year.

Table 2. Percentage of missing data for each station from 2014 to 2018.

	% of Missing Data				
	2014	2015	2016	2017	2018
Station 1	2.8%	2.1%	5.8%	10.1%	0.4%
Station 2	0.0%	2.0%	0.0%	0.0%	0.0%
Station 3	2.2%	2.6%	1.6%	0.0%	1.7%
Station 4	0.0%	0.0%	0.0%	0.0%	0.0%

Moreover, data cleaning and validation was performed. From a first analysis of the data, some values were not consistent or realistic (for example, one order of magnitude higher than the values of the previous or following hour). This required a manual adjustment of the data to reduce the error probability in the subsequent analyses. The performed adjustment was not numerical, as the figures were not modified. We simply moved one or two decimal points to make the data consistent with the previous or subsequent data.

3. Results

3.1. UHI in Bari: Meteorological Data Collection and Comparison

A first analysis involves the identification of the hourly variation patterns of air temperatures in the four stations. By aggregating all data recorded at the four stations during the five years, it has been possible to perform a statistical analysis of hourly temperatures recorded at the four stations, which is summarized in Figure 4. It can be highlighted that station 2 is the one recording the highest average values of air temperatures (18.1 °C), followed by station 1 (17.5 °C) and station 3 (17.2 °C), while station 4 is the one recording the lowest average hourly temperatures (16.4 °C). Station 3 is the one recording the absolute highest temperature (38.8 °C), while station 4 is the one recording the absolute lowest temperature (−3.2 °C). All the three stations placed in Bari have recorded during the five years of measurements hourly temperature values always above 0 °C. Furthermore, from the box plot, it can be recognized that station 3 presents the highest dispersion of values (difference between the first and third quartile of values of 11.3 °C), while station 1 recorded less dispersed values (difference between the first and third quartile of 10.1 °C).

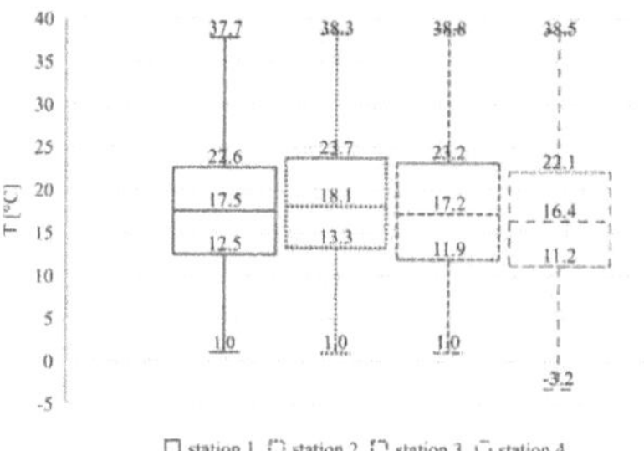

Figure 4. Statistical analysis of hourly temperatures at the four stations: maximum value, minimum value, first quartile, average, third quartile.

Therefore, it was decided that we would deepen the hourly analysis, by plotting the hourly data of air temperature during the hottest (August) and coldest (December) months for the five years of the study, as shown in Figure 5. In August, during daytime, the highest temperatures generally seem to be those recorded by station 4 in 2014 and 2018, and by station 3 in 2015-2016-2017. During night hours, station 2, in all 5 years, is the one with the highest temperatures, followed by stations 3, 1 and 4 in 2014-2017-2018, and stations 1, 3 and 4 in 2015 and 2016.

Analysing the graphs for December, during the daytime the highest temperatures remain those recorded by station 3 in 2014 and 2015. In 2016, the temperatures recorded by station 1 are higher in the first half of the month, those of station 3 in second half of the month. In 2017, station 2 records the highest daytime temperatures, while in 2018 it is station 1. With reference to the night hours, the situation is like that of August. However, in 2018, temperatures recorded at station 1 are higher, even at night.

By the analysis of the data presented in Figure 5, it is interesting to consider the patterns of temperatures recorded at weather station 1 (i.e., the one located near the sea). During the heatwave of August 2018, it can be observed that the temperatures recorded at weather station 1 are rather low compared with the previous years and with the temperatures of the other weather stations for the same year. This happens also when data are compared with the ones obtained from weather station 4, which generally records the lowest temperatures. A maximum temperature at station 1 of 28.3 °C (11 August 2018) is reached, while the temperature recorded at stations 2, 3, and 4 are, respectively, 32.8 °C (13 August 2018), 34.7 °C (11 August 2018), and 33.6 °C (8 August 2018). Moreover, the temperature peaks at station 1 are considerably lower, and the trend of maximum temperatures is more constant. This can be accounted for by the sea breeze phenomenon [21].

Figure 6 summarizes the average, maximum, and minimum monthly temperatures for each location and for the five years considered. In 2014, although with a difference of some degrees in winter (December, January, and February) where higher temperatures are recorded at station 1, in summer (between May and September), averages are roughly similar for the four stations. In the subsequent years, the minimum, maximum, and average monthly temperatures are quite distinguishable, and they rarely overlap. In general, the monthly average temperatures calculated at stations 1 and 4 remain lower than those calculated for values recorded at stations 2 and 3, with differences that may even exceed 2 °C in summer and winter. This is particularly evident in the years 2017 and 2018, and precisely in 2018, when the abnormal decrease in temperature for station 1 is observed and is mainly due to the sea breeze phenomenon.

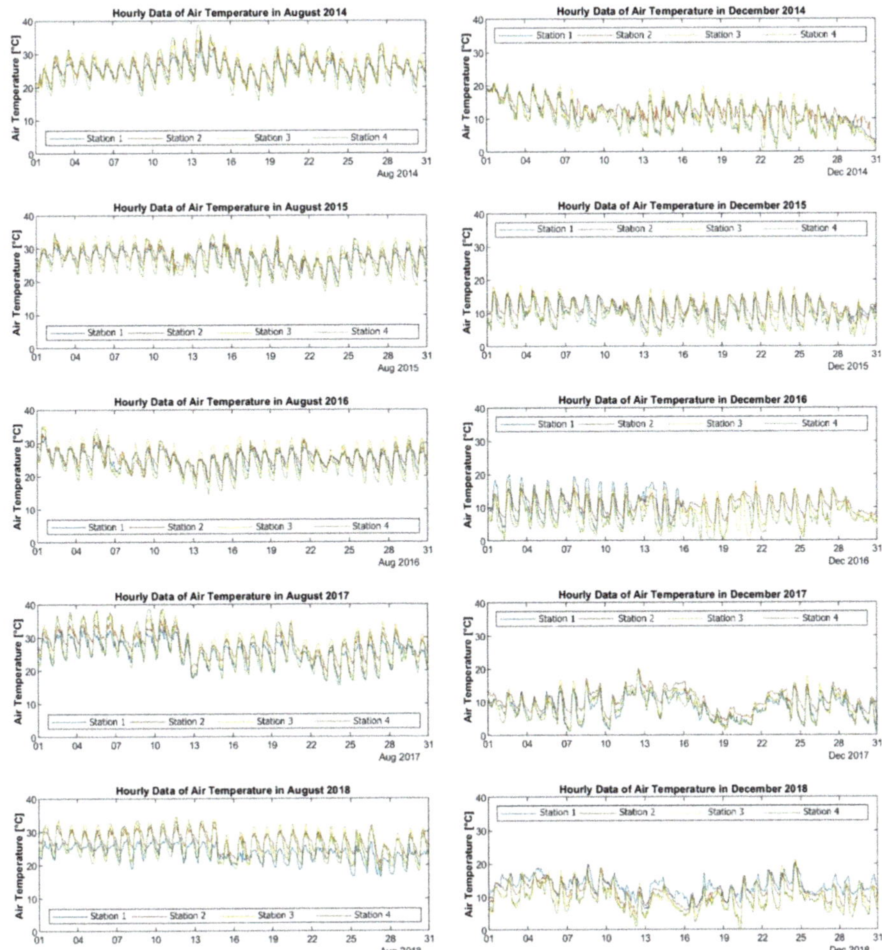

Figure 5. Hourly data of air temperature in August and December from 2014 to 2018. Blue hourly air temperature values for station 1, orange hourly air temperature values for station 2, yellow hourly air temperature values for station 3, green hourly air temperature values for station 4.

Looking at the patterns of maximum monthly temperatures, stations 3 and 4 show the highest values during the months from April to August. During the other months and, in particular, in the coldest ones, the maximum temperatures recorded by all the weather stations have values similar one to each other. As previously described, only in 2014 did the monthly maximum temperature values show a relatively small variation for the four weather stations considered.

As far as minimum temperatures are concerned, station 2 is the one that records the highest values, both in the hot and cold months. It must be noted that the minimum temperature difference between station 2 and the other stations is particularly pronounced in the cold months, from November to March.

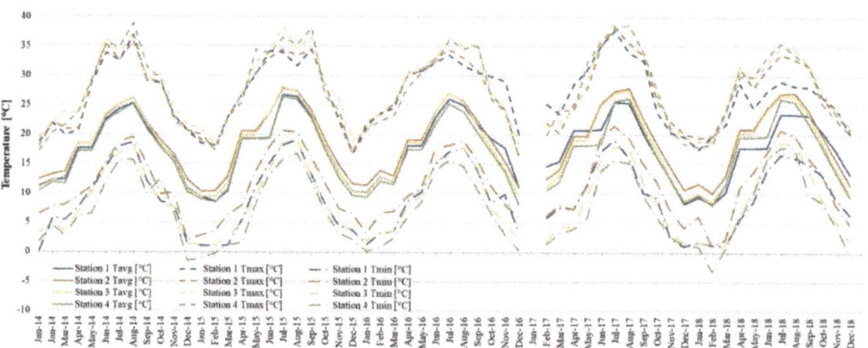

Figure 6. Monthly average temperatures, monthly maximum and minimum temperatures from 2014 to 2018.

3.2. Definition of Reference Station and Determination of Urban Heat Island Intensity

The analysis of temperatures included in the previous paragraph shows that the UHI phenomenon is present in the area considered. Therefore, in this paragraph, the classical indicator to quantitatively describe UHI effect, the Urban Heat Island Intensity (UHII), is analysed. The study illustrated in the previous paragraph was meant to understand and quantify more accurately, on the basis of air temperature data, how important the Heat Island effect is in the considered area in order to define the area where such a phenomenon is less pronounced and so as to identify the reference non-urban area.

The reference station must be selected where there is a low influence of buildings (i.e., in a flat area with high amount of vegetation). Moreover, since the influence of the wind regime is particularly relevant in the definition of the UHI phenomenon [39], for coastal cities, it is suggested that we consider a reference station at a similar distance from the coastline than the urban area under consideration [40]. In previous studies conducted in coastal cities, either airport stations [41,42] or non-urban stations located along the coastline [43–45] were selected. For all of these reasons, we had to exclude station 4 from the assessment of UHII. Therefore, among the three remaining stations, station 1 is the one that exhibits the characteristics of the non-urban area as previously defined, and therefore, has been considered as the reference station.

Consequently, several analyses on the UHII were assessed by considering:

- monthly average temperatures (UHII$_m$) and annual average temperature (UHII$_y$);
- average daily temperatures (UHII$_d$);
- maximum hourly difference of temperature recorded during each day (UHII$_{max}$);
- diurnal (UHII$_{day}$) and nocturnal (UHII$_{night}$) urban heat island intensity;

The following Tables 3 and 4 include the results of the analyses conducted, respectively, on the monthly (and annual average) and daily temperatures. In Table 3, it can be recognized a missing value for January 2017, as during that month the reference station did not record any value. For all the other analyses, any missing data for one weather station (hourly or daily), were not considered in the average also for the other weather stations, as to make the averages more consistent by using the same number of data.

Analysing the data included in Table 4, it can be observed that UHII is generally higher at weather station 2 than at weather station 3. The maximum UHII values based on average yearly data are, respectively, 2 °C and 1 °C for stations 2 and 3 and are both recorded in 2018. On a monthly basis, the highest UHII is also recorded in 2018, with both stations recoding in June monthly average temperature values 6.6 °C higher than the ones recorded at the reference station. From the analysis of the data, it can also be observed that for three out of five years (2015, 2017 and 2018), the maximum monthly UHII is observed in June. For station 2, it is also common to find maximum UHI intensities during cold months (March 2014 and January 2016), while for station 3 in 2014 and 2016, the maximum monthly UHII is found in hot months (August 2014 and July 2016).

Table 3. Values of Urban Heat Island Intensity (UHII) monthly average temperatures (UHII$_m$) and annual average temperature (UHII$_y$) at stations 2 and 3.

Station	Year	Jan	Feb	Mar	Apr	May	Jun	Jul	Aug	Sep	Oct	Nov	Dec	UHII$_y$ [°C]
						UHII$_m$ [°C]								
2	2014	1.2	1.0	1.4	0.8	0.8	0.7	0.7	1.0	0.8	0.8	0.7	1.5	1.0
	2015	0.8	1.9	2.4	1.3	1.3	4.1	1.1	1.0	1.2	0.9	1.2	1.2	1.6
	2016	1.2	1.1	1.0	1.0	1.0	1.0	1.0	0.7	0.7	−0.3	−2.3	−0.1	0.5
	2017	-	−2.1	−1.5	−1	−1	4.3	1.8	2.5	2.5	2.2	2.1	2.3	1.1
	2018	2.3	1.9	2.8	3.2	3.2	6.6	3.5	3.7	0.8	−0.5	−1.6	−1.7	2.0
3	2014	−0.2	0	0.5	0.8	0.8	0.9	0.8	1.0	0.6	0.1	−0.2	−0.2	0.4
	2015	0.0	0.8	1.4	0.7	0.7	4.1	1.4	0.8	0.6	−0.1	0.0	−0.1	0.9
	2016	−0.1	−0.1	0.1	0.5	0.5	0.9	1.1	0.5	−0.2	−1.6	−3.7	−1.5	−0.3
	2017	-	−3.5	−2.6	−1.7	−1.7	4	1.5	2	1.4	1	0.3	0.5	0.1
	2018	0.5	0.4	1.3	2.6	2.6	6.6	3.2	3.1	0.0	−1.8	−3.1	−3.3	1.0

Table 4. Maximum average daily temperatures (UHII$_d$) for stations 2 and 3.

	Year	Date	Reference [°C]	Station [°C]	UHII$_d$ [°C]
Station 2	2014	23/12	8.5	12.9	4.4
	2015	19/5	18.8	23.5	4.7
	2016	13/7	28.4	30.9	2.5
	2017	21/11	8.0	11.4	3.4
	2018	28/3	6.5	12.0	5.5
Station 3	2014	23/05	20.4	22.4	2.0
	2015	19/5	18.8	23.0	4.2
	2016	13/7	28.4	31.2	2.8
	2017	10/8	29.4	32.4	3.0
	2018	02/7	23.2	27.9	4.6

Analysing the results included in Table 4, the comparison the daily temperature differences confirms what is observed from the annual analysis. In fact, by analysing the maximum daily UHII for each year, the intensity calculated at station 2 is higher than the one calculated at station 3.

The highest intensity for both areas is recorded in 2018 and is equal to 5.5 °C for station 2, 1.3 °C higher than at station 3 on the same day (23 March 2018) and 4.6 °C for station 3, slightly higher by 0.3 °C than at station 2 on that same day (2 July 2018). Such a high difference may be correlated with the presence of the heatwave of that year, and with the fact that, as previously explained, the sea breeze effect has caused the temperatures of the reference weather station (station 1) to drop. The calculated average values were accordingly higher. In 2014, the UHII at station 2 is more than double that at station 3 on 3 December 2014, whereas on 23 May 2014, station 2 exhibits higher intensity with a difference of 0.3 °C compared to station 3. In 2015 and in 2016, the highest intensity for both the areas is recorded on the same day: 19 May 2015 and 13 July 2016. In 2015, the intensity calculated at station 2 is 0.5 °C higher than the one calculated at station 3, whereas in 2016, it reverses. This phenomenon may depend on many factors that act individually or simultaneously, whether environmental, geometric or climatic, and even on possible errors in data collection or adjustment.

In general, however, the result that shows that on some days UHII at station 3 is higher than the one at station 2 does not challenge the fact that station 2 is in the area of the city with the highest temperatures. One should consider, indeed, that the days on which this inversion occurs refer to the typically hot months (May, August) where many factors may affect the temperature values and then the calculated averages. However, the days on which UHII is higher station 2 refer to the tendentially cold months (November, December, March), except for 2015 where the day is the same for both areas. Thus, although temperatures may be lower because of the drop in temperature all over the area in the coldest months, the UHII, and accordingly the temperatures at station 2 compared with the rural reference area, are more pronounced. In the previous paragraphs, it was equally stressed that the Atmospheric UHI is more evident in cold months. Consequently, the days where UHII is higher at station 2 are deemed to be more significant, and it is stated this is precisely the area more exposed to the UHI phenomenon.

By analysing hourly values of UHII recorded at stations 2 and 3, it has been possible to identify seasonal trends in the distribution of UHI phenomenon. Figure 7 includes the seasonal distribution of the daily UHII$_{max}$, calculated as the maximum hourly difference of temperatures between the urban station and the reference one recorded during the day. From the analysis of the graphs, it can be highlighted that, although the UHI phenomenon is not strictly seasonally dependant, some differences can be recognized. For station 2, between 40% and 50% of days experience a UHII$_{max}$ between 2 °C and 4 °C. Furthermore, during the summer season (JJA), a significant percentage of days (over 30%) experience a maximum hourly UHI higher than 4 °C. For station 3, there is a more uniform distribution of values of UHII$_{max}$. It can be observed that, in comparison with station 2, station 3 shows lower values of maximum hourly UHI, with 47% to 73% of days during winter (DJF), spring (MAM), and autumn (SON) seasons showing intensities lower than 3 °C. During summer, instead, there is a high prevalence of days with maximum intensities of UHI over 3 °C (90%). In this case, the most frequent values of UHII$_{max}$ are between 4 °C and 5 °C (33% of days).

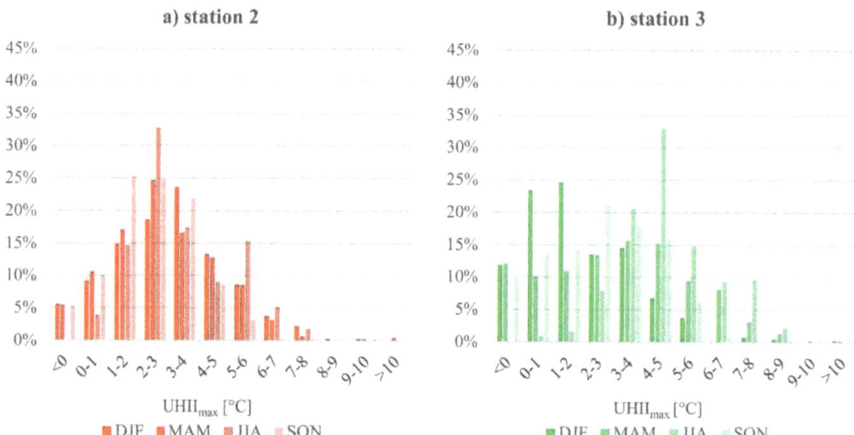

Figure 7. Seasonal distribution of daily values of maximum hourly difference of temperature recorded during each day (UHII$_{max}$) at (**a**) stations 2 and (**b**) 3.

In order to identify the diurnal and nocturnal variation in UHI intensity, a specific analysis was conducted and is summarized in Figure 8. The nocturnal variation in UHII is calculated as the monthly average of the differences between hourly temperatures collected at the weather stations during the night-time period (7 p.m.–7 a.m.), while the diurnal UHII is calculated as the monthly average of the differences between hourly temperatures collected at the weather stations during the daytime period (7 a.m.–7 p.m.). From the analysis of the data, it is also clear that the calculated UHII is higher for station 2 than for station 3. Except for the period between September 2016 and June 2017 and the last three months of 2018, the nocturnal UHII recorded at station 2 is never lower than 0.5 °C and reaches the maximum intensity in the five years (4.2 °C) in June 2018. Similarly, except for the abovementioned months, the diurnal intensity also reaches significant values, albeit lower than the nocturnal one, with a minimum recorded value of 0.2 °C and a maximum of 4.0 °C in August 2018. The average diurnal intensity calculated over the entire 5-year period for station 2 is 0.6 °C, while the nocturnal one is 1.4 °C.

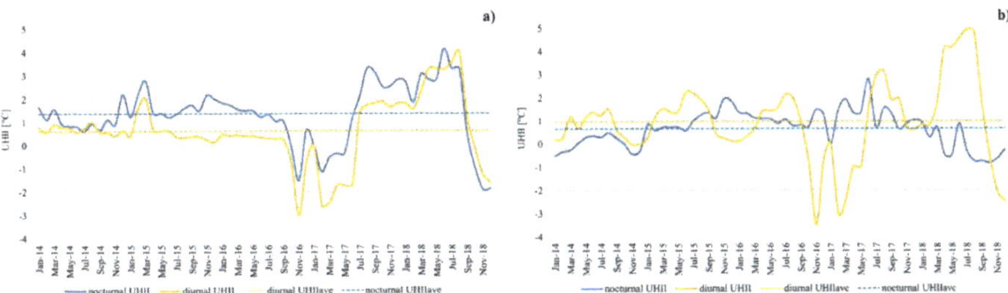

Figure 8. Diurnal and nocturnal UHII from 2014 to 2018, recorded at (**a**) station 2, and (**b**) station 3.

For station 3 (Figure 8b) the nocturnal UHII generally reaches lower values than those recorded at station 2, with a maximum of 2.8 °C in June 2017 and negative values in 2014 and 2018. The diurnal one, instead, reaches a peak of 4.8 °C in July 2018, and negative values lower than those of station 2, between October 2016 and May 2017, and in the last three months of 2018. The average diurnal UHII calculated over the entire five-years period for station 3 is 1.0 °C, while the nocturnal one is 0.7 °C.

To better understand the phenomenon, the summer months (JJA) of 2018, the last full year available, were evaluated more closely, considering the average daily temperature compared with the average monthly temperature indicated by the UNI 10349: 2016 [33] standard, maximum daily temperature and diurnal and nocturnal UHII (Figure 9).

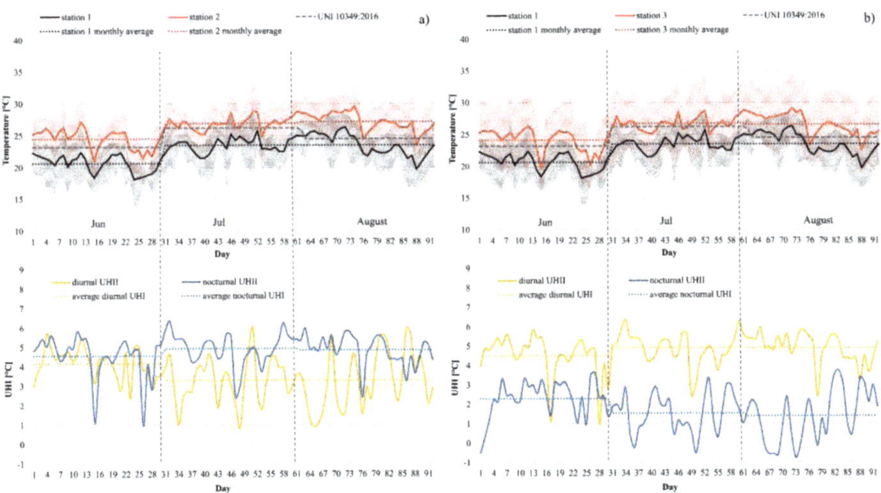

Figure 9. (**a**) Diurnal and nocturnal UHII at station 2 and comparison between stations 1 and 2 during June, July, and August; (**b**) diurnal and nocturnal UHII at station 3 and comparison between stations 1 and 3 during June, July, and August.

In addition to the comparison between the previously presented temperatures at stations 1, 2, and 3, Figure 9 allows us to relate these temperatures to those indicated by the UNI 10349: 2016 [33] standard. From the analysis of the graphs, it can be highlighted that the average temperatures at station 1 are generally lower than those set out in the standard in June, July, and in the second half of August; instead, they are closer to the ones of the standard in the first half of August. Indeed, for station 1, the average monthly temperatures are 2 °C to 3 °C lower than those set out in the standard. The temperatures recorded at station 2 are instead higher during all three months considered, except for some sporadic days and for the month of July. In this case, although temperatures are slightly higher, the average values are very similar to those of the UNI 10349 standard (27 °C recorded as monthly average at station 2 against 26.2 °C, monthly average temperature according to the standard). Overall, it can be observed that the average monthly temperatures of station 2 during the summer period of 2018 have been between 0.8 and 2.7 °C higher than standard. Temperatures recorded at station 3 show a similar trend than those recorded at station 2, but with a lower deviation from the standard. In August 2018 the average monthly temperature recorded at station 2 is 2 °C higher than the standard one, while the difference between real and standard temperatures is 0.9 °C for June and only 0.5 °C for July. The comparison between diurnal and nocturnal UHII at stations 2 and 3, shows significant differences. The nocturnal UHII intensity at station 2 is higher during the night (maximum of 6.4 °C and average of 4.8 °C) than the diurnal one (maximum of 6 °C and average of 3.6 °C). Both intensities show always a positive value, which means that on all the days considered, the temperatures recorded at station 2 are higher than those recorded at station 1. Conversely at station 3, nocturnal UHII (maximum of 3.7 °C and average of 1.7 °C) is generally lower than diurnal one (maximum of 6.4 °C, average of 4.8 °C). For station 3, sometimes the night temperatures are below the ones recorded at station 1 and, therefore, for 10 days the values of nocturnal UHII are negative. The month with higher differences of average diurnal and nocturnal UHII between stations 2 and 3 is August, when the diurnal UHII at station 2 is 1.6 °C lower than the one at station 3 and the nocturnal UHII at station 2 is 3.4 °C higher than the one at station 3.

3.3. Sample Year and Standard Comparison

The UNI 10349-1: 2016 [33] standard provides conventional climatic data for the Italian territory, useful for verifying the energy performance of buildings. The standard includes the data of the monthly average temperatures for selected locations including the city of Bari. In this section we have compared the data from the UNI 10349: 2016 standard, the one from the previous edition of the standard, UNI 10349: 1994, dated more than 20 years before the current one, and those collected during the present study in the city of Bari via fixed weather stations. This comparison was made to assess any matches between the temperatures indicated and forecast by the standard and those recorded in the area through meteorological stations. To carry out this final comparison, a reference year was created. Since the temperatures collected derive from different meteorological stations located in different areas of the city, to obtain a reference year that could be as representative as possible of the entire city and not of only one zone, the standard year was obtained by averaging the monthly temperature values of all three stations and of all the five years considered.

Figure 10a shows that the current edition of the standard indicates lower monthly temperatures than the previous one, with the exception of the summer months, between May and August, where temperatures are around 1.0 °C higher, up to a maximum of 1.5 °C in July. Figure 10b shows the monthly temperatures referring to the sample year created and the typical one suggested by legislation. Between May and August, the temperatures are very similar to each other with differences of a few tenths of a degree; the biggest difference is found in April (5.3 °C), followed by the months of February (3.7 °C) and October (3.2 °C). Instead, the differences in the remaining months are around 2.0 °C.

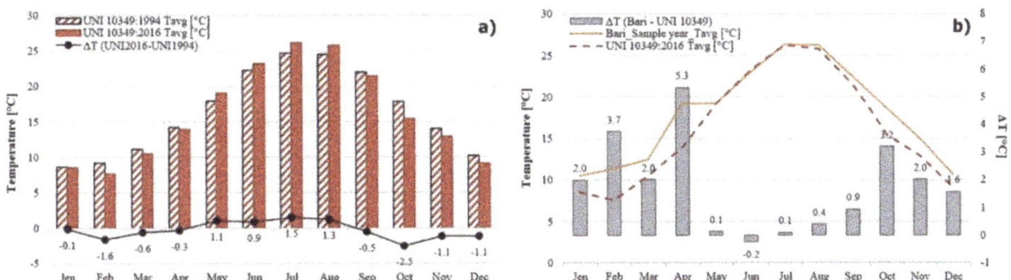

Figure 10. (**a**) Comparison between UNI 10349:1994 and UNI 10349:2016 temperature data; (**b**) Comparison between the sample year and UNI 10349: 2016 temperature data.

3.4. Frequency Distribution of Current Temperature and Future Weather Scenarios

The final aim of the study was to predict the impact of UHI in future scenarios in order to define future threats to urban liveability. Therefore, a statistical analysis of current and future air temperature values was performed.

Figure 11 includes a comparison between the frequency distribution of air temperatures recorded at the four stations and the related frequency distribution of air temperatures of two typical meteorological years. The first typical year (dashed lines in the figure) is obtained from recordings taken at the Airport in the period between 1951 and 1970 and included in the Italian Climatic data collection "Gianni De Giorgio" (IGDG) [46]. The second typical year (pointed line in the figures) is obtained from recordings taken at the Airport in the period between 1984 and 2008 and included in the version 2.0 of the International Weather for Energy Calculation (IWEC2) database managed by the American Society of Heating, Refrigerating and Air-Conditioning Engineers (ASHRAE) [47]. Therefore, both files represent typical historical years. The three shadowed areas in Figure 11 represent the envelope of the frequency distributions of air temperatures recorded in the period 2014–2018 at the three weather stations. The graphs can be analysed considering three temperature ranges: cold (below 10 °C), mild (between 10 °C and 22 °C) and warm (over 22 °C).

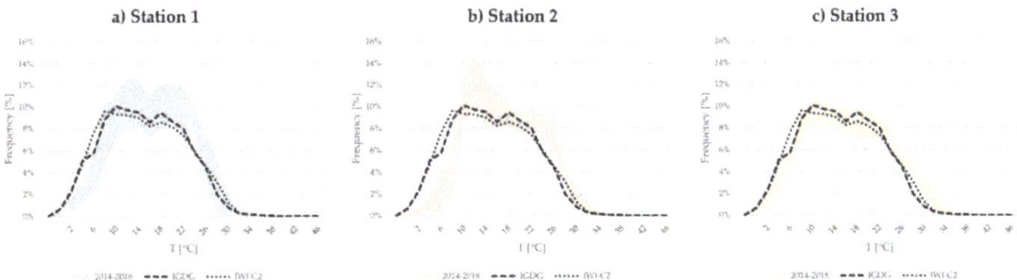

Figure 11. Frequency distribution of air temperature values recorded at stations 1, 2, and 3 and comparison with the frequency distribution of air temperatures of typical meteorological years from IGDG (1951–1970) and IWEC2 (1984–2008) databases: (**a**) frequency distribution of air temperatures at station 1, (**b**) frequency distribution of air temperatures at station 2, (**c**) frequency distribution of air temperatures at station 3.

Station 1 shows a frequency distribution of cold air temperatures with a pattern similar to the one of the two typical years but shifted of about 0 °C–2 °C towards mild temperatures. The frequency distribution of mild temperatures during the period between 2014 and 2018 is in line with the one of the two typical years. The frequency distribution of hot temperatures recorded at station 1 during the period between 2014 and 2018 shows data in line with the typical years, but with a reduced frequency of temperatures over 26 °C. This is due to the sea breeze phenomenon, particularly evident at station 1, which can counterbalance the climate change effects.

Station 2 shows a frequency distribution of cold air temperatures with a pattern similar to the one of the typical years but shifted of about 2 °C to 4 °C towards mild temperatures. Moreover, while the frequency distribution of mild temperatures is in line with the one of typical meteorological years, the frequency distribution of warm temperatures shows a shift of about 0 °C to 4 °C in comparison with the one of typical meteorological years.

Finally, station 3 shows the lower limit of the envelope of the frequency distribution of air temperatures that overlaps with the frequency distribution of air temperatures of typical meteorological years for both cold and warm temperature ranges.

Figure 12 includes an overlap of the frequency distribution of air temperatures recorded at the three weather stations with the frequency distribution of air temperatures in future weather scenarios based on the weather files obtained from the IGDG database. The future weather scenarios have been produced using the Climate Change World Weather Generator tool developed by Jentsch et al. [48] using the statistical downscaling morphing method [49]. Furthermore, the prediction is based on the scenario A2 developed by the Intergovernmental Panel for Climate Change (IPCC) [50].

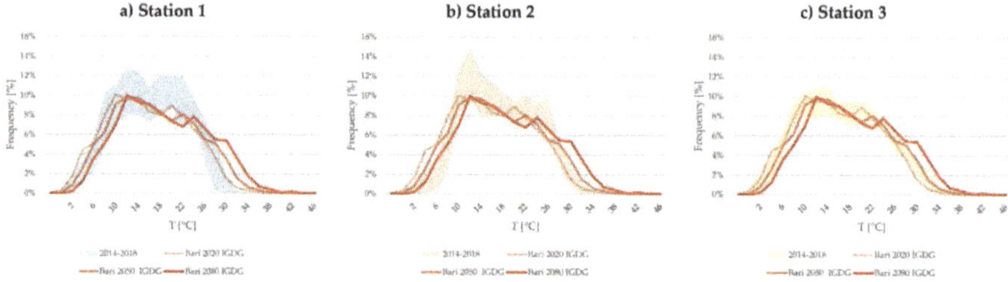

Figure 12. Frequency distribution of air temperature values recorded at station 1, 2, and 3 and comparison with the frequency distribution of air temperatures of future weather scenarios: (**a**) frequency distribution of air temperatures at station 1, (**b**) frequency distribution of air temperatures at station 2, (**c**) frequency distribution of air temperatures at station 3.

As can be observed from the graphs, the future scenario for 2020 (including the average temperatures predicted for the period 2011–2040) fits well with the current distribution of air temperatures for station 1, while it underestimates the temperatures at stations 2 and 3. It can be recognized that for both stations 2 and 3, the frequency distribution of warm air temperatures measured in the period 2014–2018 is shifted of about 0 °C–2 °C towards higher temperatures in comparison with the frequency distribution of air temperatures of the 2020 future scenario. This can be recognized as the average UHI penalty for both stations 2 and 3.

Finally, the frequency distribution of air temperatures for the scenarios of 2050 (average between 2041 and 2070) and 2080 (average between 2071 and 2100) show a further increase in average temperatures of about, respectively, 2 °C and 4 °C for the 2050 and 2080 scenarios in comparison with the 2020 one.

4. Discussion and Conclusions

Among the four selected weather stations, three of them (stations 1, 2, 3) are included in an area with a maximum distance from the coastline of less than 2 km and therefore are affected by a synergic effect of sea and land breeze, and UHI phenomenon. The fourth one (station 4) is, instead, located in a rural area far from the coastline. From the analysis of the hourly data of temperatures collected over the five years of monitoring, it is evident that station 4 shows the lowest values both in the average temperature (16.4 °C) and in the minimum one (−3.2 °C).

Nevertheless, as explained above, station 4 was excluded in the assessment of UHI intensity in Bari, since it is located too far from the coastline. For this reason, and following other studies carried out in coastal cities [44,45,51] and relevant indications from the literature [40,52], the urban weather station located in the proximity of the coastline (station 1) has been selected as the reference station.

By analysing the average monthly temperatures, we found a maximum $UHII_m$ of 6.6 °C for both stations during the summer season (JJA period). This result is in line with what has been reported for Sydney (Australia) by Santamouris et al., who did find a maximum monthly UHII between 3.7 °C and 8.5 °C and concentrated in the summer period (November to February for Sydney).

On average, the $UHII_y$ reaches the maximum value over the five years of 2 °C for station 2 and of 1 °C for station 3.

The maximum hourly values of UHII ($UHII_{max}$) also show a distribution very close to the one found in other studies in coastal Mediterranean cities reported in literature. In our study we did found that, depending on the season, $UHII_{max}$ between 2 °C and 4 °C occur in about 40% to 50% of the days at station 2, and that during the summer season over 30% of days show $UHII_{max}$ higher than 4 °C. Similar values can be found in the study of Giannaros and Melas [42]. The paper reported that in a Mediterranean coastal city (Thessaloniki, Greece), although clear seasonal variations in UHI phenomenon cannot be identified, UHI is more pronounced in the period between June and August, with maximum UHII always higher than 2 °C and sometimes approaching 3 °C.

For station 3, instead, there is a more uniform distribution of values of $UHII_{max}$. This is due to the morphology of the site, which shows a lower density and a higher percentage of greenery in comparison with the urban area monitored with station 2. Therefore, for station 3 lower values of $UHII_{max}$ are recorded, with 47% to 73% of days all seasons except the summer one showing values lower than 3 °C. Moreover, this prediction is in line with the existing literature. As an example, Papanastasiou and Kittas [53] reported for a medium coastal city (Volos, Greece) an occurrence of $UHII_{max}$ between 1 °C and 3 °C for over 90% of the time both in winter and in summer. It must be noted that the city of Volos is about half of the size of Bari in terms of population, but with a density of about 15% of the one in Bari and, therefore, with much lower expected values of UHI. A significant difference can be found, instead, for station 3 during summer period, when the most frequent $UHII_{max}$ is between 4 °C and 5 °C (33% of days).

Finally, from the analysis of the diurnal and nocturnal variation in UHII, it can be observed the difference pattern between stations 2 and 3. The site with higher density and lower percentage of greenery (station 2) shows, during the summer period, nocturnal UHII always higher than the diurnal one, with an average monthly nocturnal UHII between 4.5 °C and 4.9 °C (daily peak of 6.4 °C) and an average monthly diurnal UHII between 3.3 °C and 4.2 °C (daily peak of 6 °C). On the contrary, the site with lower density and higher percentage of greenery (station 3) shows, during the summer period, nocturnal values of UHII much lower than the diurnal ones, with an average monthly nocturnal UHII between 1.4 °C and 2.3 °C (daily peak of 3.7 °C), and an average monthly diurnal UHI between 4.5 °C and 4.9 °C (daily peak of 6.4 °C). Moreover, while at station 2 nocturnal UHII never shows negative values, at station 3, nocturnal UHII on about 10% of days is lower than zero (i.e., station 3 records temperature values during the night-time that are lower than the ones of the reference station placed in proximity to the coastline).

Based on these data and the physical characteristics of the areas considered, some considerations can be made. The choice of weather stations resulted from the need to represent different urban contexts. The recorded and collected temperature data confirm the underlying reason for our choice and they lead to some assumptions that account for the differences in temperatures:

(a) Moving away from weather station 4 to those located in Bari, by merely comparing the images of the areas, it is evident that the presence of the vegetation diminishes: weather station 4, although not being located in an isolated rural area, is surrounded by more vegetation than stations 3 and 2.

(b) Another key feature for the formation of UHI is urban geometry. Geometry influences wind flows, energy absorption, and the ability of a given surface to emit long-wave radiation back to the atmosphere. In built-up areas, where the presence of obstacles prevents the quick release of heat, surfaces, and structures become large thermal masses. Unsurprisingly, urban canyons, as it is the case of station 2, play an essential role. It is thus evident that because of urban geometry, the area in which station 2 is located is the most disadvantaged one and then subject to temperature increase. In contrast, temperatures are lower at station 3, which has a less built-up area density, and even lower at weather station 4, which is the closest representation of a non-urban area.

(c)   Anthropogenic heat contributes to the atmospheric heat islands. Thus, in general, locations with more infrastructures, like station 2, show more anthropogenic heat than those with fewer infrastructures, like stations 1 or 4.

(d)   As explained above, station 1—near the sea—is affected by the sea breeze phenomenon [16]. During the heat wave of August 2018, it recorded lower temperatures than those of the other stations (reaching a maximum temperature of 28.3 °C compared to 30 °C at the others). Moreover, compared to the temperatures recorded by the other meteorological stations, the temperature peaks for station 1 are considerably lower, and the trend of maximum temperatures is more constant. In terms of sea breeze, coastal wind is generated by the differential heating of the land and the water [54,55]. When the temperature over the land is higher than the neighbouring water, the air above it is heated and rises. Then, at lower levels, the air is replaced by cooler air flowing by advection from the adjacent sea areas. Sea breezes regularly influence coastal temperatures [56]. If there is enough moisture in the atmosphere, clouds and precipitation may form. It means that the sea breeze can modify the UHI pattern, and during summer, sea breezes can reduce and delay the Heat Island circulation (HIC). This justifies the drop in temperatures recorded at weather station 1. What we did find is a common phenomenon in coastal cities. As an example, a study carried out on a small Mediterranean town (Chania, Crete) led to a similar result. It has been verified that, in this small coastal town, subjected to rapid urbanization in recent years, when moving from the coastal line to the city centre, comfort conditions become worse [22].

(e)   The distribution of current air temperatures at the three weather stations analysed fits well with the climate change prediction for the period 2011–2040 obtained by statistically downscaling with the IPCC A2 scenario the available data on the typical meteorological year for the period 1951–1970. A further increase of air temperature of between 2 °C to 4 °C is expected to be reached in the period 2071-2100, exacerbating the current treats to urban liveability.

In conclusion, based on this analysis of the variation in temperatures in the area of Bari and its provinces, we can reasonably assert that the phenomenon of Urban Heat Island exists; in particular, it is more pronounced in the areas closer to the city centre and gradually thins out when moving away from the city centre, with specific effects in coastal areas. This difference is seemingly attributed to a number of factors, like the presence/absence of vegetation, urban geometry and, in general, to the level of urbanization that also inevitably causes a different production of anthropogenic heat, all of which actively contribute to the formation of UHI.

From the data collected, both in the first phase of our study concerning the distribution of temperatures across the territory, and the subsequent detailed analysis of the UHI intensity, it has been shown that the busiest and most urbanized areas tend to register higher temperatures, such as the areas where stations 2 and 3 are located. Indeed, if the two weather stations used may be deemed to be representative of the urban and rural areas, the area most exposed to this phenomenon is the one coinciding with the city centre: its geometric, environmental, climatic, and geographical characteristics are such that temperatures tend to remain higher during the day.

Finally, it has also been shown that the sea breeze phenomenon, particularly visible in the coastal area represented by the reference station (station 1), can mitigate temperatures and change the configuration of the phenomenon.

Author Contributions: Conceptualization, A.M., D.-D.K. and F.F.; methodology, A.M., D.-D.K. and F.F.; validation, A.M., D.-D.K. and F.F.; formal analysis, A.M. and F.F.; investigation, A.M.; writing—original draft preparation, A.M. and F.F.; writing—review and editing, D.-D.K.; supervision, F.F. and D.-D.K. All authors have read and agreed to the published version of the manuscript.

Funding: This research received no external funding.

Acknowledgments: The authors would like to acknowledge the Regional Environmental Protection Agency of Puglia (ARPA Puglia) and the Mediterranean Agronomic Institute (CIHEAM) for providing them with the environmental data required for the study.

Conflicts of Interest: The authors declare no conflict of interest.

References

1. Grimmond, C.S.B.; Ward, H.C.; Kotthaus, S. Effects of Urbanization on local and regional climate. In *The Routledge Handbook of Urbanization and Global Environmental Change*; Seto, K.C., Solecki, W.D., Griffith, C.A., Eds.; Routledge: London, UK, 2015. [CrossRef]

2. Boufidou, E.; Commandeur, T.J.F.; Nedkov, S.B.; Zlatanova, S. Mesure the climate, model the city. *Int. Arch. Photogramm. Remote Sens. Spatial Inf. Sci.* **2011**, *XXXVIII-4/C21*, 59–66. [CrossRef]

3. Santamouris, M.; Cartalis, C.; Synnefa, A.; Kolokotsa, D. On the impact of urban heat island and global warming on the power demand and electricity consumption of buildings—A review. *Energy Build.* **2015**, *98*, 119–124. [CrossRef]

4. Santamouris, M. On the energy impact of urban heat island and global warming on buildings. *Energy Build.* **2014**, *82*, 100–113. [CrossRef]

5. Santamouris, M.; Papanikolaou, N.; Livada, I.; Koronakis, I.; Georgakis, C.; Argiriou, A.; Assimakopoulos, D.N. On the impact of urban climate on the energy consuption of building. *Sol. Energy* **2001**, *70*, 201–216. [CrossRef]

6. Santamouris, M. Innovating to zero the building sector in Europe: Minimising the energy consumption, eradication of the energy poverty and mitigating the local climate change. *Sol. Energy* **2016**, *128*, 61–94. [CrossRef]

7. Sanchez-Guevara, C.; Núñez Peiró, M.; Taylor, J.; Mavrogianni, A.; Neila González, J. Assessing population vulnerability towards summer energy poverty: Case studies of Madrid and London. *Energy Build.* **2019**, *190*, 132–143. [CrossRef]

8. Santamouris, M. Recent progress on urban overheating and heat island research. integrated assessment of the energy, environmental, vulnerability and health impact synergies with the global climate change. *Energy Build.* **2019**, 109482. [CrossRef]

9. Hajat, S.; Armstrong, B.; Baccini, M.; Biggeri, A.; Bisanti, L.; Russo, A.; Paldy, A.; Menne, B.; Kosatsky, T. Impact of high temperatures on mortality: Is there an added heat wave effect? *Epidemiology* **2006**, *17*, 632–638. [CrossRef] [PubMed]

10. Baccini, M.; Biggeri, A.; Accetta, G.; Kosatsky, T.; Katsouyanni, K.; Analitis, A.; Anderson, H.R.; Bisanti, L.; D'Iippoliti, D.; Danova, J.; et al. Heat effects on mortality in 15 European cities. *Epidemiology* **2008**, *19*, 711–719. [CrossRef] [PubMed]

11. Oke, T.R. Canyon geometry and the nocturnal urban heat island: Comparison of scale model and field observations. *J. Climatol.* **1981**, *1*, 237–254. [CrossRef]

12. Oke, T.R. The energetic basis of the urban heat island. *Q. J. R. Meteorol. Soc.* **1982**, *108*, 1–24. [CrossRef]

13. Doulos, L.; Santamouris, M.; Livada, I. Passive cooling of outdoor urban spaces. The role of materials. *Sol. Energy* **2004**, *77*, 231–249. [CrossRef]

14. Rizwan, A.M.; Dennis, L.Y.C.; Liu, C. A review on the generation, determination and mitigation of Urban Heat Island. *J. Environ. Sci.* **2008**, *20*, 120–128. [CrossRef]

15. Taha, H. Urban climates and heat islands: Albedo, evapotranspiration, and anthropogenic heat. *Energy Build.* **1997**, *25*, 99–103. [CrossRef]

16. Vardoulakis, E.; Karamanis, D.; Fotiadi, A.; Mihalakakou, G. The urban heat island effect in a small Mediterranean city of high summer temperatures and cooling energy demands. *Sol. Energy* **2013**, *94*, 128–144. [CrossRef]

17. U.S. Environmental Protection Agency. Reducing Urban Heat Islands: Compendium of Strategies. Draft. 2008. Available online: https://www.epa.gov/heat-islands/heat-island-compendium (accessed on 21 April 2020).

18. Zhou, X.; Okaze, T.; Ren, C.; Cai, M.; Ishida, Y.; Watanabe, H.; Mochida, A. Evaluation of urban heat islands using local climate zones and the influence of sea-land breeze. *Sustain. Cities Soc.* **2020**, *55*. [CrossRef]

19. Bauer, T.J. Interaction of urban heat island effects and land–sea breezes during a new york city heat event. *J. Appl. Meteorol. Climatol.* **2020**, *59*, 477–495. [CrossRef]

20. Papanastasiou, D.K.; Melas, D.; Bartzanas, T.; Kittas, C. Temperature, comfort and pollution levels during heat waves and the role of sea breeze. *Int. J. Biometeorol.* **2010**, *54*, 307–317. [CrossRef]

21. Freitas, E.D.; Rozoff, C.M.; Cotton, W.R.; Silva Dias, P.L. Interactions of an urban heat island and sea-breeze circulations during winter over the metropolitan area of São Paulo, Brazil. *Bound.-Layer Meteorol.* **2007**, *122*, 43–65. [CrossRef]

22. Kolokotsa, D.; Psomas, A.; Karapidakis, E. Urban heat island in southern Europe: The case study of Hania, Crete. *Sol. Energy* **2009**, *83*, 1871–1883. [CrossRef]

23. Salvati, A.; Coch Roura, H.; Cecere, C. Assessing the urban heat island and its energy impact on residential buildings in Mediterranean climate: Barcelona case study. *Energy Build.* **2017**, *146*, 38–54. [CrossRef]

24. Geletič, J.; Lehnert, M.; Savić, S.; Milošević, D. Inter-/intra-zonal seasonal variability of the surface urban heat island based on local climate zones in three central European cities. *Build. Environ.* **2019**, *156*, 21–32. [CrossRef]

25. Sailor, D.J.; Fan, H. Modeling the diurnal variability of effective albedo for cities. *Atmos. Environ.* **2002**, *36*, 713–725. [CrossRef]

26. Salvati, A.; Monti, P.; Coch Roura, H.; Cecere, C. Climatic performance of urban textures: Analysis tools for a Mediterranean urban context. *Energy Build.* **2019**, *185*, 162–179. [CrossRef]

27. Paramita, B.; Matzarakis, A. Urban morphology aspects on microclimate in a hot and humid climate. *Geogr. Pannonica* **2019**, *23*, 398–410. [CrossRef]

28. Voogt, J.A. Urban Heat Island. In *Encyclopedia of Global Environmental Change*; Munn, T., Ed.; Wiley: Chichester, UK, 2002; Volume 3, pp. 660–666.

29. Offerle, B.; Grimmond, C.S.B.; Fortuniak, K.; Kłysik, K.; Oke, T.R. Temporal variations in heat fluxes over a central European city centre. *Theor. Appl. Climatol.* **2006**, *84*, 103–115. [CrossRef]

30. Hinkel, K.M.; Nelson, F.E. Anthropogenic heat island at Barrow, Alaska, during winter: 2001–2005. *J. Geophys. Res. Atmos.* **2007**, *112*. [CrossRef]

31. Oke, T.R. Street design and urban canopy layer climate. *Build. Environ.* **1988**, *11*, 103–113. [CrossRef]

32. Oke, T.R. The distinction between canopy and boundary-layer urban heat Islands. *Atmosphere* **1976**, *14*, 268–277. [CrossRef]

33. UNI. 10349-1: 2016. Riscaldamento e Raffrescamento Degli Edifici—Dati Climatici. Parte 1: Medie Mensili per la Valutazione Della Prestazione Termo-Energetica Dell'edificio e Metodi per Ripartire L'irradianza Solare Nella Frazione Diretta e Diffusa e per Calcolare L'irradianza Solare su di una Superficie Inclinata. 2016. Available online: https://nt24.it/2016/02/riscaldamento-e-raffrescamento-degli-edifici-dati-climatici-in-inchiesta-pubblica/ (accessed on 21 April 2020).

34. Italian National Institute of Statistics. Statistical Data. Available online: https://www.istat.it/ (accessed on 6 April 2020).

35. World Meteorological Organization. Available online: https://worldweather.wmo.int/en/home.html (accessed on 29 May 2020).

36. ARPA Puglia. Official Website. Available online: www.arpa.puglia.it (accessed on 21 April 2020).

37. CIHEAM. Official Website. Available online: https://www.iamb.it (accessed on 21 April 2020).

38. Stewart, I.D.; Oke, T.R. Local climate zones for urban temperature studies. *Bull. Am. Meteorol. Soc.* **2012**, *93*, 1879–1900. [CrossRef]

39. Giridharan, R.; Kolokotroni, M. Urban heat island characteristics in London during winter. *Sol. Energy* **2009**, *83*, 1668–1682. [CrossRef]

40. Sakakibara, Y.; Owa, K. Urban–Rural temperature differences in coastal cities: Influence of rural sites. *Int. J. Climatol.* **2005**, *25*, 811–820. [CrossRef]

41. Acero, J.A.; Arrizabalaga, J.; Kupski, S.; Katzschner, L. Urban heat island in a coastal urban area in northern Spain. *Theor. Appl. Climatol.* **2013**, *113*, 137–154. [CrossRef]

42. Giannaros, T.M.; Melas, D. Study of the urban heat island in a coastal Mediterranean City: The case study of Thessaloniki, Greece. *Atmos. Res.* **2012**, *118*, 103–120. [CrossRef]

43. Anjos, M.; Lopes, A. Urban Heat Island and Park Cool Island intensities in the coastal city of Aracaju, North-Eastern Brazil. *Sustainability* **2017**, *9*, 1379. [CrossRef]

44. Santamouris, M.; Haddad, S.; Fiorito, F.; Osmond, P.; Ding, L.; Prasad, D.; Zhai, X.; Wang, R. Urban heat island and overheating characteristics in Sydney, Australia. An analysis of multiyear measurements. *Sustainability* **2017**, *9*, 712. [CrossRef]

45. Yun, G.Y.; Ngarambe, J.; Duhirwe, P.N.; Ulpiani, G.; Paolini, R.; Haddad, S.; Vasilakopoulou, K.; Santamouris, M. Predicting the magnitude and the characteristics of the urban heat island in coastal cities in the proximity of desert landforms. The case of Sydney. *Sci. Total Environ.* **2020**, *709*. [CrossRef] [PubMed]

46. Weather Data Sources. Available online: https://energyplus.net/weather/sources (accessed on 5 June 2020).

47. ASHRAE. International Weather for Energy Calculations 2.0 (IWEC Weather Files) DVD. 2012. Available online: https://www.ashrae.org/technical-resources/bookstore/ashrae-international-weather-files-for-energy-calculations-2-0-iwec2 (accessed on 21 April 2020).

48. Jentsch, M.F.; James, P.A.B.; Bourikas, L.; Bahaj, A.S. Transforming existing weather data for worldwide locations to enable energy and building performance simulation under future climates. *Renew Energy* **2013**, *55*, 514–524. [CrossRef]

49. Belcher, S.E.; Hacker, J.N.; Powell, D.S. Constructing design weather data for future climates. *Build. Serv. Eng. Res. Technol.* **2005**, *26*, 49–61. [CrossRef]

50. IPCC. Climate Change 2014: Syntesis Report. In *Contribution of Working Groups I, II and III to the Fifth Assessment Report of the Intergovernmental Panel on Climate Change*; IPCC: Geneva, Switzerland, 2014; p. 151.

51. Livada, I.; Synnefa, A.; Haddad, S.; Paolini, R.; Garshasbi, S.; Ulpiani, G.; Fiorito, F.; Vassilakopoulou, K.; Osmond, P.; Santamouris, M. Time series analysis of ambient air-temperature during the period 1970–2016 over Sydney, Australia. *Sci. Total Environ.* **2019**, *648*, 1627–1638. [CrossRef]

52. Oke, T.R. *Initial Guidance to Obtain Representative Meteorological Observations at Urban Sites*; Report prepared for the World Meteorological Organization, Instruments and observing methods, Report no. 81; World Meteorological Organization (WMO): Geneva, Switzerland, 2004.

53. Papanastasiou, D.K.; Kittas, C. Maximum urban heat island intensity in a medium-sized coastal Mediterranean city. *Theor. Appl. Climatol.* **2012**, *107*, 407–416. [CrossRef]

54. Estoque, M.A. A theoretical investigation of the sea breeze. *Q. J. R. Meteorol. Soc.* **1961**, *87*, 136–146. [CrossRef]

55. Rotunno, R. On the linear theory of the land and sea breeze. *J. Atmos. Sci.* **1983**, *40*, 1999–2009. [CrossRef]

56. Simpson, J.E.; Mansfield, D.A.; Milford, J.R. Inland penetration of sea-breeze fronts. *Q. J. R. Meteorol. Soc.* **1977**, *103*, 47–76. [CrossRef]

Review

Urban Overheating and Cooling Potential in Australia: An Evidence-Based Review

Komali Yenneti [1,2], **Lan Ding** [3], **Deo Prasad** [3], **Giulia Ulpiani** [4,*], **Riccardo Paolini** [3], **Shamila Haddad** [3] and **Mattheos Santamouris** [3]

[1] School of Architecture and Built Environment, Faculty of Science and Engineering, University of Wolverhampton, West Midlands WV1 1LY, UK; komaliy@wlv.ac.uk

[2] Australia India Institute, University of Melbourne, Parkville, VIC 3010, Australia

[3] High Performance Architecture Research Cluster, Faculty of Built Environment, UNSW Sydney, Sydney, NSW 2052, Australia; lan.ding@unsw.edu.au (L.D.); d.prasad@unsw.edu.au (D.P.); r.paolini@unsw.edu.au (R.P.); s.haddad@unsw.edu.au (S.H.); m.santamouris@unsw.edu.au (M.S.)

[4] Industrial Engineering and Mathematical Sciences Department, Università Politecnica delle Marche, via Brecce Bianche 1, 60131 Ancona, Italy

* Correspondence: giulia.ulpiani@sydney.edu.au

Received: 21 September 2020; Accepted: 31 October 2020; Published: 4 November 2020

Abstract: Cities in Australia are experiencing unprecedented levels of urban overheating, which has caused a significant impact on the country's socioeconomic environment. This article provides a comprehensive review on urban overheating, its impact on health, energy, economy, and the heat mitigation potential of a series of strategies in Australia. Existing studies show that the average urban heat island (UHI) intensity ranges from 1.0 °C to 13.0 °C. The magnitude of urban overheating phenomenon in Australia is determined by a combination of UHI effects and dualistic atmospheric circulation systems (cool sea breeze and hot desert winds). The strong relation between multiple characteristics contribute to dramatic fluctuations and high spatiotemporal variabilities in urban overheating. In addition, urban overheating contributes to serious impacts on human health, energy costs, thermal comfort, labour productivity, and social behaviour. Evidence suggest that cool materials, green roofs, vertical gardens, urban greenery, and water-based technologies can significantly alleviate the UHI effect, cool the ambient air, and create thermally balanced cities. Urban greenery, especially trees, has a high potential for mitigation. Trees and hedges can reduce the average maximum UHI by 1.0 °C. The average maximum mitigation performance values of green roofs and green walls are 0.2 °C and 0.1 °C, respectively. Reflective roofs and pavements can reduce the average maximum UHI by 0.3 °C. In dry areas, water has a high cooling potential. The average maximum cooling potential using only one technology is 0.4 °C. When two or more technologies are used at the same time, the average maximum UHI drop is 1.5 °C. The mitigation strategies identified in this article can help the governments and other stakeholders manage urban heating in the natural and built environment, and save health, energy, and economic costs.

Keywords: urban heat; Australia; UHI effect; mitigation; climate change

---

1. Introduction

The history of urbanisation is often defined as the history of human development. In the past two centuries, the urban population increased more than 100 times [1]. Today, more than 50% of the world's population lives in cities and forecasts suggest that this number will rise to 70% by 2050 [2]. The burgeoning urban population growth and subsequent urban expansion will greatly affect local and regional climates, urban environmental quality, and public life [3]. Worse, dark coloured building surfaces, roads, pavements, vehicle emissions, and reduced urban green spaces are already contributing to increased atmospheric heat, extreme temperatures, frequent and extended heat spells, and thermal stress.

In Australia, urban overheating has become an increasingly important issue, and urban residents often suffer from excess heat and frequent heatwaves [4,5]. Urban overheating is generally the consequence of the urban heat island (UHI) effect, a local phenomenon caused by city characteristics (urban density, structure, form, and land use), building and paving materials, anthropogenic heat released by vehicle exhausts and building energy use, and the loss of natural features (green areas, water) [5].

Evidence on the UHI effect is available for almost all Australian cities [4]. However, urban overheating in Australia is often triggered by the self-amplifying mechanism of synoptic weather conditions combined with the UHI effect [6]. The significant co-existence of the dualistic atmospheric systems of cool sea breeze from the ocean and hot winds from the inland desert makes the spatiotemporal characteristics of urban overheating highly variable and heterogeneous. As a result, the analysis of the behaviour and formation of urban overheating is very challenging.

Urban overheating and frequent, extreme, and extended heatwaves have significant impact on energy [7], health [8,9], thermal comfort [10], environment [11], and the economy [12]. Advanced technologies and strategies have been developed to mitigate the UHI effect and manage urban heat. The implementation of mitigation techniques and strategies, such as urban greening, green roofs, vertical gardens, cool roofs, and cool pavements, can provide a path for sustainable urban development.

Against this background, the aim of this article was to provide an in-depth evidence-based review on the characteristics of urban overheating, its impacts on human health, energy and economy, and the potential of appropriate mitigation technologies and strategies in Australia.

The analysis of the information presented in this paper is based on data collected from all scientific articles published in peer-reviewed journals, original articles published in accredited media, and original reports published by government agencies and credible research institutions over the last 15 years. The journal articles were searched with a combination of key words such as 'UHI', 'Australia', 'heat mitigation', 'health', 'energy', and 'economy'. Papers with at least a section including the relevant information required for this paper were only considered for final inclusion. All articles and publications that provided spatiotemporal data were included in the study.

This work focuses on canopy layer ground basis observations of the near surface air temperature. Therefore, all publications on boundary layer heat island and surface heat islands using remote sensing, subsurface, and non-urban heat island were excluded. The studies on the UHI effect were separately analysed based on two criteria: (a) the experimental protocol used (standard equipment and weather monitoring stations, mobile traverses around the selected area, and non-standard equipment) and (b) the form and type of the UHI intensity reported. In addition, relevant information on UHI impacts and the potential of mitigation measures were extracted from government reports and other secondary data pathways. The data collected was processed, analysed, and interpreted to provide a meaningful understanding on the UHI effects, impacts, and mitigation in Australia.

After the introduction, the structure of this article is as follows: Section 2 discusses the dynamic characteristics of urban overheating in Australian cities. Section 3 provides evidence on the impact of urban overheating on energy, health, and the economy. Section 4 analyses the potential of different urban heat mitigation strategies. Finally, the article ends with conclusions and policy implications.

2. The Magnitude and Characteristics of Urban Overheating

The UHI effect is a key phenomenon of local climate change, wherein the temperatures in inner cities are usually higher than the sorrounding rural areas. It is exhibited when "a significant difference in temperature can be observed within a city or between a city and its suburbia and/or its sorrounding rural areas, and areas of maximum temperature can expectedly be found within the densest part of the urban area" [13] (p. 73). Evidence on the intensity of UHIs is available for almost all major cities in Australia (Table 1).

Table 1. The intensity of the urban heat island (UHI) effect in Australian cities and regions.

No.	City	Population Density (People/sq.km: 2016)	Intensity of the Heat Island (°C)	Details of Data Sources	Reference
1	Melbourne	17,506	annual average: 1.4	one urban and one rural weather stations	[14]
			annual average (depending on summer or winter): 0.5–2	one central business district (CBD) and three sorrounding non-CBD area weather stations	[15]
			average mean maximum intensity: 4	mobile traverse from the western fringe, approximately 2 km south of the city center, through the CBD to the northern fringe	[16]
			annual average: 1.4	two urban and two rural reference stations	[17]
2	Sydney	1171	maximum intensity: 11	six meteorological stations distributed across the city	[4]
			maximum intensity: 13	eight different stations within the city	[6]
3	Alice Springs	85	UHI is evident at night. Average intensity: 4.1	ten sensors installed within the city center	[5]
4	Camperdown	4362	average intensity: 1.2	mobile transect from a position in the rural area through town center to a rural area on the other side of the town	[16]
5	Colac	520			
6	Hamilton	480			
7	Hobart	131	maximum intensity: 5.7	mobile sensors	[18]
8	Darwin	703	maximum intensity: 2	weather station at airport	[19]
9	Perth	317	UHI is evident at night. maximum intensity: 0.8	one urban, two urban fringe and three rural stations	[20]
10	Adelaide	400	UHI is evident at night. maximum intensity: 1.3	one urban, two urban fringe and two rural stations	[20]

It should be noted that different measurement methods capture different magnitudes of the urban-reference temperature differences (Figure 1). The average intensity using standard measuring methods (weather monitoring stations) varies between 1.0 °C and 13.0 °C, while the average intensity using non-standard methods (e.g., mobile measurements, micro-scale sensor based measurements) is between 1.0 °C and 7.0 °C. The magnitude of the UHI effect using mobile transects or other non-standard methods is higher than that by using standard measurement stations. Mobile transects and non-standard measurement methods are commonly used to measure UHI in densely populated urban areas, while fixed weather monitoring stations are used in thermally undisturbed areas.

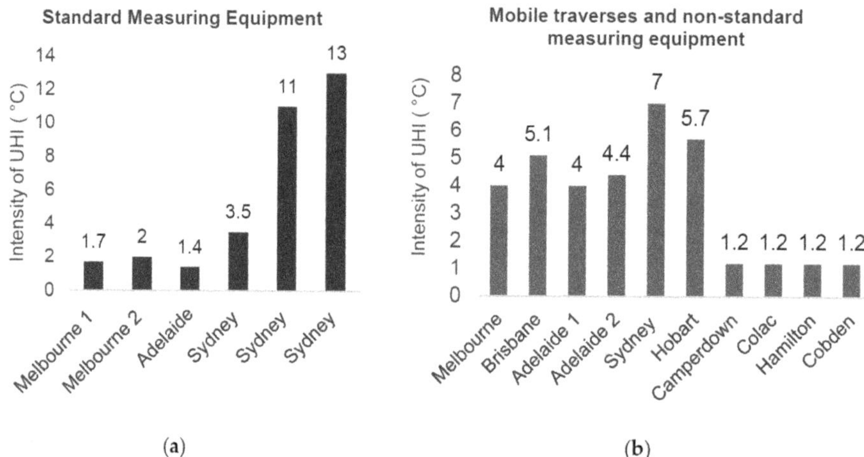

Figure 1. The intensity of the UHI effect in Australian cities and regions (**a**) using standard measuring equipment; and (**b**) nonstandard measuring equipment.

Considering the relatively short duration of the studies based on mobile traverses and non-standard measuring equipment, the UHI intensity reported usually is the maximum temperature difference measured during the entire experiment. However, studies using standard measurement equipment obtained multi-year measurement results on either the annual average, annual average maximum, or absolute maximum UHI intensity or a combination of them (Table 1). The quality and accuracy of the UHI intensity provided by the studies depends on many parameters, such as the duration of the experiment, the number of stations used, the selected experimental protocols, and the accuracy of the measuring equipment [21].

In the next few paragraphs, the characteristics of overheating in Australian cities are discussed. In most parts of the world, the intensity of urban overheating is mainly determined by the UHI effect, a phenomenon caused by city-specific factors (urban expansion, land use, dense built environment, urban layout), anthropogenic heat released by buildings and vehicles, extensive alteration of urban natural spaces (green spaces, water bodies), and the presence of heat sources and sinks [22,23]. However, the magnitude of urban overheating in Australian cities is influenced by both the UHI effect and synoptic weather conditions [21,24,25]. As a result, the spatiotemporal characteristics of urban overheating in Australia, especially in coastal cities, are highly variable and heterogeneous. Moreover, the analysis of the behaviour and formation of urban overheating becomes very challenging.

A good example to demonstrate the variability of urban overheating in Australia is Greater Sydney. Despite being located along the coast and close to the ocean, Sydney's average UHI intensity is much higher than other Australian cities (Figure 1). Further, the behaviour of urban overheating changes with time and space. In Sydney, a high degree of thermal imbalances exists between different parts of the city, with temperatures changing daily, monthly, and yearly [4,6].

Results from multiple numerical and experimental studies indicate that the dramatic fluctuations in the intensity of urban overheating in Sydney are related to two major large-scale atmospheric circulation systems: (a) sea breeze and (b) hot desert winds [4,6,24]. The significant co-existence of advective and convective processes related to local weather conditions, along with the UHI effect itself, can affect the spatiotemporal pattern of urban overheating in Sydney [4,6].

Spatially, the absolute intensity of urban overheating increases with increasing distance from the coastline (Figure 2). The average intensity of overheating in the inland western suburbs is at least 2–5 °C higher than the coastal areas in the eastern suburbs [6].

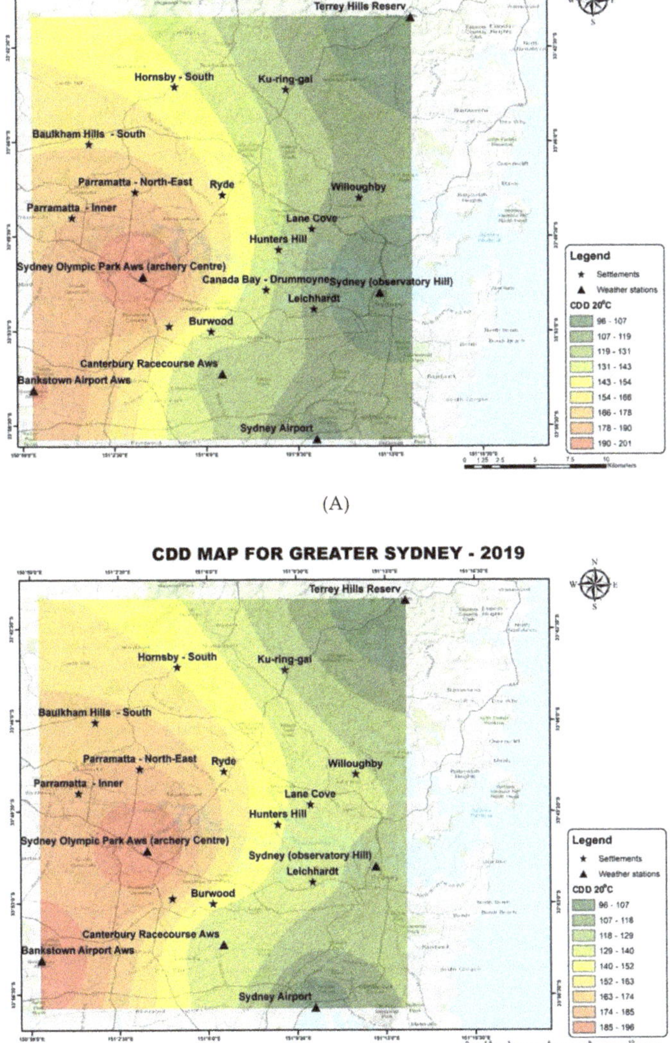

Figure 2. Patterns of UHI intensity in Greater Sydney, based on cooling degree days (CDD) (**A**) 2018 (**B**) 2019.

Sydney is strategically positioned on the South Pacific coast, but is also located close to one of the largest desert areas in the world, known as "the Australian arid biome". This inland desert is a cradle of strong hot winds and a massive heat source [6,26]. The sea breeze inhibits hot air advection as it interacts with the UHI circulation and contributes to the cooling effect in the eastern suburbs [6]. This is a pattern commonly observed in other Australian cities, such as Adelaide [27] and Melbourne [21]. On the other side, the stagnation region in the city canyons, coupled with the warm desert winds leads to weaker penetration of cool winds in the western part of the city [4,5].

Temporally, the intensity of urban overheating in the western region is lower at night. As a result, the average night time ambient temperature in western Sydney is lower than in eastern areas [4]. This is particularly due to the high intensity of the nocturnal oasis effect [28], a phenomenon triggered by radiation and convection cooling processes (low density, reduced solar heat gain, and nearness to vegetation areas) [29,30]. However, during the hot summer months, the degree of night-time oasis phenomenon reduces in the western suburbs due to strong warm winds from the inland areas [28]. At the same time, the cooling effect of the sea breeze from the ocean alleviates the heat island effect and lowers the temperature in the eastern suburbs [4]. These two climatic processes (heat island effect and oasis effect) are the main influencing factors of the discrepancies in temperature distribution in summer [4,6].

The constant dynamic struggle between the sea breeze (cooling mechanism) and the warm dry air (heating mechanism) from inland during the summer season further complicate urban overheating characteristics in Sydney [31,32]. Compared with other hot arid cities around the world (e.g., Ghardaia in Algeria) [33], the characteristics of urban overheating in Sydney are unique and undergo significant changes in space and time; the city center is cooler during the day than in the fringe areas of the city, and warmer at night [29].

Alice Springs is another city that represents Australia's hot desert climate. Urban overheating in Alice Springs presents a constant pattern and is mainly governed by city-specific variables [5]. Existing evidence shows that the inner city area is warmer than the sorrounding desert environment at night and cooler in the morning [5]. Similar to Alice Springs, the night-time UHI was observed in other cities (Adelaide and Perth). Clearly, a presence of 'daytime cool island' effect has been observed in the city, which can be explained by the shading of buildings, vegetation and trees (especially in the eastern parts of city areas), and shading of solar radiation. Urban cool island (UCI) effect has been also observed in other cities located in similar hot dry climates [30]. Moreover, a higher intensity of UHI during the day has been observed from afternoon to evening. The large amount of heat stored in the urban fabric, unobstructed solar radiation contributing to diurnal heat island and a desert mass inland are the main reasons for the UHI increase in the afternoon in Alice Springs [5]. The delay in urban cooling via the slow heat release from the urban fabric and the long-wave radiation loss from the urban canopy may be related to higher nocturnal UHI amplitude [5].

The intensity of urban overheating in other Australian cities (such as Darwin, Adelaide, and Melbourne) can also be explained by city-specific variables. In Darwin, heat retention within the urban canopy, high humidity, low wind speed, and lack of sea breezes exacerbate urban overheating [34]. The form, layout, structure, morphology, and anthropogenic heat greatly influence the advection rate. Compared with other areas of the city, the low-rise open layout areas have lower diurnal temperatures [34].

Last but not least, evidence from recent research confirms that global climate change further exacerbates urban overheating [24,35]. Over the past 100 years, the average temperature of the earth has risen by about 0.7 °C, while the average temperature of Australia has risen by about 0.9 °C [36]. Interestingly, a large part of this temperature increase (close to 0.7 °C) occurred after 1950. Further, the hottest years in Australian modern history (since 1910) occurred in the past two decades [37]. Climate change projections also indicate that under 1990 baseline conditions and business as usual (BAU), by 2030 and 2100, the temperature rise in Australia will be 1.5 °C and 4.5 °C, respectively [36]. Even under strong carbon emission reduction scenarios, the average temperature increase will be about

2 °C by 2100 [38]. These changes are expected to increase the average and maximum temperatures in summer, the frequency of heatwaves, hot days, and warm nights [25,37,39].

Heatwaves and extreme high-temperature events are some of the most significant effects of El Niño Southern Oscillation (ENSO) and global climate change. Thus, the increasing frequency of heatwaves is an important issue in Australia. There is no consensus on the definition of heatwaves because people in different regions have different climatic adaptive capacities [40–42]. However, in Australia, a heatwave is pronounced if the absolute maximum temperature threshold (35 °C) is exceeded for two to five consecutive days [43].

The combined effect of heatwaves and local climate change is argued to increase summer temperatures in Australia [24,35,44]. Yet, in-depth studies on changes in the intensity of overheating during heatwaves and the possible synergies between these two phenomena are very limited [24]. For example, recent research in Sydney argues that there is a strong relationship between heatwaves and the degree of overheating: the difference between the maximum average UHI intensity during heatwaves and non-heatwave period was found to be at least 8 °C [24]. In addition, the UHI effect is enhanced and more pronounced during the day (noon). Moreover, in Sydney, the synergies between overheating and heatwaves can also be attributed to specific synoptic weather conditions in the city [24]. The advective heat flux from desert winds, as well as anthropogenic and sensible heat fluxes, can be considered as key reasons for overheating during heatwaves [24].

These results are consistent with studies carried out in other coastal cities [45,46], although there are large differences between heatwave and non-heatwave periods in Sydney. Further, the results are different from studies based on non-coastal cities [47,48], where no changes or reductions in overheating patterns were found during heatwaves. In studies comparing rural and non-coastal areas, the phenomenon of urban heating was more pronounced at night [49–51].

Among other weather extremes, Australia is particularly susceptive to and whose frequency is being exacerbated by climate change are bushfires. A recent publication [52] demonstrated that long-lasting bushfire seasons may alter the overheating pattern in the city of Sydney. In the study, the authors compared the UHI intensity during the bushfire event in 2019/2020 to that recorded during the same period over the previous 20 years. Results from the study show that bushfires were responsible for the disappearance of cool island events and the exacerbation of UHI events over the median. The interlacement between UHI and urban pollution is indeed very intimate and expected to deteriorate in the future [53].

3. Impacts of Urban Overheating

Urban overheating is a major local climate change phenomenon in Australia. As mentioned in the previous section, the average UHI in Australian cities is as high as 4–6 °C, and in some metropolitan cities, it exceeds 10 °C. Consequently, this local climate anomaly may seriously affect urban sustainability and human well-being, and the interrelationship between urban overheating and its impact on various aspects of human life has been documented for major Australian cities. This section provides a comprehensive review of the impact of urban overheating on public health, energy and the economy.

3.1. Health and Well-Being

In Australia, overheating in cities seriously threatens public health. The Australian Emergency Management Agency and other government organisations have recognised that overheating poses a serious threat to health and well-being. Long-term exposure to extreme temperatures and heat may cause cardiovascular, respiratory, and thermoregulation (cramps, rashes, and heat stroke) related problems, and affect cognitive and emotional abilities [54]. The most at-risk groups include the elderly, children, pregnant women, patients with chronic diseases, people with physical and mental disabilities, and low-income communities.

A considerable body of literature has demonstrated that local climate change and higher urban temperatures, especially during heatwaves, can amplify heat-related mortality and morbidity [38,55–57]. In particular, existing works have found that when the ambient temperatures rise above a certain threshold, mortality and morbidity significantly increase. For example, recent evidence indicates that people living in warmer areas of Western Sydney have a 6% higher risk of heat-related death than those living in colder areas of East Sydney [58]. Furthermore, a 2 °C rise in the maximum threshold temperature (27 °C) can increase the average mortality rate by 5.3% Similar findings were also observed by [59]. In Perth, a degree rise in the temperature threshold of 30 °C increased the mortality rate of patients with cardiovascular-diseases by at least 10% [60]. Studies of a similar nature have also demonstrated a presence of strong synergies between overheating and increase in heat-related mortality in many other Australian cities, such as Brisbane [61] and Adelaide [56].

In terms of morbidity, some comprehensive studies conducted in all capital cities in Australia found that a degree rise in temperature can increase the emergency hospitalisation rate of heart-disease patients by an average of 10%, when the maximum ambient temperature is considered to be 30 °C [60,62]. However, other evidence suggests that results may vary depending on the research methods, spatial, socioeconomic, and climatic variabilities, as well as public adaptation [58]. For example, recent research has shown that Sydney's unique overheating phenomenon can be a major cause of higher morbidity in the western parts of the city [58]. It is further estimated that the incidence of all-cause heat morbidity is between 0.05% and 4.6%, and that during heatwaves, this value is between 1% and 11% [58]. Moreover, a 1 °C increase in daily maximum temperature can increase the incidence of heat-related morbidity by 1.1% to 4.6%, when the threshold temperature is regarded as 27 °C. Other studies conducted in Sydney [63] and Brisbane [64] also observed similar results.

Overall, it is evident that the risk of heat-related mortality and morbidity rises significantly with the rise in threshold temperature and during heatwaves. However, the risk gradient may depend on a variety of factors, such as local climate, age, outdoor and indoor environments, thermal quality of the housing, physiological characteristics of the population, demographic and socioeconomic factors, adaptation, and infrastructure [38,65–67]. For example, for low-income people living in poor and warm parts of the city with poor-quality housing, and lack of resources to maintain thermal comfort (air conditioning), the health risks are very high [68,69]. As a result, low-income people may spend more energy than others, or even live in uncomfortable indoor environmental conditions that may affect their health and well-being [58]. Despite these risks, limited research has explored the relative impact of urban overheating on low-income communities in Australia [70].

3.2. Energy Consumption and Demand

Urban overheating has severely affected the energy consumption and peak electricity demand in Australian cities. Many studies have explored the relationship between urban overheating and energy consumption, and found a positive correlation between the two [71].

Santamouris [7] found that urban energy consumption per person-year increases by 0.73 ± 0.64 kWh/m^2/°C, or 78 ± 47 kWh/°C, while peak electricity demand increases by 0.45–12.3%/°C, depending on AC penetration and setpoint temperature. Evidence from a recent experimental study conducted in Sydney indicated that urban overheating can increase indoor overheating levels by 56% and cooling energy demand by 16% per year [58]. It was further found that the cooling penalties of residential and commercial buildings were 6.4% and 15.6% per year, respectively, or about 1.8 kWh/m^2/°C and 6.7 kWh/m^2/°C per year, respectively. However, the distinct overheating phenomenon in Sydney (Section 2) can have a differential effect on the city's cooling energy demand.

According to a parametric study of the Sydney metropolis, the buildings in western Sydney consume three times as much energy as eastern Sydney [71]. Moreover, the annual cooling energy demand in western Sydney was as high as 140.2 kWh/person/°C, while the cooling penalties for residential and commercial buildings were 45.1 kWh/person/°C and 95.1 kWh/person/°C per year, respectively [58]. The higher energy penalties imposed by the commercial sector can be attributed to the increased use of commercial energy and the relative smaller population in the western region.

Similarly, in the desert city of Alice Springs, the heat island effect (Section 2) also significantly affected the city's cooling energy demand and building consumption. The energy demand, measured in cooling degree days (CDDs), was between 923–475, when the base temperature ranged between 23 °C and 27 °C [5]. This finding indicates that Alice Spring manifests three times the energy demand of Sydney [4].

In general, studies on the impact of urban overheating on energy determined that for every degree rise in threshold temperature (18 °C) increase, the average cooling energy demand will increase by 0.45% to 4.6% [72], while the annual average energy consumption will increase by 0.5% to 8.5% [65]. Worse, the cooling energy demand of urban buildings will be at least 13% higher than similar rural buildings [7]. Considering that Australia's average summer temperature is higher than 27 °C and 90% of the country is urbanised, these figures may seriously affect thermal comfort and well-being [15,73].

3.3. Economy and Productivity

Urban overheating may pose a serious threat to the Australian economy by reducing labour productivity [74]. Yet, there is not sufficient evidence on the synergies and interdependencies between local climate change and the economy. Most of the existing work on local climate change has focused on the relationship between overheating during heatwaves and indoor workplace productivity [38,74] and outdoor workers productivity [75,76].

For example, a study has shown that urban overheating will cause the Australian gross national product (GNP) to drop at least 1.3% per year [77]. A recent study highlighted that due to heat stress, 7% of the Australian population did not go to work at least one day in the year 2013/14 [74]. The study further emphasised that 70% of the population did go to work, but they felt inefficient, and on average, people were exposed to heat stress for at least 10 days a year and lost about 27 work hours per year. If the sample is extrapolated to the entire working population in Australia, the annual productivity loss from thermal stress is $7.92 billion [74]. Existing findings on the economic costs of extreme temperature events varies widely, with estimates ranging from $1.8 billion to $7 billion [74]. These economic and productivity losses make the cost of heat stress comparable to the cost of chronic health problems.

Extreme temperatures and urban overheating, especially during heatwaves [15,78], severely impact other sectors of the economy, such as transport, construction, agriculture, and tourism, in addition social behaviour (e.g., domestic violence, burglary, assault) [15]. However, evidence on the synergies between urban overheating and other sectors of the economy is also inadequate.

4. Impact of Mitigation Strategies on Cooling Cities

To mitigate urban overheating and offset its impact on cities, appropriate mitigation techniques and strategies are available. These measures create a thermally balanced city by increasing the reflectivity of urban areas, reducing anthropogenic heat, and dissipating excess urban heat. In this context, this section comprehensively reviews the progress of research on urban heat mitigation in Australia.

A number of studies have been attempted to estimate the potential of various strategies in urban heat mitigation in Australia. The list and the performance of each study are given in Table 2. The reported studies are based on the following mitigation techniques: (a) reducing the absorption of solar radiation and keeping urban surfaces cool (e.g., cool materials); and (b) increasing evapotranspiration in urban environments (e.g., urban greenery, green infrastructure, and water-based systems). The rest of this section provides a discussion on these strategies and their mitigation performance in Australia.

Table 2. Performance of mitigation strategies reported in Australian cities.

S.No	Mitigation Strategy	Location	Maximum UHI Mitigation Potential	Reference
	Urban Green Spaces			
1	Urban greenery	Sydney	1.4	[71]
2	Urban parks	Melbourne	0.3	[79]
3	Urban vegetation	Melbourne	1	[80]
4	Urban vegetation	Melbourne	2	[81]
5	Urban vegetation	Brisbane	1.08	[82]
6	Urban greenery	Alice Springs	0	[5]
7	Urban parks (trees)	Gold Coast (Brisbane)	1.2	[83]
8	Urban parks (grass)	Gold Coast (Brisbane)	0.7	[83]
9	Urban greenery	Adelaide	2	[84]
	Green roofs			
10	Green roofs	Adelaide	0.06	[85]
13	Green roofs	Melbourne	1.4	[86]
14	Green roofs	Canberra	0.4	[87]
15	Green roofs	Sydney	0.5	[71]
	Green Walls			
16	Green wall	Adelaide	0.25	[85]
17	Living wall	Adelaide	1.5	[88]
	Reflective Materials			
18	Cool streets	Sydney	1.4	
19	Cool pavements	Sydney	0.5	[71]
20	Cool roofs	Sydney	0.6	
21	Cool roofs	Melbourne	0.5	[80]
	Water			
23	Water sprinklers	Alice Springs	0	[5]
	Shading			
24	Street shading	Alice Springs	0	[5]
	Combination			
25	Greenery and Reflective materials		0.95	[89]
26	Water and Shading		3	
27	Reflective materials (roofs, pavements, streets)	Sydney	3	[71]
28	Urban vegetation and Cool roofs	Melbourne	0.82	[80]
29	Trees, Reflective materials (roofs and pavements), Evaporative cooling systems and Shading	Alice Springs	1.1	[5]
30	Trees and Green roofs	Melbourne	2.4	[86]
31	Trees and Grass	Canberra	0.8	[87]
32	Reflective materials (buildings and pavement)	Alice Springs	0.9	[5]
33	Reflective materials (roofs, pavements) and trees	Sydney	1.3	[90]

4.1. Use of Water

For centuries, people have been using water as an important strategy to minimise heat stress and cool the sorrounding environment. A waterbody helps regulate temperatures and act as a thermal buffer by reducing heat convection to air above and evaporation [89,91], where absorbed thermal energy converts sensible heat to latent heat with the production of water vapour [92] (p. 1047). Natural water bodies such as lakes, rivers, and wetlands, in addition to artificial water bodies, can reduce the UHI effect and contribute to the UCI effect [93]. Along with natural waterbodies, for decorative and climatic reasons, passive water systems such as small artificial lakes, ponds, and swimming pools have been widely used in public places. Similarly, active or hybrid systems such as evaporation towers, sprinklers, fountains, and water misting technologies are now widely used in public places around the world [89].

A water body is capable of lowering the UHI effect by 1–2 °C, and sorrounding local environments by 2–6 °C [93]. Unfortunately, previous studies in Australia have not fully assessed the possible impact of water bodies on urban temperatures, especially beyond their sorroundings (Figure 3). A trivial number of existing studies demonstrate positive correlation between water and UHI reduction, while the corresponding temperature drop at the local scale is between 4–13 °C [5,94]. For example, in Alice Springs, the use of water sprinklers technology in the CBD had no major effect on sorrounding temperatures, while it contributed to a local maximum mitigation of 13 °C [5]. Similarly, a local maximum reduction of 3.9 °C was observed in Darwin by using evaporative cooling systems [34].

Notwithstanding, the ability of water to influence the UHI effect and the sensible cooling effectiveness (of both dynamic and static water systems) depends on the urban area, the physical and geometric characteristics of the system, its inherent properties, the net effect of radiation balance, atmospheric advection, the climate variables (humidity, wind velocity) that contribute to sensible to latent heat conversion, and its interactions with sorrounding climatic conditions [92,94]. Furthermore, despite their intense local impact, there is little experimental information about the performance of blue installations in Australia.

4.2. Urban Green Technologies and Strategies

Urban greenery is one of the key elements of a sustainable city. Greenery can alleviate urban temperatures, cool the sorrounding environment, and influence the urban microclimate through (a) shading the building surfaces, deflecting solar radiation, reducing the heat convection to the air above occupied spaces, thereby reducing space cooling energy, and any resulting anthropogenic heat emissions that have the potential to increase thermal energy released back into the urban climate; (b) evapotranspiration, a process where water absorbed through roots of plants is evaporated into the air through their leaves by absorbing energy from solar radiation, which keeps themselves cool through the photosynthesis and the sorrounding air by latent heat absorption; and (c) acting as wind shields and contributing to wind pattern changes [95–97]. In addition, greenery can improve thermal comfort and human health, promote psychological balance, and make cities more attractive [89]. Urban greenery may be part of urban landscapes, parks, streets, hedges, open spaces, and integrated into buildings through green roofs [98] and green walls/vertical gardens [99].

Many studies have been conducted on the mitigation potential of various types of urban greenery in Australian cities. The full list and the mitigation potential evaluated in each study is provided in Table 2. In total, 11 studies evaluated the potential of increased tree canopy cover, 7 studies analysed the potential cooling effect of green and planted roofs and walls, and 2 studies identified the performance of combinations of various types of greenery. In general, different forms of urban greenery can provide a mitigation potential in the range of 0.3–2.5 °C, with an average value of 1.0 °C (Figure 4). A further discussion on the mitigation potential of each of the urban greenery strategies is provided in the following paragraphs.

4.2.1. Urban Green Spaces

In tropical and subtropical climate regions, such as Australia, increasing the number of trees and hedges is a cost-effective mitigation strategy [100]. Based on a comparative analysis of five selected cities in different climate zones, Brown et al. [101] concluded that shading and canopy cover are by far the most effective cooling strategies in Australia. Evidence from existing research show that increasing the tree canopy cover can reduce the UHI by 0.3–2 °C, with an average value of 1.07 °C. However, the temperature differences vary both spatially and temporally.

Spatially, a detailed study conducted in Melbourne found that increasing canopy coverage from 27% to 40% can reduce the UHI by 1 °C [80], while increasing the percentage of outdoor pavements integrating greenery has a maximum cooling potential of 1.4 °C in Sydney [71]. By reviewing the existing studies in Melbourne [79] and Brisbane [83], it can be summarised that urban parks are 0.3–1.2 °C cooler than a sorrounding non-green area.

The performance of urban greenery at the local level is comparable to the global mitigation potential. Findings from Alice Springs revealed that increasing tree canopy cover can contribute to a maximum local temperature drop by 1 °C [5], while shading on all main streets in the city center can reduce the local temperature by 1.3 °C in Darwin [34].

Temporally, experimental results from different cities establish that a high percentage of tree canopy cover correlates strongly with cooler nocturnal temperatures in Australia (Figure 3). Research from Melbourne show that when tree canopy coverage was 5–10%, there was no large temperature difference between day and night, while the difference increases by 0.6 °C at 40% vegetation cover [80]. Similar results were also observed in Brisbane [82] and Adelaide [84].

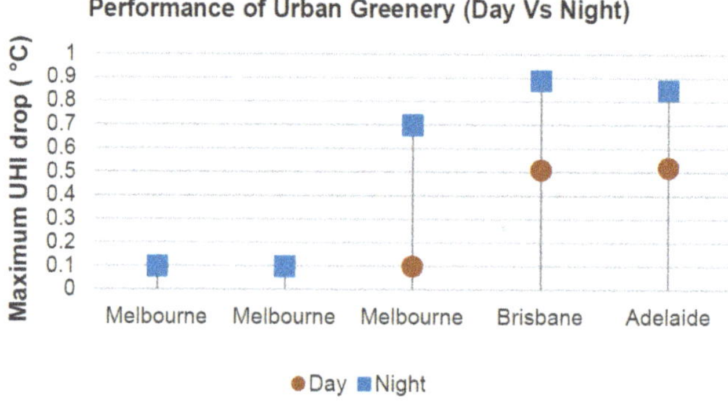

Figure 3. Performance of urban greenery (day vs. night).

While trees and hedges have significant potential in urban heat mitigation, their performance is highly localised and is affected by complex factors, such as vegetation type, canopy density, height of leaves above the ground, leaf area index (LAI), water content, water availability, evapotranspiration, distance between buildings and trees, thermal balance in sorrounding areas, urban density, and local weather conditions [97,102]. Similarly, the performance of urban parks depends on factors such as the characteristics of the area, the features of urban park, types of trees, and wind speed. According to [103], the impact of urban parks is limited to one park width, and the cooling gradient outside the park varies from 0.1 to 1.5 °C/100 m. To design large urban green spaces, such as parks, that have the greatest cooling effect in hot summer weather, landscape architects and city planners need to understand the relative impact of various design interventions.

4.2.2. Green Roofs

A green roof is a building roof with fully or partially covered vegetation. Green roofs are usually categorised into three types: intensive roofs (with small trees and shrubs), semi-intensive roofs (with small herbaceous plants, ground covers, grasses, and shrubs), and extensive roofs (covered with thin vegetation layers) [96,104]. Green roofs can mitigate urban heat, improve urban environmental quality, and have many other environmental and economic benefits. However, it must be recognised that green roofs (and also green walls) have disadvantages, such as high initial investment, high maintenance cost, and requirement of more structural strength to support extra load [105].

Evidence from existing research demonstrates that green roofs have a mitigation potential of 0.06–1.4 °C (Figure 4). According to an experimental survey of a typical urban area in Adelaide, covering 30% of the total roof area of all buildings with green roofs during a typical warm summer can reduce UHI by 0.06 °C and the daily energy consumption by 2.57 W/m^2/day [85]. Similarly, [86] documented that the cooling potential of green roofs in Melbourne is 1.4 °C. [71] reported that green roofs can reduce the UHI in Sydney by 0.5 °C, while the heat mitigation potential of green roofs in Canberra is 0.4 °C [87]. Given the small number of studies, more research on the thermal performance of green roofs is warranted.

4.2.3. Green Walls

Green walls have been around for a long time as hanging gardens or climbing plants. Today, green walls, also known as living walls, vertical gardens, and bio-walls, are an important category of natural urban sustainability solutions.

Green walls can be divided into two categories: (i) green façades and (ii) living wall systems [106]. A green façade can be designed to be direct or indirect, wherein a direct green façade is a type of traditional green wall with evergreen or deciduous climbers connected directly to the building's surface, while an indirect green façade consists of a vertical structure supported by trellis or steel cables for climbing plants [107]. On the other hand, a living wall is a modern vertical greening system and requires complex planting boxes, pre-vegetation, and pre-fabricated support structures to promote plant growth [108]. The application of modular panels in living walls helps plants obtain sufficient nutrients to survive. The success of a green wall depends on several factors, such as the choice of plants and vegetation (local and non-local), an irrigation system, wall orientation, and design conditions.

Despite their environmental and economic benefits, there is little information about the performance of green walls in Australia. Thermal performance analysis of a living wall in Adelaide show that, in summer, a living wall can mitigate UHI by 1.5 °C [88], while another study found that green walls has a maximum UHI mitigation performance of 0.25 °C [85]. Nevertheless, Australia's green wall industry is still in its infancy, and new case studies are needed to resolve many research gaps.

4.3. Use of Reflective Materials

Increase in albedo will significantly reduce the UHI effect and extreme temperatures. Advanced materials are commercially available with high emissivity and high reflectivity. These can be applied on roofs, exterior walls of buildings, and outdoor urban spaces (such as pavements). Cool roofs, cool facades, and cool pavements help mitigate UHI, reduce cooling energy consumption in air-conditioned buildings, improve thermal comfort in non-air-conditioned buildings, and improve outdoor air quality and comfort [109,110].

4.3.1. Cool Roofs and Façades

Cool roofs and exterior walls are building components with high solar reflectance and high emissivity coefficient materials. The reflective materials commonly used in buildings are white and can be a single layer or liquid. Typical liquid products are white coatings, elastomers, acrylic, or polyurethane coatings, while single-layer products are EPDM (ethylene propydiene tetrolymer membrane), CPE (chlorinated polyethylene), PVC (polyvinyl chloride), TPO (thermoplastic polyolefin), and CPSE (chlorosulfonated polyethylene) [23]. Akbari and Kolokotsa [96] provide an extensive list of existing cool materials for cool roofs and exterior walls. Furthermore, extensive breakthrough research has been conducted to develop coloured thermochromic materials that become more reflective at higher temperatures [111]. However, more research is needed to develop thermochromic agents as viable and economical cooling materials.

Reflective materials used in buildings can be divided into four categories: (i) natural materials with high solar reflectivity (e.g., white marble), (ii) white synthetic coatings with high reflectivity, (iii) coloured coatings with high reflectivity in the infrared solar spectrum, and (iv) smart coatings such as thermochromic coatings and materials with enhanced optical and thermal properties [65].

Although some studies have been conducted to determine the effect of cool roofs on urban heat, outdoor and indoor comfort, and building energy consumption, research in Australia is still very limited. As such, [71] found that increasing the albedo of all roofs in Sydney can reduce the UHI by 0.6 °C, while [80] reported that an increase in the albedo of 60% of Melbourne's rooftops can lead to a cooling potential of 0.5 °C.

Furthermore, the thermal performance of a cool roof depends on many factors, such as local climatic conditions (solar radiation intensity, humidity, wind speed, and cloud cover), the solar reflectance and thermal emittance of roof materials, heat capacity, and U-value of the roof [65]. Moreover, ageing can remarkably reduce the solar reflectance—and thus the cool roof direct and indirect benefits—even by more than 20% during the first years for cool roofing materials with initial albedo equal to 80%, as documented with natural exposure programs conducted in the US, Europe, Brazil, and Japan [112].

The design, construction and materials of cool roofs are guided by cool roof standards, building codes, grades, and labels. Different building energy efficiency standards including *ASHRAE 90.1 and 90.2* and *International Energy Conservation Code*, have adopted cool roof requirements [109]. In many developed countries, cool roof committees (e.g., the American Cool Roof Ratings Council and the European Cool Roof Council) have been established to promote and standardise cool materials. Although cool roofs can provide important opportunities for Australia to save energy and mitigate urban heat, the lack of cool roof councils or the lack of building regulations with cool roof credits or requirements can pose challenges.

4.3.2. Cool Pavements

Urban pavements, streets, driveways, parking lots, squares, and sports fields cover a large proportion of urban structures, and are mainly composed of highly heat-absorbing surfaces (such as asphalt and concrete). Higher surface temperatures will increase the ambient temperature and exacerbate the UHI effect. In contrast, cool materials can lower the surface temperature of pavements and help alleviate urban overheating [96].

The standard reflective paving materials used are fly ash (concrete additives), chip seal, slurry coatings (also known as "micro-surface layers", "fog coatings", "overlays"), reflective synthetic adhesives and light colours coatings [65]. Important research has been carried out to develop extremely high reflective materials for pavements. Important research has been conducted to develop extremely high reflectivity materials for pavements. These include water-retaining or permeable materials, infrared reflective coatings, heat reflective coatings, colour-changing coatings, nanotechnology additives (for example, emerald coatings), and photovoltaic pavements [113]. Further, Ref. [96,113] provide a list of materials used for pavement and their reflectance values.

In Australia, little work has been conducted on the performance of cool pavements. As such, Ref. [71] found that an increase in the albedo of all streets and pavements in Sydney can mitigate the UHI by 1.4 °C and 0.5 °C, respectively. More analysis is needed to determine the economic feasibility and thermal performance of reflective pavements. In addition, current building standards, public information plans, or incentive plans do not consider reflective surfaces.

4.4. Combined Mitigation Strategies

Evidence from existing research suggests that that the combined use of different mitigation strategies has higher mitigation potential than the contribution of each technology. The combined use of greenery and reflective materials can reduce the maximum UHI by 0.82–0.95 °C [80,89], while the combination of cool roofs and cool pavements can reduce the maximum UHI by 0.9–3 °C [5,71]. Further, the combination of trees and grass [87], as well as trees and green roofs [86], can reduce the maximum UHI by 0.8 °C and 2.4 °C, respectively. The combined use of trees, cool materials, evaporative cooling, shading [5], and water and solar control [89] can reduce the maximum UHI by 1.1 and 3, respectively.

In summary, the above mitigation strategies can significantly offset the impact of UHI and local climate change. The average maximum UHI reduction using only one technology is close to 0.42 °C. When two or more technologies were used simultaneously, the maximum UHI relief increased by an average of 1.59 °C. Urban greenery, especially trees, has a high potential for mitigation. Trees and hedges can reduce the average maximum UHI by 1.08 °C. The average maximum mitigation performance values of green roofs and green walls are 0.26 °C and 0.19 °C, respectively. Reflective roofs and pavements, instead can reduce the average maximum UHI by 0.33 °C. Both green roofs and cool roofs have been found to present high mitigation potential in tropical cities, such as Darwin [19]. However, cool roofs may have a higher mitigation potential compared to green roofs in tropical climates as vegetation can add to latent flux due to evapotranspiration and needs high maintenance and extra load capacity [114]. Advanced water technologies (e.g., misting systems) have a high heat reduction potential in dry and hot climatic areas [94]. In the literature [94], the average maximum temperature reduction is 8 °C, which qualify mist cooling as a tremendous asset against urban overheating at local scale. The highest temperature reductions (>10–15 °C) were reported for hot desert and hot-summer Mediterranean climates [94]. Spray cooling finds a fertile ground for investigation and implementation not just in dry hot climates where evaporation is spontaneously enhanced, but in warm, temperate and humid climates too.

Figure 4 shows the mitigation potential for all reported individual strategies and their combinations.

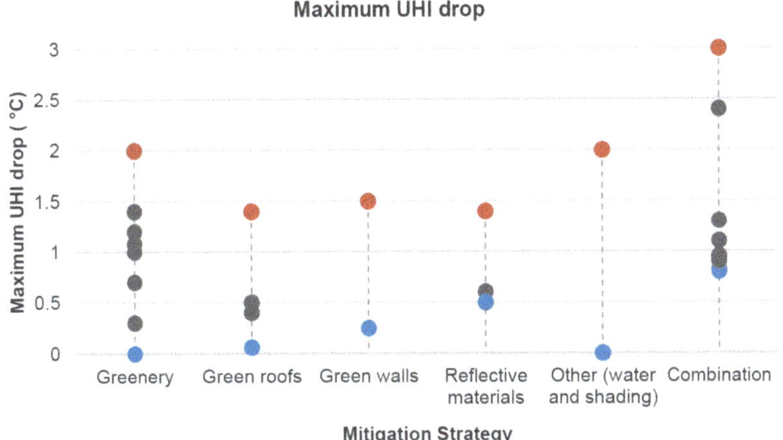

Figure 4. Range of the maximum UHI reduction for all reported mitigation strategies. Blue colour: reported lowest maximum UHI drop value; red colour: reported highest maximum UHI drop value; grey colour: range of maximum UHI drop value between blue and red.

5. Conclusions

Australian cities are warming faster than sorrounding rural areas. The average maximum temperature of the last century has been recorded in the past two decades [36]. Even without global warming, cities are already facing the urban heat island (UHI) effect, where urban areas have become hotter than sorrounding rural areas.

In Australian cities, the intensity of UHI is very significant. The current average intensity varies from 1.0 °C to 13.0 °C. The UHI amplitude changes functionally with the measurement technique adopted. The average intensity using standard method varies between 1.0 °C and 13.0 °C, while the average intensity using non-standard methods is between 1.0 °C and 7.0 °C. Mobile transects and other non-standard measurement methods seem to capture higher UHI intensity as they are commonly used in densely populated urban areas. On the other hand, fixed measuring stations installed in the sorrounding rural areas seem to capture lower UHI intensity as they are used in thermally undisturbed areas.

Urban overheating has different characteristics in different cities and regions of Australia. Urban expansion and reduction of green coverage often lead to the UHI effect. However, the synoptic weather conditions have a greater influence on urban overheating than the UHI itself in many Australian cities. In Sydney, the absolute amplitude of overheating increases as the distance from the coastline increases. The average ambient summer temperatures in the inland western suburbs are at least 2–5 °C higher than the coastal areas of the eastern suburbs [6]. This dramatic fluctuation in the intensity of overheating in Sydney is related to synoptic weather conditions, i.e., sea breeze (cooling mechanism) and hot desert winds (heating mechanism). The sea breeze inhibits hot air advection as it interacts with the UHI circulation and contribute to the cooling effect in the eastern suburbs [6], while the stagnation region in the city canyons, coupled with warm desert winds lead to weaker penetration of cool winds in the western part of the city [4,5]. The intensity of urban overheating in other Australian cities (such as Darwin, Adelaide, Melbourne, and Alice Springs) can be explained by city-specific variables (form, layout, structure, morphology, and anthropogenic heat). Moreover, global climate change and heatwaves have exacerbated overheating in Australian cities [24,35,44].

Urban overheating has caused damage to human health and severely affected energy demand, the economy, and the overall urban sustainability. Between 1993 and 2014, extreme heat has caused more deaths in Australia than floods, hurricanes, lighting, wildfires, and earthquakes combined [37]. Increased temperatures and UHI effects can also harm public health through heat stress and other heat-related diseases. The most vulnerable to overheating are the elderly, young children, chronically ill, mentally ill, outdoor workers, and low-income or socially isolated residents. Existing evidence from different cities in Australia suggests that when the threshold temperature rises by a certain degree, mortality and morbidity will increase. Overall, a 1 °C rise in the threshold temperature (27 °C) can increase the incidence of heat-related morbidity from 1.1% to 4.6% [58,59,64,75].

In addition to causing public health problems, urban overheating also increases urban energy consumption and demand. The overheating of cities will lead to increased energy consumption to meet higher cooling requirements, which will increase greenhouse gas emissions (GHG) and air pollutants. Existing research documented that the average cooling energy can increase by 0.45% to 4.6% per degree rise in threshold temperature of 18 °C [7]. The increase in energy demand will greatly increase the financial burden on governments and may also affect thermal comfort. Worse, urban overheating can affect social behaviour, work and labour productivity, thereby affecting urban development and economic growth [74,76].

Local governments can respond to the impact of urban overheating through emergency plans, outreach activities, and resilient building. However, emergency response and adaptation actions alone cannot save the most vulnerable people. The emergency plans fail to address other interrelated aspects of urban overheating, such as energy disruptions and decreased workplace productivity. Long-term mitigation strategies must be adopted in the natural and built environment to keep residents, buildings and communities cool while also saving energy, health and economic costs.

This article analysed various studies on urban heat mitigation in Australia to support government actions. There is evidence that cool materials, green roofs, vertical gardens, urban greenery, and water-based technologies can significantly alleviate the UHI effect, cool the ambient air, and create a thermally balanced city. Urban greenery, especially trees, has a high potential for mitigation. Trees and hedges can reduce the average maximum UHI by 1.0 °C. The average maximum mitigation performance values of green roofs and green walls are 0.2 °C and 0.1 °C, respectively. Reflective roofs and pavements can reduce the average maximum UHI by 0.3 °C. Water has high heat reduction potential in dry areas [94]. The combined use of greenery and reflective materials can reduce the maximum UHI by 0.8–0.9 °C [80,89], while the combination of cool roofs and cool pavements has a cooling effect in the range of 0.9–3 °C [5,71]. The combination of trees and grass [87], and trees and green roofs [86], can mitigate the maximum UHI by 0.8 °C and 2.4 °C, respectively. The combined use of trees, cool materials, evaporative cooling, shading [5], and water and solar control [89] can reduce the maximum UHI by 1.1 °C and 3 °C, respectively. The average maximum UHI reduction using only one technology is close to 0.4 °C, and the combined use of multiple strategies can reduce the UHI by 1.59 °C, while providing many co-benefits.

The results of this paper can be useful to urban planners and policy makers in reducing urban heat. Governments can use the comprehensive evidence compiled in this article to compare, analyse, and determine the best cooling strategy. Through the use of analytics and a multi-criteria decision-making process, local governments can compare the different mitigation strategies available and determine the most appropriate mitigation strategy. Effective use of evidence in planning and policy is essential to manage urban heat and guide sustainable urban development.

Author Contributions: Conceptualization, K.Y. and L.D.; methodology, K.Y., M.S., and L.D.; software, K.Y.; validation, K.Y., M.S., and L.D.; formal analysis, K.Y.; investigation, K.Y., M.S., L.D., and G.U.; resources, K.Y., L.D. and D.P.; data curation, K.Y., G.U., and R.P.; writing—original draft preparation, K.Y.; writing—review and editing, K.Y., M.S., L.D., R.P., S.H., D.P., and G.U.; visualization, K.Y.; supervision, L.D. and D.P.; project administration—K.Y. and L.D. All authors have read and agreed to the published version of the manuscript.

Funding: This research received no external funding.

Acknowledgments: We would like to acknowledge Salghuna N Nair for her help in generating maps using ArcGIS.

Conflicts of Interest: The authors declare no conflict of interest.

References

1. Schell, L.M.; Smith, M.T.; Bilsborough, A. *Human Biological Approaches to the Study of Third World Urbanism*; Cambridge University Press: New York, NY, USA, 1993.
2. United Nations, Department of Economic and Social Affairs (UN-DESA). *World Urbanization Prospects*; United Nations, Department of Economic and Social Affairs: New York, NY, USA, 2014.
3. Emmanuel, R.; Krüger, E. Urban heat island and its impact on climate change resilience in a shrinking city: The case of Glasgow, UK. *Build. Environ.* **2012**, *53*, 137–149. [CrossRef]
4. Santamouris, M.; Haddad, S.; Fiorito, F.; Osmond, P.; Ding, L.; Prasad, D.; Zhai, X.; Wang, R. Urban Heat Island and Overheating Characteristics in Sydney, Australia. An Analysis of Multiyear Measurements. *Sustainability* **2017**, *9*, 712. [CrossRef]
5. Haddad, S.; Ulpiani, G.; Paolini, R.; Synnefa, A.; Santamouris, M. Experimental and Theoretical analysis of the urban overheating and its mitigation potential in a hot arid city—Alice Springs. *Arch. Sci. Rev.* **2019**, *63*, 425–440. [CrossRef]
6. Yun, G.Y.; Ngarambe, J.; Duhirwe, P.N.; Ulpiani, G.; Paolini, R.; Haddad, S.; Vasilakopoulou, K.; Santamouris, M. Predicting the magnitude and the characteristics of the urban heat island in coastal cities in the proximity of desert landforms. The case of Sydney. *Sci. Total Environ.* **2020**, *709*, 136068. [CrossRef] [PubMed]
7. Santamouris, M. On the energy impact of urban heat island and global warming on buildings. *Energy Build.* **2014**, *82*, 100–113. [CrossRef]
8. Taylor, J.; Wilkinson, P.; Davies, M.; Armstrong, B.; Chalabi, Z.; Mavrogianni, A.; Symonds, P.; Oikonomou, E.; Bohnenstengel, S.I. Mapping the effects of urban heat island, housing, and age on excess heat-related mortality in London. *Urban Clim.* **2015**, *14*, 517–528. [CrossRef]
9. Chien, L.-C.; Guo, Y.; Zhang, K. Spatiotemporal analysis of heat and heat wave effects on elderly mortality in Texas, 2006–2011. *Sci. Total Environ.* **2016**, *562*, 845–851. [CrossRef] [PubMed]
10. Salata, F.; Golasi, I.; Petitti, D.; Vollaro, E.D.L.; Coppi, M.; Vollaro, A.D.L. Relating microclimate, human thermal comfort and health during heat waves: An analysis of heat island mitigation strategies through a case study in an urban outdoor environment. *Sustain. Cities Soc.* **2017**, *30*, 79–96. [CrossRef]
11. Sarrat, C.; Lemonsu, A.; Masson, V.; Guédalia, D. Impact of urban heat island on regional atmospheric pollution. *Atmos. Environ.* **2006**, *40*, 1743–1758. [CrossRef]
12. Chapman, L.; Azevedo, J.A.; Prieto-Lopez, T. Urban heat & critical infrastructure networks: A viewpoint. *Urban Clim.* **2013**, *3*, 7–12.
13. O'Malley, C.; Piroozfarb, P.A.E.; Farr, E.R.P.; Gates, J. An Investigation into Minimizing Urban Heat Island (UHI) Effects: A UK Perspective. *Energy Procedia* **2014**, *62*, 72–80. [CrossRef]
14. Chen, D.; Ren, Z.; Wang, C.H.; Thatcher, M.; Wang, X. *Urban Heat Island on Australian Housing Energy Consumption*; Healthy Buildings: Brisbane, Queensland, Australia, 2012.
15. van Raalte, L.; Nolan, M.; Thakur, P.; Xue, S.; Parker, N. *Economic Assessment of the Urban Heat Island Effect*; AECOM Australia Pty Ltd.: Melbourne, Australia, 2012.
16. Torok, S.J.; Morris, C.J.G.; Skinner, C.; Plummer, N. Urban heat island features of southeast Australian towns. *Aust. Meteorol. Mag.* **2001**, *50*, 1–13.
17. Erell, E.; Williamson, T.J. Intra-urban differences in canopy layer air temperature at a mid-latitude city. *Int. J. Climatol.* **2007**, *27*, 1243–1255. [CrossRef]
18. Nunez, M. *The Urban Heat Island*; University of Tasmania: Hobart, Australia, 1979; p. 46.

19. Haddad, S.; Paolini, R.; Ulpiani, G.; Synnefa, A.; Hatvani-Kovacs, G.; Garshasbi, S.; Fox, J.; Vasilakopoulou, K.; Nield, L.; Santamouris, M. Holistic approach to assess co-benefits of local climate mitigation in a hot humid region of Australia. *Sci. Rep.* **2020**, *10*, 1–17. [CrossRef] [PubMed]

20. Rogers, C.D.; Gallant, A.J.; Tapper, N.J. Is the urban heat island exacerbated during heatwaves in southern Australian cities? *Theor. Appl. Climatol.* **2019**, *137*, 441–457. [CrossRef]

21. Santamouris, M. Analyzing the heat island magnitude and characteristics in one hundred Asian and Australian cities and regions. *Sci. Total Environ.* **2015**, *512–513*, 582–598. [CrossRef] [PubMed]

22. O'Malley, C.; Piroozfar, P.; Farr, E.R.P.; Pomponi, F. Urban Heat Island (UHI) mitigating strategies: A case-based comparative analysis. *Sustain. Cities Soc.* **2015**, *19*, 222–235. [CrossRef]

23. Santamouris, M. Cooling the cities—A review of reflective and green roof mitigation technologies to fight heat island and improve comfort in urban environments. *Sol. Energy* **2014**, *103*, 682–703. [CrossRef]

24. Khan, H.S.; Paolini, R.; Santamouris, M.; Caccetta, P. Exploring the Synergies between Urban Overheating and Heatwaves (HWs) in Western Sydney. *Energies* **2020**, *13*, 470. [CrossRef]

25. Livada, I.; Synnefa, A.; Haddad, S.; Paolini, R.; Garshasbi, S.; Ulpiani, G.; Fiorito, F.; Vassilakopoulou, K.; Osmond, P.; Santamouris, M. Time series analysis of ambient air-temperature during the period 1970–2016 over Sydney, Australia. *Sci. Total Environ.* **2019**, *648*, 1627–1638. [CrossRef]

26. Byrne, M.; Yeates, D.; Joseph, L.; Kearney, M.; Bowler, J.; Williams, M.; Cooper, S.; Donnellan, S.; Keogh, J.S.; Leys, R. Birth of a biome: Insights into the assembly and maintenance of the Australian arid zone biota. *Mol. Ecol.* **2008**, *17*, 4398–4417. [CrossRef]

27. Masouleh, Z.P.; Walker, D.J.; McCauley Crowther, J. A Long-Term Study of Sea-Breeze Characteristics: A Case Study of the Coastal City of Adelaide. *J. Appl. Meteorol. Climatol.* **2019**, *58*, 385–400. [CrossRef]

28. Rajagopalan, P.; Lim, K.C.; Jamei, E. Urban heat island and wind flow characteristics of a tropical city. *Sol. Energy* **2014**, *107*, 159–170. [CrossRef]

29. Lazzarini, M.; Molini, A.; Marpu, P.R.; Ouarda, T.B.; Ghedira, H. Urban climate modifications in hot desert cities: The role of land cover, local climate, and seasonality. *Geophys. Res. Lett.* **2015**, *42*, 9980–9989. [CrossRef]

30. Potchter, O.; Goldman, D.; Kadish, D.; Iluz, D. The oasis effect in an extremely hot and arid climate: The case of southern Israel. *J. Arid. Environ.* **2008**, *72*, 1721–1733. [CrossRef]

31. Alonso, M.; Fidalgo, M.; Labajo, J. The urban heat island in Salamanca (Spain) and its relationship to meteorological parameters. *Clim. Res.* **2007**, *34*, 39–46. [CrossRef]

32. Camilloni, I.; Barrucand, M. Temporal variability of the Buenos Aires, Argentina, urban heat island. *Theor. Appl. Climatol.* **2012**, *107*, 47–58. [CrossRef]

33. Rchid, A. The Effects of Green Spaces (Palme Trees) on the Microclimate in Arides Zones, Case Study: Ghardaia, Algeria. *Energy Procedia* **2012**, *18*, 10–20.

34. Haddad, S.; Paolini, R.; Ulpiani, G.; Synnefa, A.; Hatvani-Kovacs, G.; Garshasbi, S.; Fox, J.; Vasilakopoulou, K.; Neild, L.; Santamouris, M. Holistic approach towards urban sustainability: Co-benefits of urban heat mitigation in a hot humid region of Australia. *Sci. Rep.* **2020**, in press. [CrossRef]

35. Hatvani-Kovacs, G.; Belusko, M.; Pockett, J.; Boland, J. Can the Excess Heat Factor Indicate Heatwave-Related Morbidity? A Case Study in Adelaide, South Australia. *EcoHealth* **2016**, *13*, 100–110. [CrossRef] [PubMed]

36. Commonwealth Scientific and Industrial Research Organisation (CSIRO). *The Report—State of the Climate 2014*; Commonwealth Scientific and Industrial Research Organisation (CSIRO): Canberra, Australia, 2015.

37. Bureau of Meteorology (BoM). *Annual Climate Statement 2015*; Bureau of Meteorology, Commonwealth of Australia: Melbourne, Australia, 2016.

38. Bambrick, H.J.; Dear, K.; Woodruff, R.; Hanigan, I.; McMichael, A. *The Impacts of Climate Change on Three Health Outcomes: Temperature-Related Mortality and Hospitalisations, Salmonellosis and other Bacterial Gastroenteritis, and Population at Risk from Dengue*; Australian Government: Canberra, Australia, 2008; p. 47.

39. Adams, M.; Duc, H.; Trieu, T. *Impacts of Land-Use Change on Sydney's Future Temperatures*; State of New South Wales and Office of Environment and Heritage: Sydney, Australia, 2015; p. 25.

40. Tong, S.; Fitzgerald, G.; Wang, X.-Y.; Aitken, P.; Tippett, V.; Chen, D.; Wang, X.; Guo, Y. Exploration of the health risk-based definition for heatwave: A multi-city study. *Environ. Res.* **2015**, *142*, 696–702. [CrossRef]

41. Anderson, G.B.; Bell, M.L. Heatwaves in the United States: Mortality risk during heatwaves and effect modification by heatwave characteristics in 43 US communities. *Environ. Health Perspect.* **2011**, *119*, 210–218. [CrossRef]

42. Barnett, A.; Hajat, S.; Gasparrini, A.; Rocklöv, J. Cold and heatwaves in the United States. *Environ. Res.* **2012**, *112*, 218–224. [CrossRef] [PubMed]

43. Perkins, S.E.; Alexander, L.V. On the measurement of heatwaves. *J. Clim.* **2013**, *26*, 4500–4517. [CrossRef]
44. Lam, C.K.C.; Gallant, A.J.; Tapper, N.J. Perceptions of thermal comfort in heatwave and non-heatwave conditions in Melbourne, Australia. *Urban Clim.* **2018**, *23*, 204–2186. [CrossRef]
45. Founda, D.; Santamouris, M. Synergies between urban heat island and heatwaves in Athens (Greece), during an extremely hot summer (2012). *Sci. Rep.* **2017**, *7*, 1–11. [CrossRef]
46. Ao, X.; Wang, L.; Zhi, X.; Gu, W.; Yang, H.; Li, D. Observed synergies between urban heat islands and heatwaves and their controlling factors in Shanghai, China. *J. Appl. Meteorol. Climatol.* **2019**, *58*, 1955–1972. [CrossRef]
47. Brázdil, R.; Budíková, M. An urban bias in air temperature fluctuations at the Klementinum, Prague, The Czech Republic. *Atmos. Environ.* **1999**, *33*, 4211–4217. [CrossRef]
48. Ramamurthy, P.; Bou-Zeid, E. Heatwaves and urban heat islands: A comparative analysis of multiple cities. *J. Geophys. Res. Atmos.* **2017**, *122*, 168–178. [CrossRef]
49. Rizvi, S.H.; Alam, K.; Iqbal, M.J. Spatio-temporal variations in urban heat island and its interaction with heatwave. *J. Atmos. Sol.-Terr. Phys.* **2019**, *185*, 50–57. [CrossRef]
50. Li, D.; Sun, T.; Liu, M.; Yang, L.; Wang, L.; Gao, Z. Contrasting responses of urban and rural surface energy budgets to heatwaves explain synergies between urban heat islands and heatwaves. *Environ. Res. Lett.* **2015**, *10*, 054009. [CrossRef]
51. Li, D.; Bou-Zeid, E. Synergistic interactions between urban heat islands and heatwaves: The impact in cities is larger than the sum of its parts. *J. Appl. Meteorol. Climatol.* **2013**, *52*, 2051–2064. [CrossRef]
52. Ulpiani, G.; Ranzi, G.; Santamouris, M. Experimental evidence of the multiple microclimatic impacts of bushfires in affected urban areas: The case of Sydney during the 2019/2020 Australian season. *Environ. Res. Commun.* **2020**, *2*, 065005. [CrossRef]
53. Ulpiani, G. On the linkage between urban heat island and urban pollution island: Three-decade literature review towards a conceptual framework. *Sci. Total Environ.* **2020**, *751*, 141727. [CrossRef]
54. Williams, S.; Nitschke, M.; Weinstein, P.; Pisaniello, D.L.; Parton, K.A.; Bi, P. The impact of summer temperatures and heatwaves on mortality and morbidity in Perth, Australia 1994–2008. *Environ. Int.* **2012**, *40*, 33–38. [CrossRef]
55. Loughnan, M.E.; Tapper, N.J.; Phan, T.; Lynch, K.; McInnes, A.J. *A Spatial Vulnerability Analysis of Urban Populations during Extreme Heat Events in Australian Capital Cities*; National Climate Change Adaptation Research Facility: Gold Coast, Australia, 2013; p. 128.
56. Williams, S.; Nitschke, M.; Sullivan, T.; Tucker, G.R.; Weinstein, P.; Pisaniello, D.L.; Parton, K.A.; Bi, P. Heat and health in Adelaide, South Australia: Assessment of heat thresholds and temperature relationships. *Sci. Total Environ.* **2012**, *414*, 126–133. [CrossRef]
57. Tong, S.; Ren, C.; Becker, N. Excess deaths during the 2004 heatwave in Brisbane, Australia. *Int. J. Biometeorol.* **2010**, *54*, 393–400. [CrossRef]
58. Santamouris, M.; Paolini, R.; Haddad, S.; Synnefa, A.; Garshasbi, S.; Hatvani-Kovacs, G.; Gobakis, K.; Yenneti, K.; Vasilakopoulou, K.; Feng, J. Heat mitigation technologies can improve sustainability in cities. An holistic experimental and numerical impact assessment of urban overheating and related heat mitigation strategies on energy consumption, indoor comfort, vulnerability and heat-related mortality and morbidity in cities. *Energy Build.* **2020**, *217*, 110002.
59. Vaneckova, P.; Beggs, P.J.; De Dear, R.J.; McCracken, K.W.J. Effect of temperature on mortality during the six warmer months in Sydney, Australia, between 1993 and 2004. *Environ. Res.* **2008**, *108*, 361–369. [CrossRef]
60. Inglis, S.C.; Clark, R.A.; Shakib, S.; Wong, D.T.; Molaee, P.; Wilkinson, D.; Stewart, S. Hot summers and heart failure: Seasonal variations in morbidity and mortality in Australian heart failure patients (1994–2005). *Eur. J. Heart Fail.* **2008**, *10*, 540–549. [CrossRef]
61. Tong, S.; Wang, X.Y.; Barnett, A.G. Assessment of Heat-Related Health Impacts in Brisbane, Australia: Comparison of Different Heatwave Definitions. *PLoS ONE* **2010**, *5*, e12155. [CrossRef]
62. Loughnan, M.E.; Nicholls, N.; Tapper, N.J. The effects of summer temperature, age and socioeconomic circumstance on Acute Myocardial Infarction admissions in Melbourne, Australia. *Int. J. Health Geogr.* **2010**, *9*, 41. [CrossRef]
63. Gosling, S.N.; McGregor, G.R.; Páldy, A. Climate change and heat-related mortality in six cities Part 1: Model construction and validation. *Int. J. Biometeorol.* **2007**, *51*, 525–540. [CrossRef]

64. Tong, S.; Wang, X.Y.; Yu, W.; Chen, D.; Wang, X. The impact of heatwaves on mortality in Australia: A multicity study. *BMJ Open* **2014**, *4*, e003579. [CrossRef]

65. Santamouris, M. Regulating the damaged thermostat of the cities—Status, impacts and mitigation challenges. *Energy Build.* **2015**, *91*, 43–56. [CrossRef]

66. Burton, A.; Bambrick, H.; Friel, S. If you don't know how can you plan? Considering the health impacts of climate change in urban planning in Australia. *Urban Clim.* **2015**, *12*, 104–118. [CrossRef]

67. Yu, W.; Mengersen, K.; Hu, W.; Guo, Y.; Pan, X.; Tong, S. Assessing the relationship between global warming and mortality: Lag effects of temperature fluctuations by age and mortality categories. *Environ. Pollut.* **2011**, *159*, 1789–1793. [CrossRef]

68. Commonwealth Scientific and Industrial Research Organisation (CSIRO). *Pathways to Climate Adapted and Healthy Low Income Housing, Final Report*; CSIRO, National Climate Change Adaptation Research Facility: Canberra, Australia, 2013.

69. Byrne, J.; Matthews, T.; Ambrey, C. Comment: Why Poorer Suburbs Are More at Risk in Warming Cities. Available online: http://www.sbs.com.au/topics/science/earth/article/2016/10/18/comment-why-poorer-suburbs-are-more-risk-warming-cities (accessed on 25 April 2019).

70. Zografos, C.; Anguelovski, I.; Grigorova, M. When exposure to climate change is not enough: Exploring heatwave adaptive capacity of a multi-ethnic, low-income urban community in Australia. *Urban Clim.* **2016**, *17*, 248–265. [CrossRef]

71. Santamouris, M.; Haddad, S.; Saliari, M.; Vasilakopoulou, K.; Synnefa, A.; Paolini, R.; Ulpiani, G.; Garshasbi, S.; Fiorito, F. On the energy impact of urban heat island in Sydney: Climate and energy potential of mitigation technologies. *Energy Build.* **2018**, *166*, 154–164. [CrossRef]

72. Guan, H.; Soebarto, V.; Bennett, J.; Clay, R.; Andrew, R.; Guo, Y.; Gharib, S.; Bellette, K. Response of office building electricity consumption to urban weather in Adelaide, South Australia. *Urban Clim.* **2014**, *10*, 42–55. [CrossRef]

73. Jamei, E.; Rajagopalan, P. Urban development and pedestrian thermal comfort in Melbourne. *Sol. Energy* **2017**, *144*, 681–698. [CrossRef]

74. Zander, K.K.; Botzen, W.J.; Oppermann, E.; Kjellstrom, T.; Garnett, S.T. Heat stress causes substantial labor productivity loss in Australia. *Nat. Clim. Chang.* **2015**, *5*, 647–651. [CrossRef]

75. Hanna, E.G.; Kjellstrom, T.; Bennett, C.; Dear, K. Climate Change and Rising Heat: Population Health Implications for Working People in Australia. *Asia Pac. J. Public Health* **2011**, *23*, 14S–26S. [CrossRef] [PubMed]

76. Kjellstrom, T.; Gabrysch, S.; Lemke, B.; Dear, K. The 'Hothaps' programme for assessing climate change impacts on occupational health and productivity: An invitation to carry out field studies. *Glob. Health Action* **2009**, *2*, 2. [CrossRef]

77. Garnaut, R. *The Garnaut Climate Change Review—Update 2011: Australia in the Global Response to Climate Change Summary*; Cambridge University Press: Melbourne, Australia, 2011.

78. Chhetri, P.; Hashemi, A.; Basic, F.; Manzoni, A.; Jayatilleke, G. *Bushfire, Heatwave and Flooding Case Studies from Australia*; School of business IT and Logistics, RMIT University: Melbourne, Australia, 2012.

79. Chen, D.; Wang, X.; Khoo, Y.B.; Thatcher, M.; Lin, B.B.; Ren, Z.; Wang, C.; Barnett, G. (Eds.) *Assessment of Urban Heat Island and Mitigation by Urban Green Coverage*; Springer: Berlin/Heidelberg, Germany, 2013.

80. Jacobs, S.J.; Gallant, A.J.; Tapper, N.J.; Li, D. Use of Cool Roofs and Vegetation to Mitigate Urban Heat and Improve Human Thermal Stress in Melbourne, Australia. *J. Appl. Meteorol. Climatol.* **2018**, *57*, 1747–1764. [CrossRef]

81. Chen, D.; Thatcher, M.; Wang, X.; Barnett, G.; Kachenko, A.; Prince, R. Summer cooling potential of urban vegetation—A modeling study for Melbourne, Australia. *Cities* **2015**, *27*, 35–38. [CrossRef]

82. Chapman, S.; Thatcher, M.; Salazar, A.; Watson, J.E.; McAlpine, C.A. The Effect of Urban Density and Vegetation Cover on the Heat Island of a Subtropical City. *J. Appl. Meteorol. Climatol.* **2018**, *57*, 2531–2550. [CrossRef]

83. Zhang, J. A Study of Parks' Thermal Performances and Their Influential Factors in Urban Heat Island Mitigation for the City of Gold Coast. Master's Thesis, Griffith University, Queensland, Australia, 2019.

84. Soltani, A.; Sharifi, E. Daily variation of urban heat island effect and its correlations to urban greenery: A case study of Adelaide. *Front. Arch. Res.* **2017**, *6*, 529–538. [CrossRef]

85. Razzaghmanesh, M.; Beecham, S.; Salemi, T. The role of green roofs in mitigating Urban Heat Island effects in the metropolitan area of Adelaide, South Australia. *Urban For. Urban Green.* **2016**, *15*, 89–102. [CrossRef]

86. Bruse, M.; Skinner, C.J. Rooftop greening and local climate: A case study in Melbourne. In Proceedings of the International Conference on Urban Climatology & International Congress of Biometeorology, Sydney, Australia, 8 November 1999.

87. Mitchell, V.G.; Cleugh, H.A.; Grimmond, C.S.; Xu, J. Linking urban water balance and energy balance models to analyse urban design options. *Hydrol. Process. An Int. J.* **2008**, *22*, 2891–2900. [CrossRef]

88. Razzaghmanesh, M.; Razzaghmanesh, M. Thermal performance investigation of a living wall in a dry climate of Australia. *Build. Environ.* **2017**, *112*, 45–62. [CrossRef]

89. Santamouris, M.; Ding, L.; Fiorito, F.; Oldfield, P.; Osmond, P.; Paolini, R.; Prasad, D.; Synnefa, A. Passive and active cooling for the outdoor built environment—Analysis and assessment of the cooling potential of mitigation technologies using performance data from 220 large scale projects. *Sol. Energy* **2017**, *154*, 14–33. [CrossRef]

90. Qi, J.; Ding, L.; Lim, S. Planning for cooler cities: A framework to support the selection of urban heat mitigation techniques. *J. Clean. Prod.* **2020**, *275*, 122903. [CrossRef]

91. Oke, T.R. *Boundary Layer Climates*; Routledge: Abingdon-on-Thames, UK, 2002.

92. Gunawardena, K.; Wells, M.; Kershaw, T. Utilising green and bluespace to mitigate urban heat island intensity. *Sci. Total Environ.* **2017**, *584*, 1040–1055. [CrossRef]

93. Manteghi, G.; Bin Limit, H.; Remaz, D. Water Bodies an Urban Microclimate: A Review. *Mod. Appl. Sci.* **2015**, *9*, 1–10. [CrossRef]

94. Ulpiani, G. Water mist spray for outdoor cooling: A systematic review of technologies, methods and impacts. *Appl. Energy* **2019**, *254*, 113647. [CrossRef]

95. Razzaghmanesh, M.; Beecham, S.; Kazemi, F. Impact of green roofs on stormwater quality in a South Australian urban environment. *Sci. Total Environ.* **2014**, *470–471*, 651–659. [CrossRef]

96. Akbari, H.; Kolokotsa, D. Three decades of urban heat islands and mitigation technologies research. *Energy Build.* **2016**, *133*, 834–842. [CrossRef]

97. Armson, D.; Stringer, P.; Ennos, A. The effect of tree shade and grass on surface and globe temperatures in an urban area. *Urban For. Urban Green.* **2012**, *11*, 245–255. [CrossRef]

98. Niachou, A.; Papakonstantinou, K.; Santamouris, M.; Tsangrassoulis, A.; Mihalakakou, G. Analysis of the green roof thermal properties and investigation of its energy performance. *Energy Build.* **2001**, *33*, 719–729. [CrossRef]

99. Afshari, A. A new model of urban cooling demand and heat island—Application to vertical greenery systems (VGS). *Energy Build.* **2017**, *157*, 204–217. [CrossRef]

100. Narita, K.-I.; Mikami, T.; Sugawara, H.; Honjo, T.; Kimura, K.; Kuwata, N. Cool-island and Cold Air-seeping Phenomena in an Urban Park, Shinjuku Gyoen, Tokyo. *Geogr. Rev. Jpn.* **2004**, *77*, 403–420. [CrossRef]

101. Brown, R.D.; Vanos, J.; Kenny, N.; Lenzholzer, S. Designing urban parks that ameliorate the effects of climate change. *Landsc. Urban Plan.* **2015**, *138*, 118–131. [CrossRef]

102. Gill, S.; Rahman, M.; Handley, J.; Ennos, A. Modelling water stress to urban amenity grass in Manchester UK under climate change and its potential impacts in reducing urban cooling. *Urban For. Urban Green.* **2013**, *12*, 350–358. [CrossRef]

103. Skoulika, F.; Santamouris, M.; Kolokotsa, D.; Boemi, N. On the thermal characteristics and the mitigation potential of a medium size urban park in Athens, Greece. *Landsc. Urban Plan.* **2014**, *123*, 73–86. [CrossRef]

104. Vijayaraghavan, K. Green roofs: A critical review on the role of components, benefits, limitations and trends. *Renew. Sustain. Energy Rev.* **2016**, *57*, 740–752. [CrossRef]

105. Bianchini, F.; Hewage, K. How "green" are the green roofs? Lifecycle analysis of green roof materials. *Build. Environ.* **2012**, *48*, 57–65. [CrossRef]

106. Cuce, E. Thermal regulation impact of green walls: An experimental and numerical investigation. *Appl. Energy* **2017**, *194*, 247–254. [CrossRef]

107. Pérez, G.; Rincón, L.; Vila, A.; González, J.M.; Cabeza, L.F. Green vertical systems for buildings as passive systems for energy savings. *Appl. Energy* **2011**, *88*, 4854–4859. [CrossRef]

108. Nugroho, A.M. The Impact of Living Wall on Building Passive Cooling: A Systematic Review and Initial Test. *IOP Conf. Ser. Earth Environ. Sci.* **2020**, *448*, 012120. [CrossRef]

109. Akbari, H.; Matthews, H.D. Global cooling updates: Reflective roofs and pavements. *Energy Build.* **2012**, *55*, 2–6. [CrossRef]

110. Santamouris, M.; Yun, G.Y. Recent development and research priorities on cool and super cool materials to mitigate urban heat island. *Renew. Energy* **2020**, *161*, 792–807. [CrossRef]

111. Synnefa, A.; Santamouris, M.; Apostolakis, K. On the development, optical properties and thermal performance of cool colored coatings for the urban environment. *Sol. Energy* **2007**, *81*, 488–497. [CrossRef]

112. Paolini, R.; Terraneo, G.; Ferrari, C.; Sleiman, M.; Muscio, A.; Metrangolo, P.; Poli, T.; Destaillats, H.; Zinzi, M.; Levinson, R. Effects of soiling and weathering on the albedo of building envelope materials: Lessons learned from natural exposure in two European cities and tuning of a laboratory simulation practice. *Sol. Energy Mater. Sol. Cells* **2020**, *205*, 110264. [CrossRef]

113. Santamouris, M. Using cool pavements as a mitigation strategy to fight urban heat island—A review of the actual developments. *Renew. Sustain. Energy Rev.* **2013**, *26*, 224–240. [CrossRef]

114. Yang, J.; Kumar, D.L.M.; Pyrgou, A.; Chong, A.; Santamouris, M.; Kolokotsa, D.; Lee, S.E. Green and cool roofs' urban heat island mitigation potential in tropical climate. *Sol. Energy* **2018**, *173*, 597–609. [CrossRef]

Publisher's Note: MDPI stays neutral with regard to jurisdictional claims in published maps and institutional affiliations.

Article

The Spatial and Temporal Characteristics of Urban Heat Island Intensity: Implications for East Africa's Urban Development

Xueqin Li [1,*], **Lindsay C. Stringer** [1] and **Martin Dallimer** [2]

[1] Department of Environment and Geography, University of York, York YO10 5NG, UK; lindsay.stringer@york.ac.uk

[2] School of Earth and Environment, University of Leeds, Woodhouse Lane, Leeds LS2 9JT, UK; M.Dallimer@leeds.ac.uk

[*] Correspondence: xl3053@york.ac.uk

Abstract: Due to the combination of climate change and the rapid growth in urban populations in Africa, many urban areas are encountering exacerbated urban heat island (UHI) effects. It is important to understand UHI effects in order to develop suitable adaptation and mitigation strategies. However, little work has been done in this regard in Africa. In this study, we compared surface UHI (SUHI) effects between cities located in different climate zones in East Africa, investigating how they change, both spatially and temporally. We quantified the annual daytime and night-time SUHI intensities in the five most populated cities in East Africa in 2003 and 2017, and investigated the links to urban area size. We consider the possible drivers of SUHI change and consider the implication for future development, highlighting the role of factors such as topography and building/construction materials. We suggest that UHI mitigation strategies targeting East African cities may benefit from more comprehensive analyses of blue and green infrastructure as this offers potential opportunities to enhance human comfort in areas where UHI effects are highest. However, this needs careful planning to avoid increasing associated issues such as disease risks linked to a changing climate.

Keywords: urbanisation; climate; cities; densification; mitigation; population; temperature

Citation: Li, X.; Stringer, L.C.; Dallimer, M. The Spatial and Temporal Characteristics of Urban Heat Island Intensity: Implications for East Africa's Urban Development. *Climate* **2021**, *9*, 51. https://doi.org/10.3390/cli9040051

Academic Editors: Giulia Ulpiani and Michele Zinzi

Received: 3 March 2021
Accepted: 24 March 2021
Published: 28 March 2021

Publisher's Note: MDPI stays neutral with regard to jurisdictional claims in published maps and institutional affiliations.

1. Introduction

More than half of the world's population was residing in urban areas in 2018, with 4.2 billion people living in urban settlements. As the world continues to urbanise, this proportion is expected to increase from 55% to 68% between 2018 and 2050 [1]. One major challenge in coping with urbanisation is the urban heat island (UHI) effect: a phenomenon in which urban areas are warmer than their adjacent rural areas [2]. Air UHI (AUHI) and surface UHI (SUHI) are two major indicators used in UHI analysis. AUHI is based on air temperature measurements above or below the roof level, whereas SUHI is derived from remotely sensed surface temperatures. Here, we focus on the SUHI. The UHI effect is caused by differences in urban and rural energy balance [3]. Compared to rural areas, impervious surfaces in urban areas have lower albedo, and higher thermal capacity and thermal conductivity [4–6]. Urban structures and their configurations can also raise the surface roughness, trapping radiation in built-up areas, and changing overall air movement [5], with precinct ventilation performance further influencing the UHI and thermal comfort [7].

The UHI effect brings about numerous direct and indirect impacts on urban inhabitants, primarily affecting their health [5,8,9]. In particular, the UHI effect exposes urban inhabitants to additional heat stress, generating extra thermal discomfort in many cities under particular urban development patterns and geographic conditions, and increasing heat-related health risks [10]. Generally, the UHI effect prolongs and intensifies heat waves, which increases morbidity and mortality during the heatwave events [11,12].

Warming intensified by urban land use and land cover change has been identified in many cities, which implies that heat stresses will be further exacerbated in the future [5,13], as the urbanisation trend continues. For example, the UHI effect became more prominent in areas of rapid urbanisation in the Pearl River Delta of China [14]; SUHI was found to be influenced by urban land use in Rotterdam in the Netherlands [15]; and amplified temperatures detected in Noida, India, were mainly due to an increase in impervious areas, associated with urbanisation [16]. The impacts caused by the UHI effect will be exacerbated when taking climate change into consideration [5,8,17]. The interactions of UHI effect and climate change are non-linear [5], and are not simply the sum of the combined contributions [18].

Enhancement of UHI intensity has been detected in various cities around the world. Planners have devised models at different scales for conceptualising UHI effect and urban growth. However, the major focus of these studies is on expansion forms of urban growth, rather than densification [19]. In contrast, urban densification targets higher urban density. This approach can lead to economic efficiency and resource conservation as services and infrastructure are more easily integrated into more densely built cities. Urban expansion is usually associated with urban sprawl, which is a kind of rapid urban land expansion with non-contiguous and unplanned development. Urban sprawl is exemplified via unlimited outward expansion, and rapid growth toward suburban areas [20]. Urban sprawl brings numerous challenges for urban development, including unsustainable land use problems, pollution, and environmental degradation [21]. In developing countries of Africa and Asia, rapid urban expansion has become a critical public policy issue in recent years [22]. Urban expansion mainly contributes to the broadening of the area affected by the UHI, while densification is likely to exacerbate the intensity of the UHI effect [23]. Although urban expansion has become the most common form of urban development, some African countries have adopted policies that aim to promote urban densification [24–26].

Many countries with the fastest estimated and projected rates of urbanisation are located in Africa [1], where rapid and often unplanned urbanisation is exacerbating the impacts of a wide range of natural and anthropogenic disasters. The East Africa region is one of Africa's fastest urbanising areas and its growth rate is higher than the average for Africa. Estimates from the United Nations (UN) show that 24% of East Africa's population lives in urban areas in 2018, compared to only 7% in 1960; this represents an increase from 3 million people in 1960 to 65 million urban inhabitants in 2018 [1]. As East Africa continues to urbanise, the proportion of urban populations is expected to increase from 24% to 39% by 2050. The urban poor who live in informal settlements are highly susceptible to heat stress, and are particularly vulnerable to sustained high temperatures due to their location, urban design and poverty levels. As urban area growth and climate change intensify, hundreds of millions of inhabitants will be under heat stress. For example, over 70 million people in Kenya and Uganda experienced 20 to 25 days of extreme heat per year in 2018, yet it could exceed 125 days by 2090, putting far more people at risk [27].

Climate change leads to higher temperatures and longer, more severe, and more frequent heat waves, causing UHI affected areas to bear the brunt of these harsher heat events. East Africa has been identified as one of the most vulnerable regions in Africa in the face of climate change [28]. Most of the East African region has experienced a significant temperature increase from the beginning of the early 1980s [29]. The Famine Early Warning Systems Network (FEWS NET) also reported temperature increases in many areas of Ethiopia, Kenya, South Sudan, and Uganda over the last 50 years [30]. Temperatures in large parts of East Africa are projected to rise during the current century, and the maximum and minimum temperatures will be higher than the baseline period (1961 to 1990) [28]. Climate model projections under the SRES A2 and B1 scenarios indicate temperature increases in all seasons across the whole of Ethiopia, which may lead to a higher frequency of heat waves as well as higher rates of evaporation and evapotranspiration [31].

The UHI effect has become one of the clearest examples of how urbanisation affects the local and regional climate [19]. Therefore, essential measures should be taken to reduce the negative effects of UHI, especially in the coming decades, where the frequency and intensity of extreme heat events are anticipated to become more severe due to the interactions between urban climate, heatwaves, global climate change, and anticipated rapid urbanisation [5]. Globally, the UHI effect has been extensively studied, however most research has been conducted in cities located in Europe, North America and China, while there has been little research in those regions most vulnerable to climate change, such as cities in Africa [19]. Most UHI studies also tend to focus on a single city. Variability in methodological approaches can make direct comparisons across studies (and therefore cities) difficult, making it harder to build a picture of regional changes. This is problematic because climate conditions are important factors, affecting UHI in different ways [32]. To understand more about the importance of regional climate conditions and patterns of urbanisation in determining the extent and distribution of UHI effects, it is important to undertake analyses across multiple cities simultaneously using a common methodology. Thus far, however, few such studies exist.

Given that the combined impacts of urbanisation and climate change will potentially have large impacts on future urban temperatures [19] and exacerbate existing heat stress in East African cities, understanding the UHI effect in these cities is important if UHI is to be mitigated. This study compares SUHI between East African cities located in different climate zones and assess how it changes spatially and temporally. This allows us to understand the possible driving factors, and consider the implications for future urban development. We address the following questions: (1) How does SUHI vary in different climate zones in East Africa? (2) What are the spatial and temporal variations of SUHI in these cities and what are the possible driving factors? (3) What are the implications for future urban development in the region?

2. Materials and Methods

2.1. Study Area

Five cities spanning the four climate zones in the region were selected as study areas based on the following criteria [33]: (i) being the largest or capital city; (ii) being the main economic and commercial centre of their country; and (iii) experiencing rapid urbanisation with the highest population in their respective countries. The location of these cities and the corresponding climatic information (en.climate-data.org) and recorded urban development pattern are provided in Table 1, Table 2 and Figure 1. Climate classification data was taken from 1 km resolution Köppen-Geiger maps [34].

Table 1. Geographic and climatic information of the study cities.

City	Population (2015)	Longitude/ Latitude	Altitude (Meters about Sea Level)	Annual Rainfall (mm)	Temperature (°C)		
					Annual	Maximun	Minimum
Khartoum (Sudan)	5,128,000	32.527° E 15.522° N	386	70	29.9	34.3 (May)	22.7 (Jaunuary)
Dar es Salaam (Tanzania)	5,116,000	39.103° E −6.807° S	16	1114	26.1	27.9 (February)	24.4 (July)
Nairobi (Kenya)	3,914,000	36.858° E −1.181° S	1669	674	18.8	17 (July)	20.5 (February or March)
Addis Ababa (Ethiopia)	3,871,000	38.565° E 8.884° N	2350	1874	15.6	17.2 (April)	14.1 (December)
Kampala (Uganda)	2,577,000	32.596° E 0.291° N	1224	1747	21.4	22.7 (February)	20.7 (July)

Table 2. Climate zones and urban development information of the study cities.

City	Climate Zone	Climate Condition	Urban Development Pattern
Khartoum (Sudan)	Warm desert climate	There are three seasons in the city, a dry season from November to February, hot season from March to May, and a wet season from June to October [35].	The development of the city is tightly linked to the evolution of the Blue and White Nile Rivers, and is also influenced by increasing land exploitation for intensive and extensive cultivation [36].
Dar es Salaam (Tanzania)	Tropical savannah climate	The climate of the city is greatly influenced by the northeast monsoon which prevails from March to October. The southeast monsoon dominates between October and March. The climate is also greatly influenced by the sea due to proximity of the city to the Indian Ocean [37].	Due to its location along the coast and the existence of the four main roads entering the city, Dar es Salaam has been observed to expand in a radial structure [37]. The city is experiencing multiple development projects involving partnerships between the government and UN HABITAT. These projects aim to reduce the area of unplanned settlements [37].
Nairobi (Kenya)	Subtropical oceanic highland climate	The city experiences a short rainy period in November/December and a heavy rainy season from March till the beginning of June [38].	With functionalism as the main planning principle, the city has grown in concentric zones [33,39]. More than half the population reside in informal settlements [40]. The city is served by a highly polluted river (the Nairobi River) and exhibits ahigh proportion of open vegetated spaceswithin the urban area [40].
Addis Ababa (Ethiopia)	Humid subtropical climate/Subtropical oceanic highland climate	The short rainy season extends from March to May and the main rainy season from June to September, followed by a dry season from October to February [41,42].	The city is experiencing rapid urbanisation, with increasing built-up area (impervious surfaces), and is characterized by mixed land uses dominated by informal settlements located close to urban growth centers [33].
Kampala (Uganda)	Tropical rainforest climate	There are two wet seasons (March-May and September-November). Torrential rains are often observed from March to May and July is normally the driest month [43].	The city has expanded in all directions with growth primarily concentrated along main roads. Between 1989 and 2010 the total built-up area increased exponentially, a result of both natural increase and migration [43,44].

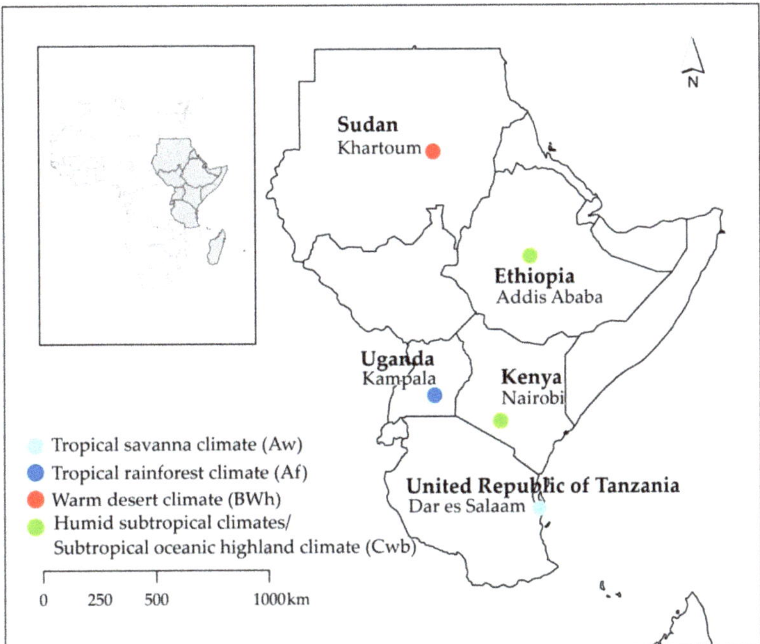

Figure 1. The location of study area and Africa.

2.2. Data and Methodology

The definition of urban and non-urban for SUHI estimation is based on data from Landscan urban extent (Natural Earth; https://www.naturalearthdata.com/, (accessed on 1 May 2020). City centres were defined as the centre of the most populated areas in each city. Urban units are closed polygons around contiguous urban agglomerations. An advantage of using this dataset is that it is based on a consistent algorithm implemented on the MODIS land use satellite. It also consistently bounds global hotspots of human habitation. This approach is also more appropriate than using administrative boundaries, as the urban areas of many developing countries have expanded rapidly and do not always map directly onto administrative jurisdictions. Administrative boundaries are also not comparable across East Africa's cities, and often do not include the full extent of the urban areas [33].

We used the Global Surface UHI Explorer to quantify SUHI. The Global Surface UHI Explorer is based on the Simplified Urban Extent (SUE) algorithm [45], which defines the SUHI as the average LST difference between the urban and non-urban pixels, as classified from spectral reflectance data, within an urban agglomeration or city [45].

The SUE uses the MODIS-derived LST data from TERRA (MOD11A2) and AQUA (MYD11A2). After selecting clear-sky pixels with average LST error of less than or equal to 3 K for quality control, data estimated the LST at four local times: 01:30, 10:30, 13:30, and 22:30. Day time and night-time SUHI varies according to different climate conditions, so we focus on annual daytime and night-time SUHI here. Based on Global Multi-resolution Terrain Elevation Data 2010 (GMTED2010), the SUE eliminated the influence of the elevation differences, by filtering to only include those clusters with a mean elevation difference of less than 50 m [45].

Data from the TERRA platform is available from February 2000 until the present, and data from AQUA is available from July 2002 to the present. As data in the Global Surface UHI Explorer version 3 was only available from 2003 to 2017, we chose the years 2003 and 2017 for observation [45].

The Welch's Test (unequal variance) was conducted to see if two years of UHI are significantly different. Welch's Test is a modification of the Student's *t*-test that performs better than the Student's *t*-test whenever sample sizes and variances are unequal between groups, and gives the same result when sample sizes and variances are equal [46].

$$x = \frac{\overline{X_1} - \overline{X_2}}{\sqrt{\frac{S_1^2}{N_1} + \frac{S_2^2}{N_2}}} \tag{1}$$

The degrees of freedom of the Welch *t*-test is estimated as follows:

$$df = \frac{\left(\frac{S_1^2}{N_1} + \frac{S_2^2}{N_2}\right)^2}{\left(\frac{S_1^4}{N_1^2(N_1-1)} + \frac{S_2^4}{N_2^2(N_2-1)}\right)} \tag{2}$$

$\overline{X_j}$ = sample mean,
$S_{j=}$ = sample deviation,
N_j = sample size, $j \in \{1, 2\}$

We set a null hypothesis that 15 years of urban development did not change the UHI intensity. We used Hedges' g (the unbiased version of Cohen's d) to compute the effect size for different years' UHI with different sample sizes (n) [47], by adjusting the calculation of the pooled standard deviation with weights for the sample sizes.

$$Hedges' g = \frac{M_1 - M_2}{SD^*_{pooled}} \tag{3}$$

$M_1 - M_2$ = difference in means,
SD^*_{pooled} = *pooled and weighted standard deviation,*

Overall this approach is identical to Cohen's d with a correction of a positive bias in the pooled standard deviation [48,49]. The result of Hedges' g is represented by gHedge in the subtitle of each plot in Tables 4 and 5 [50]. We draw on the suggestions of Cohen (1988) for interpreting the magnitude of effect sizes, whereby intervals for Cohen's d under 0: negative effect; 0 to 0.2: no effect; 0.2 to 0.5: small effect; 0.5 to 0.8: intermediate effect; 0.8 and higher: strong effect [48]. All the statistics analyses were undertaken in RStudio using the function ggbetweenstats from R package ggpubr [51].

3. Results

3.1. Annual SUHI

The results of SUHI intensity in 2003 and 2017 are shown in Table 3, Figures 2 and 3. The highest SUHI intensity in 2003 across the five cites was found in Kampala (tropical rainforest climate zone), followed by Nairobi (subtropical oceanic highland climate), Dar es Salaam (tropical savannah climate) and Addis Ababa (subtropical oceanic highland climate). Notably, Khartoum (warm desert climate) has a daytime SUHI of below zero, indicating that the city was cooler than its surrounds. This SUHI sequence demonstrated a slight change in 2017, when Kampala (tropical rainforest climate) still exhibited the highest annual UHI intensity but Dar es Salaam (tropical savannah climate) rose to second place followed by Nairobi (subtropical oceanic highland climate). The annual SUHI intensity for Addis Ababa (subtropical oceanic highland climate) had become the lowest except for Khartoum which was still showing to be a UCI in 2017, with a below zero daytime SUHI.

Table 3. SUHI intensity during daytime and night-time in the study cities in 2003 and 2017. The n value is the number of UHI pixels.

City	Annual Daytime SUHI (°C)				Annual Night-Time SUHI (°C)			
	2003		**2017**		**2003**		**2017**	
	n	**SUHI Intensity**	**n**	**SUHI Intensity**	**n**	**SUHI Intensity**	**n**	**SUHI Intensity**
Khartoum	446	−0.44	640	−0.34	446	1.01	640	1.1
Addis Ababa	131	0.17	254	0.42	131	0.25	254	−0.13
Kampala	240	2.21	285	1.93	240	1.02	285	0.82
Nairobi	42	1.73	242	0.57	42	0.93	242	0.43
Dar es Salaam	131	0.89	173	1.85	131	0.26	173	0.3

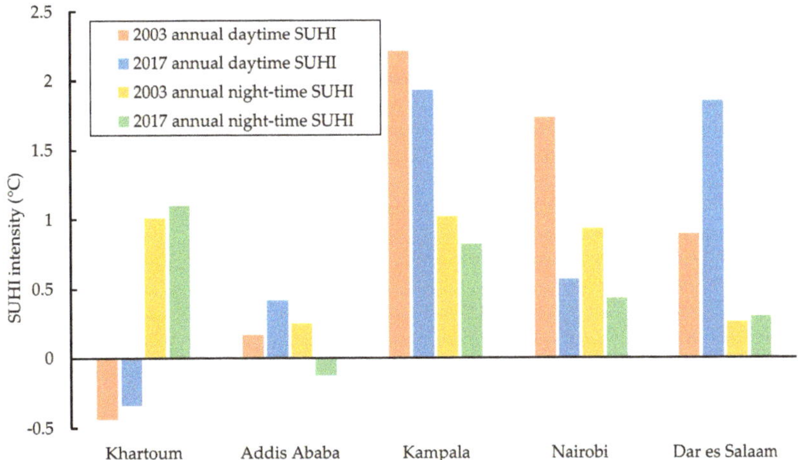

Figure 2. Daytime and night-time SUHI spatial distribution of each city in 2003 and 2017.

For annual night-time SUHI intensity in 2003, Kampala (tropical rainforest climate) still had the highest SUHI intensity. This was followed closely by Khartoum (warm desert climate), Nairobi (subtropical oceanic highland climate), Dar es Salaam (tropical savannah climate) and Addis Ababa (subtropical oceanic highland climate). In 2017, Khartoum (warm desert climate) showed the highest night-time SUHI intensity, followed by Kampala (tropical rainforest climate), Nairobi (subtropical oceanic highland climate), Dar es Salaam (tropical savannah climate) and then Addis Ababa (subtropical oceanic highland climate).

For all five cities, SUHI intensity and magnitude varied with time. In 2003, annual daytime UHI intensity observations were −0.44 °C in Khartoum, 0.17 °C in Addis Ababa, 2.21 °C in Kampala, 1.73 °C in Nairobi and 0.89 °C in Dar es Salaam. Annual night-time SUHI intensity in 2003 was −1.01 °C in Khartoum, −0.25 °C in Addis Ababa, 1.02 °C in Kampala, 0.93 °C in Nairobi and 0.26 °C in Dar es Salaam. In 2017, annual daytime SUHI intensity observations from GEE were −0.34 °C in Khartoum, 0.42 °C in Addis Ababa, 1.93 °C in Kampala, 0.57 °C in Nairobi and 1.85 °C in Dar es Salaam. Annual night-time SUHI intensity in 2017 was 1.1 °C in Khartoum, −0.13 °C in Addis Ababa, 0.82 °C in Kampala, 0.43 °C in Nairobi and 0.30 °C in Dar es Salaam.

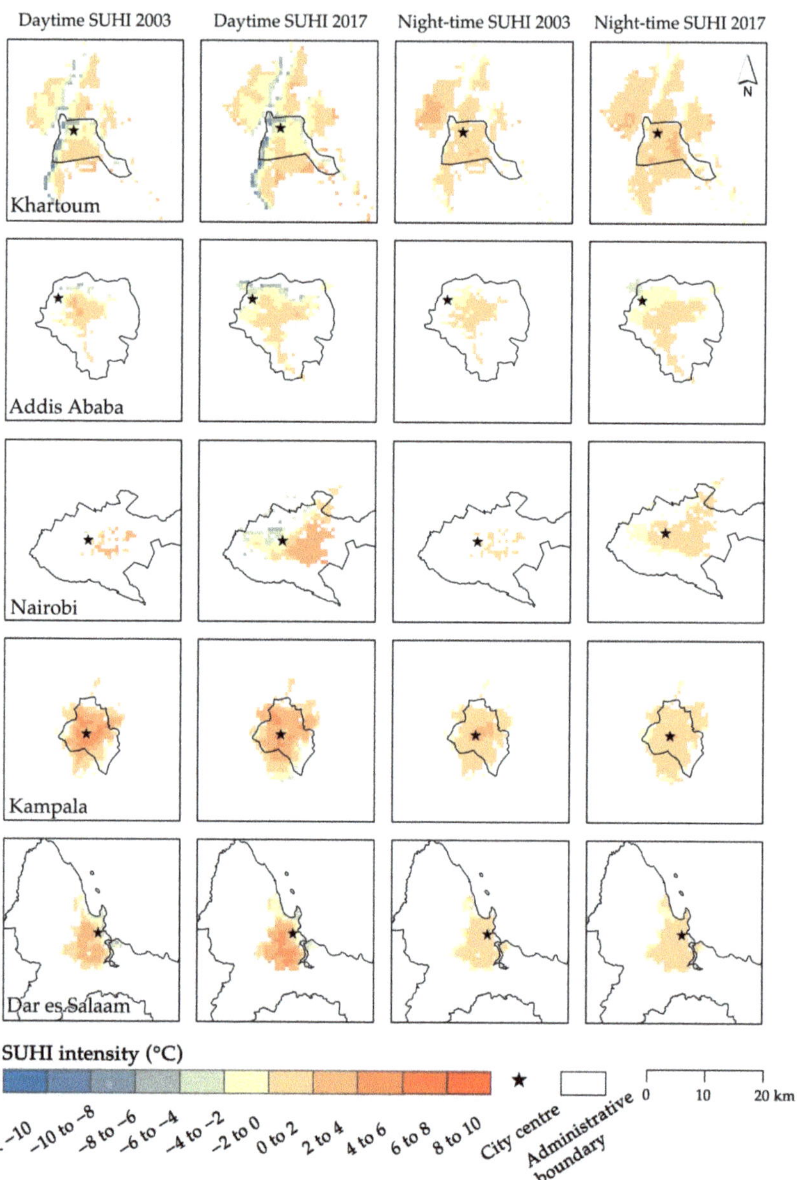

Figure 3. Daytime and night-time SUHI spatial distribution in each city in 2003 and 2017.

Addis Ababa, Kampala, Nairobi and Dar es Salaam demonstrate positive daytime SUHI intensity in 2003 while Khartoum, located within the warm desert climate, was found to demonstrate a surface urban cool island (SUCI) effect. In 2017, Kampala, Nairobi and Dar es Salaam still demonstrate a positive daytime SUHI intensity, while Addis Ababa shows a SUCI, as does Khartoum. All cities had positive night-time SUHI in 2003. This was true in 2017, except for Addis Ababa. The urban areas located in the humid subtropical climate changed from displaying a SUHI to SUCI. In 2003 and 2017, Kampala presents the highest daytime SUHI intensity and Khartoum demonstrates the lowest daytime SUCI.

In contrast, Khartoum shows the highest SUHI intensity for the night-time while Addis Ababa has the lowest night-time SUHI intensity in 2003 and 2017.

Figure 3 shows the distribution of daytime and night-time SUHI intensity and magnitude in 2003 and 2017 for each of the five study areas. Compared to 2003, the urban area affected by SUHI has increased. For Khartoum, this area increased by 43.5%, from 44,600 ha to 64,000 ha. In Addis Ababa, the SUHI-affected area rose from 13,100 ha in 2003 to 25,400 ha in 2017, presenting a 93.9% increase. The largest change in the aerial extent of SUHI affected areas was for Nairobi, growing from 4,200 ha to 24,200 ha; an increase of 476.9%. Changes were smallest in Kampala (24,000 ha to 28,500 ha; 18.8%) and Dar es Salaam (6.8% from 16,200 ha to 17,300 ha).

From 2003 to 2017, the strength of SUCI in central areas of Khartoum, Nairobi and Dar es Salaam increased and the SUHI affected areas gradually expanded from the urban centre to the surrounding area. In Addis Ababa and Kampala, the SUHI intensity strength in central urban areas decreased and the affected areas reduced. In particular, in Kampala, the annual daytime SUHI intensity centre fragmented. There were two subsidiary SUHI centres in 2003, which became four separated SUHI centres, surrounding the urban area centre.

Annual night-time SUHI in Khartoum was positive and stronger than in other study cities. In 2003, the SUHI centre was located in the north west, and then moved toward the east, and in 2017 there were two SUHI centres in Khartoum located on the east side and middle west, divided by White Nile river. In 2003, the major SUHI centres in Addis Ababa and Nairobi were located south east of the cities, and their SUHI affected areas expanded and intensified in 2017. Night-time SUCI effect areas also expanded and intensified in 2017. Addis Ababa's SUCI expanded toward the urban north west, the same as the daytime SUHI expansion trajectory. However, Nairobi's night-time SUCI mostly expanded to the west, differing from the SUCI and daytime SUHI which appeared in north west side of the city. Compared to 2003, the area of the central city experiencing an annual night-time SUHI in Kampala shrank in 2017. However, as a whole a larger area of the city was affected by night-time SUHI. Night-time SUHI affected a larger area in Dar es Salaam in 2017, in a similar way to changes in daytime SUHI.

3.2. Summary Statistics

Figures 4 and 5 present daytime and night-time SUHI value distributions for each city in 2003 and 2017. A summary of annual mean, Welch's Test, and Hedges' g is provided in Tables 4 and 5. For Khartoum's urban areas located in a warm desert climate, the annual daytime SUHI was less intense than the night-time SUHI in 2003 and 2017. Daytime SUHI intensity increased significantly in Khartoum's urban areas over the last decade and a half, even though the expansion of urban areas over time would amplify the SUCI [39]. For Addis Ababa and Nairobi, both located in a humid subtropical climate, the annual daytime SUHI intensity was less than the night-time intensity in 2003, and the annual daytime SUHI intensity was more pronounced than the night-time intensity in 2017. For Nairobi, also located in a humid subtropical climate zone, the annual daytime and night-time SUHI intensities were greater than the night-time intensity both in 2003 and 2017. For Kampala and Dar es Salaam, annual daytime SUHI intensity was greater than the night-time intensity in 2003, and the annual daytime surface UHI intensity is stronger than the night-time intensity in 2017.

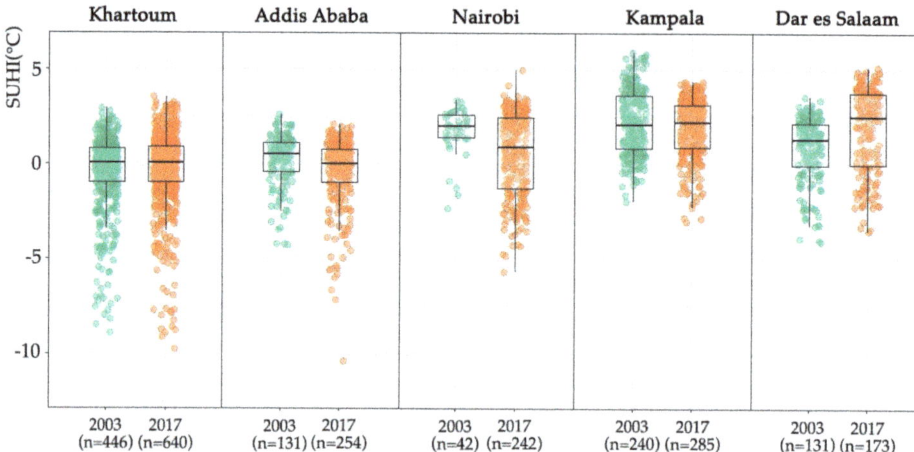

Figure 4. Daytime SUHI value distributions for each city in 2003 and 2017. The boxplots show the distribution of daytime SUHI in 2003 and 2017 in the five cities. The rectangle showing the ends of the first and third quartiles and central line the median. The n value under the x-axis is the number of SUHI pixels.

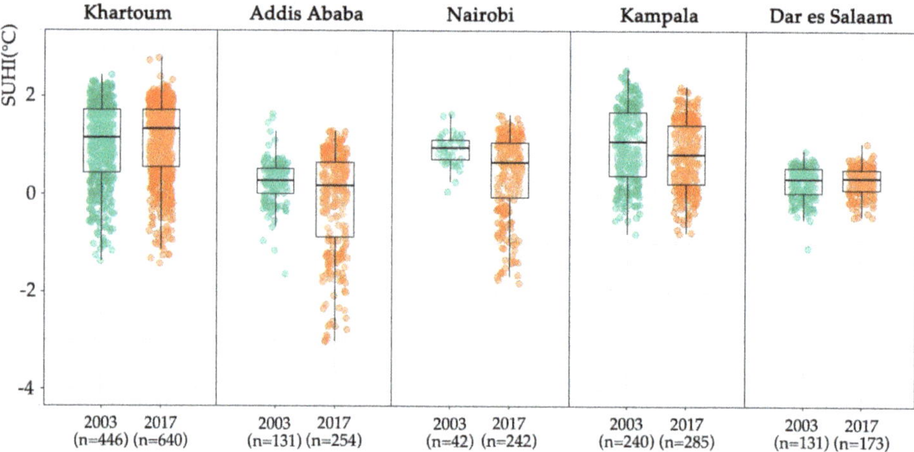

Figure 5. Night-time SUHI value distributions for each city in 2003 and 2017. The boxplots show the distribution of night-time SUHI in 2003 and 2017 in the five cities. The rectangle showing the ends of the first and third quartiles and central line the median. The n value under the x-axis is the number of SUHI pixels.

Table 4. Summary of annual daytime SUHI intensity statistics.

City	Daytime SUHI Intensity (°C)						Welch's Test		Hedges' g
	Maximum		Minimum		Annual Mean				
	2003	2017	2003	2017	2003	2017			
Khartoum	2.94	3.51	−8.91	−9.76	−0.44	−0.34	−0.79	$p = 0.43$	−0.05
Addis Ababa	2.6	2.08	−4.28	−10.39	0.17	−0.42	3.45	$p = 0.001$	0.36
Kampala	5.9	4.38	−1.93	−2.99	2.21	1.93	1.93	$p = 0.054$	0.17
Nairobi	3.35	4.96	−2.36	−5.68	1.73	0.57	4.86	$p \leq 0.001$	0.65
Dar es Salaam	3.59	5.15	−4.07	−3.49	0.89	1.85	−4.39	$p \leq 0.001$	−0.48

Table 5. Summary of annual night-time SUHI intensity statistics.

City	Night-Time SUHI Intensity (°C)						Welch's Test		Hedges' g
	Maximum		Minimum		Annual Mean				
	2003	2017	2003	2017	2003	2017			
Khartoum	2.43	2.77	−1.38	−1.42	1.01	1.1	−1.63	$p = 0.103$	−0.1
Addis Ababa	1.64	1.29	−1.64	−3.03	0.25	−0.13	4.92	$p \leq 0.001$	0.47
Kampala	2.55	2.21	−0.82	0.79	1.02	0.82	3.12	$p = 0.002$	0.27
Nairobi	1.64	1.63	0.06	−1.82	0.93	0.43	6.91	$p < 0.001$	0.8
Dar es Salaam	0.91	1.05	−1.09	−0.46	0.26	0.30	−1.22	$p = 0.225$	−0.13

From Table 4, Welch *t*-tests confirm that there was no statistically significant difference between 2003 and 2017 in terms of annual daytime UHI intensity for Khartoum and Kampala ($p > 0.05$). However, there is highly statistically significant difference between 2003 and 2017 annual daytime SUHI intensity in Addis Ababa ($p = 0.001$), Nairobi ($p \leq 0.001$) and Dar es Salaam ($p \leq 0.001$). The negative Hedges' g indicates that 15 years of land use change caused a negative effect in SUHI growth in Khartoum (Hedges' g = −0.05) and Dar es Salaam (Hedges' g = −0.48). Different positive Hedges' g values indicate no size effect in Kampala (Hedges' g = 0.36), a small effect in Addis Ababa (Hedges' g = 0.36) and an intermediate effect size in Nairobi (Hedges' g = 0.65).

For night-time SUHI (Table 5), Welch *t*-tests indicate that there is no statistically significant difference between 2003 and 2017 annual night-time SUHI intensity for Khartoum and Dar es Salaam ($p > 0.05$). However, the difference between the 2003 and 2017 annual night-time SUHI intensity is statistically significant ($p < 0.05$) for Kampala. There is a highly statistically significant difference between the 2003 and 2017 annual daytime SUHI intensity in Addis Ababa ($p \leq 0.001$) and Nairobi ($p \leq 0.001$). The negative Hedges' g indicates that 15 years of land use change negatively affected SUHI growth in Khartoum (Hedges' g = -0.1) and Dar es Salaam (Hedges' g = -0.13). Different positive Hedges' g values indicate there is a small effect in Addis Ababa (Hedges' g = 0.47) and Kampala (Hedges' g = 0.27), and an intermediate effect size in Nairobi (Hedges' g = 0.8).

4. Discussion

4.1. Climate Zones

The spatial and temporal heterogeneity of the SUHI effect in East Africa is explicitly delineated in this study. Generally, the climate effects on SUHI could be explained by the differences in soil moisture [3,52]. The hot-wet cities (i.e., Kampala and Dar es Salaam) normally present higher soil moisture contents compared with desert cities (i.e., Khartoum) [53] and exhibit larger urban–rural differences in soil moisture, resulting in higher daytime and lower night-time LST differences between urban and surrounding rural areas (i.e., UHI), thus amplifying SUHI differences between day and night. In addition, climate indirectly affects UHI intensities through regulating vegetation conditions, surface albedo, and anthropogenic heat emissions [54]. This partly explains why Khartoum (warm desert climate) presented a SUCI effect during the daytime, but a positive SUHI at night. Vegetation in the Khartoum urban area compared to bare land in the rural buffer is a possible contributor to the negative annual daytime SUHI [55]. The reversal of the SUHI diurnality in desert cities has been reported elsewhere, e.g., in Las Vegas [56] and Abu Dhabi [57]. During 2003 and 2017, almost all the annual mean SUHI intensities of the cities fit the distribution of decreases with latitudinal increases, except Khartoum and Dar es Salaam. Chakraborty similarly notes latitudinal variation in surface UHI for daytime and night-time [45].

While previous authors [45] used a global analysis algorithm to analyse multiple cities in East Africa in an attempt to develop a standardized data collection approach our research offers additional insight through our statistical analysis. This allows us to extend existing work. Generally, our results in East Africa show the arid urban areas present the highest SUHI intensity for night-time, consistent with [45] and [58]. Peng [58] identified the range of annual SUHI intensity in Africa as being from -0.2 (°C) to 2.0 (°C) for daytime, and 0.4 (°C) to 1.4 (°C) for night-time. Our results fall outside of this range, spanning -0.44 (°C) to 2.21 (°C) for day-time and -0.13 (°C) to 1.1 (°C) for night-time. These differences in findings may be down to the selection of different study periods and different land cover data, LandScan from Natural Earth was used by [43] while MODIS Global Land Cover Map was used by [56].

Similar to findings in previous investigations, mean annual daytime SUHI intensities were higher than mean night-time SUHI intensities in most cities. A study in South America found that the mean daytime SUHI intensities was higher than the mean night-time SUHI intensities in all areas except for the arid climate zone [59]. The southeast of China also presents stronger SUHI in the daytime than night-time [60], but the annual mean SUHI intensity was apparently weaker in arid northwest China's cities [6]. The differences between daytime SUHIs and night-time SUHIs are probably because the mechanism of daytime heat island formation is different from that at night. The daytime SUHI was generally explained as the result of an increase in sensible heat flux and a reduction in latent heat flux due to widespread vegetated and evaporating soil surfaces being encroached by impervious surfaces [6], while the release of more stored energy in the urban areas compared to adjacent areas major contribute to night-time SUHI [6]. Compared to 2003, some parts of the cities had a stronger SUHI while some places had a stronger SUCI effect in 2017. This means the temperature differences of all the cities were becoming stronger compared to in the past. Given projected climate changes for East Africa [28], temperatures in large parts of East Africa will continue to rise during the 21st century. This suggests that these already intensified SUHIs will become more complicated. The important role of climate conditions and their diurnal variations in SUHI intensity should not be ignored [6,61] but the changes are also inadequately explained by a focus on climate zones alone.

4.2. Urbanisation Pattern

SUHI intensity and magnitude change is also associated with urban development patterns [19], with built-up areas being one of the most important land-cover types in urban areas, and expected to have a significant effect on SUHI [60].

For the case study cities, the pattern of urban expansion and densification explains the change of SUHI intensity to some extent [19]. In Dar es Salaam, there is a highly statistically significant difference (Welch's Test, $p \leq 0.001$) in annual daytime SUHI between 2003 and 2017, with a negative size effect (Hedges' g $= -0.48$). The statistical results could be explained by the densification of Dar es Salaam. Previous research has also pointed out that Tanzania's urban built-up densities and informal settlements have been growing persistently in recent decades [62].

The urban growth of Addis Ababa and Nairobi reflect an urban expansion development pattern. From 2003 to 2017, the SUHI influenced areas of Addis Ababa expanded from 13,100 ha to 25,400 ha, with a 93.9% increase. The mean annual daytime and night-time SUHI intensity of Addis Ababa presented a highly statistically significant decrease (Welch's Test, $p \leq 0.001$) from 2003 to 2017, but the maximum annual SUHI was higher and the minimum SUHI became lower in 2017 compared to 2003. Similarly, Nairobi is experiencing fast urban expansion. The annual SUHI magnitude of Nairobi saw the greatest changes among the five cities, growing from 4200 ha to 24,200 ha with an increase of 476.9%. The mean annual daytime and night-time SUHI also presented a highly statistically significant decrease (Welch's Test, $p \leq 0.001$) between 2003 and 2017. In particular, the Hedges' g values for Nairobi indicate an intermediate size effect in the daytime (Hedges' g $= 0.65$) and strong size effect (Hedges' g $= 0.8$) in the night-time. In other words, these two cities were experiencing rapid expansion with some places getting hotter and some places getting cooler. Researchers have already pointed out that the built-up area of Addis Ababa and its surrounding towns is expanding into the peri-urban region [63]. This development not only influences the SUHI intensity but also leads to high losses of farmland, directly influencing food production for the urban populations [64]. The urban growth in Nairobi was at the cost of bare land (land without any cover as well as deserts), showing a star-shaped pattern along the main transport routes from the city centre [65].

The statistical analysis could indicate the urban development of Kampala and Khartoum followed a more comprehensive pattern, with urban expansion accompanied with densification. For these two cities, there is no significant difference in annual daytime SUHI between 2003 and 2017, with no effect or negative size effect (both Hedges' g < 0.2). There is no highly significant difference in annual night-time SUHI between 2003 and 2017, with no effect or a small size effect (Hedges' g < 0 or 0.2 < Hedges' g < 0.5).

Which kind of urban development pattern is best for the development of cities to reduce any increase in SUHI intensity is still unclear. From this study, we suggest that a compact city will face enhanced SUHI, while an overexpanding city will mitigate mean annual SUHI but cause local thermal anomalies. In Africa, urbanisation has not frequently been associated with economic development and improvements in wellbeing [66], although urbanisation can offer many development opportunities for countries if well managed. Consequently, it is important that further urban development occurs in a way that is beneficial to people, and which does not exacerbate risks associated with climate change, such as increased exposure to temperature extremes.

Policy makers and city planners should pay more attention to one of the main variables influencing rapid urbanisation in Africa: natural rural-urban migration [67]. People migrate to urban areas primarily in response to perceived better employment and economic opportunities available there (pull factors); they also migrate to escape unfavourable conditions exacerbated by rural challenges and climate change (push factors) [68]. The IPCC pointed out that climate change could influence the size and characteristics of human settlements in Africa owing to the scale and type of rural-urban migration are partially driven by climate change [69].

Fast urbanisation sometimes implies a shift of poverty levels from rural areas to urban areas, which is why we see poverty characteristics often align with settlement patterns in African cities, with informal settlements showing particular growth. In Dar es Salaam, over 70% of the estimated 2.4 million inhabitants lived in informal settlements in 2000 [70], with this number projected to increase to 80% in the coming decades [71]. Such settlements often experience elevated heat exposure [45,72], as well as other environmental risks, such as greater exposure to pollution [44]. In informal settlements in Nairobi, temperatures regularly exceed temperatures at the central, non-slum monitoring station by several degrees or more [73]. The extra heat exposure experienced by residents of informal settlements, is mainly due to their dense population, tin housing and little vegetation, showing the importance of building materials in contributing to UHI effects. In Kampala, millions of urban inhabitants will live in informal areas by 2030 with high potential for them to suffer from epidemiological diseases related to unsanitary conditions [74]. It is also important to note that urban areas in Africa contain a large proportion of the people who are migrating from rural areas because of climate change impacts, in combination with other factors. Policymakers should consider designing UHI adaptation strategies to minimise its impacts on the most socioeconomically vulnerable populations in informal settlements, who may be less equipped to adapt to environmental stressors. Promoting green space in lower-income neighbourhoods is one appropriate urban strategies that policymakers may adopt to ameliorate some of UHI's inequitable burden on economically disadvantaged inhabitants [72], while encouraging use of building materials with higher heat reflectance offers another strategy (this is explored further below). Improving support for climate change adaptation in the rural source areas of urban immigrants may also be a suitable strategy for policymakers to pursue.

4.3. Climate Change and Urbanisation

Urban areas consume over two-thirds of the world's energy and account for more than 70% of global CO2 emissions [75]. Moreover, 90% of urban growth will take place in less developed regions in the coming decades [76], where urbanisation is generally unplanned, fuelling the continuous growth of mostly poor informal settlements in areas such as East Asia, South Asia and sub-Saharan Africa [76]. One billion urban inhabitants are projected to live in informal settlements, where the impact of climate change is most acute [77]. The current urban development deficiency in a form of infrastructure shortfalls, brings opportunities to "leapfrog" to low or zero-emission systems and structures [77]. This implies that a major reconceptualisation of countries' approaches to urban development could still be undertaken. As the World Migration Report [78] notes, about two-thirds of the investments in urban infrastructure to 2050 in Africa have yet to be made, offering an opportunity to manage the urbanisation trends in a way that could harness the benefits of rapid urbanisation while reducing the risks. Combining multiple UHI mitigation and adaptation strategies may effectively negate the localised temperature increasing due to both UHI and climate change [72,79].

Many researchers have noticed the relationship between urban morphology and the UHI intensity [7,80]. The morphological characteristics of the city such as building height, building density, road width and orientation may affect the heat balance in urban canyons [80]. In Dar es Salaam, scattered settlements provided better temperature regulation services than other residential configurations due to their urban morphology characteristics, since former are associated with relatively large proportions of green structures [81]. The negative relationship between distance from the city centre and SUHI intensity was detected in Addis Ababa [82]. As noted above, the characteristics of housing materials also influence the SUHI intensity, with different housing types providing different insulation properties and protecting from heat and cold to different extents, even directly impacting on temperature-related mortality [83]. Analysis suggests that traditionally constructed housing provides more protection from heat than formal low-cost housing [83]. This should not be ignored in future African cities development, where the focus is often on upgrading unplanned or informal settlements from traditionally constructed housing [81].

There were various synergies between heat wave and UHI as the varied spatially based on urban development patterns and geographic conditions [84]. Many countries and region have developed other kinds of UHI mitigation strategies for urban development, and beyond looking at entire cities' development, it is important to take the climate condition and climate projections into account at the local scale. For example, in a predominantly alpine climate country Austria, the ADAPT-UHI project developed tools by mapping the potential UHI risk index, modelling different adaptation scenarios, and evaluating and rating cities' quality of existing green and blue infrastructure [81]. Researchers help city planners make decisions on climate change adaptation and minimize the effects of UHI in the future. In the hot desert climate city of Phoenix, Arizona, researchers found that a warming climate was detected to aggravate the interaction between UHI effect and heat waves, while the amplified temperature-exacerbated UHI effect can be almost completely offset by adopting a widespread green roof strategy [85]. In Kampala, a project "Promoting Green Urban Development in Africa: Enhancing the Relationship Between Urbanization, Environmental Assets and Ecosystem Services" conducted under the leadership of the World Bank [86], seeks to conserve the environment as urbanisation proceeds. Addis Ababa is trying to incorporate green infrastructure areas for "multifunctional uses", in order to address stormwater management, UHI mitigation, air filtration, and preservation of urban agriculture [87]. These various approaches offer opportunities for East African cities but require proactive policies and a multi-sector approach to city development [86–88]. Studies of projects that pursue climate compatible development (CCD) in Malawi, South Africa, Tanzania, India and Kenya have revealed such a framework that advances triple-wins across development, mitigation and adaptation within the developing world is feasible [89–92].

Anticipated additional climate variability and change could further exacerbate disease incidence and spread in East Africa [69]. In East Africa's highlands above 2000 m altitude, environmental conditions currently do not support malaria transmission. Due to climate change, these areas will experience increased malaria [93]. The changing weather conditions may increase the spread of Rift Valley fever (RVF) virus and could also introduce the RVF virus into domestic animal populations [94]. When considering the use of blue green infrastructure within city development to mitigate UHI effects, local governments should rationally allocate vegetation and water configurations to avoid creating habitats suitable for viral vectors. In addition, for specific climate zones, such as the desert climate zone, understanding its UCI effect (or oasis effect) in urban areas located is important [55]. More than 90% of the global population residing in deserts and dryland are located in developing countries with high rates of rural to urban migration, indicating that rapid growth in desert cities is likely [95], suggesting further research in this area is urgently needed.

4.4. Uncertainty, Limitations and Further Considerations

Our research has focused only on the annual UHI. This decision was made primarily due to the different classification of seasons in Africa between study countries. For comparability we therefore chose to present the annual UHI trends in our study cities. We recognise that there may be specific climate features of each selected study year, and our results can only represent the differences between those two selected years. This provides important insights but from the analysis undertaken, we cannot make robust claims about long-term change [45].

Topography, morphology and biodiversity, are well known to influence UHI effects but were not directly measured in this research [96]. We have nevertheless identified their potential contributions when explaining our results. As with all UHI modelling approaches, the model output's reliability is strongly reliant on the definition and boundary-setting of urban areas. In particular, there remain challenges in integrating data from the Global Surface UHI Explorer with the global-scale Landscan urban extent dataset, as in terms of missing data for certain months or years, limiting the ability to explore dynamic changes in SUHI associated with urbanisation.

Despite these limitations, this paper has contributed novel empirical information and applied a method that enables comparison across different East African cities, offering an approach that is widely applicable. The method can be used to identify changes in smaller African cities as well as the capital cities considered in this research. Such insights would be particularly valuable in smaller cities where urbanisation is most rapidly accelerating [1].

5. Conclusions

The ongoing effects of UHI in rapidly urbanising cities, further exacerbated by global climate change, have intensified heat stress for countless urban inhabitants, especially those who are most vulnerable during heatwave events. We investigated the SUHI in five East African study cities across various climate zones in 2003 and 2017. SUHI intensities differed between day and night, and mean annual daytime SUHI intensities were higher than the mean night-time SUHI intensities in most cities. Results revealed that the differences in climate conditions and urban development patterns had a significant impact on the changes to the SUHI. Policy makers and city planners should consider UHI mitigation strategies, taking into account climate conditions, projected climate changes and urban development patterns, as well as the implications of informal settlement growth. Blue and green infrastructure, in general, constitutes an essential means for mitigating urban heat and enhancing human comfort, but this needs to be carefully planned to avoid negative impacts linked to human wellbeing and the spread of disease.

Author Contributions: Conceptualization, X.L., L.C.S. and M.D.; methodology, X.L., L.C.S. and M.D.; software, X.L.; formal analysis, X.L.; writing—original draft preparation, X.L.; writing—review and editing, L.C.S. and M.D.; supervision, L.C.S. and M.D.; All authors have read and agreed to the published version of the manuscript.

Funding: This research was funded by UK government's Natural Environment Research Council (NERC), grant number NE/R002681/1. X.L. was funded by a scholarship (Fujian-Taiwan Joint Innovative Centre for Germplasm Resources and Cultivation of Crop (grant No. 2015-75. Fujian 2011 Program, China).

Data Availability Statement: Original SUHI data used in this study have been obtained from https://yceo.users.earthengine.app/view/uhimap (accessed on 3 May 2020) and city centre data on Figure 3 are from https://www.naturalearthdata.com/ (accessed on 30 May 2020).

Acknowledgments: Thank you to the reviewers for their helpful comments in improving this manuscript.

Conflicts of Interest: The authors declare no conflict of interest. The funders had no role in the design of the study; in the collection, analyses, or interpretation of data; in the writing of the manuscript, or in the decision to publish the results.

References

1. DESA. World Population Prospects 2019 Highlights. Available online: https://population.un.org/wpp/Publications/Files/WPP2019_Highlights.pdf (accessed on 15 December 2020).
2. Howard, L. *The Climate of London: Deduced from Meteorological Observations Made in the Metropolis and at Various Places around It*; Harvey and Darton: London, UK, 1833.
3. Oke, T.R. The energetic basis of the urban heat island. *Q. J. R. Meteorol. Soc.* **1982**, *108*, 1–24. [CrossRef]
4. Oke, T.R. City size and the urban heat island. *Atmos. Env.* **1973**, *7*, 769–779. [CrossRef]
5. Li, Y.; Schubert, S.; Kropp, J.P.; Rybski, D. On the influence of density and morphology on the Urban Heat Island intensity. *Nat. Commun.* **2020**, *11*, 2647. [CrossRef]
6. Zhou, D.; Zhao, S.; Liu, S.; Zhang, L.; Zhu, C. Surface urban heat island in China's 32 major cities: Spatial patterns and drivers. *Remote Sens. Environ.* **2014**, *152*, 51–61. [CrossRef]
7. He, B.-J.; Ding, L.; Prasad, D. Relationships among local-scale urban morphology, urban ventilation, urban heat island and outdoor thermal comfort under sea breeze influence. *Sustain. Cities Soc.* **2020**, *60*, 102289. [CrossRef]
8. Grimm, N.B.; Faeth, S.H.; Golubiewski, N.E.; Redman, C.L.; Wu, J.; Bai, X.; Briggs, J.M. Global change and the ecology of cities. *Sci. Am. Assoc. Adv. Sci.* **2008**, *319*, 756–760. [CrossRef]
9. Holloway, T.; Patz, J.A.; Campbell-Lendrum, D.; Foley, J.A. Impact of regional climate change on human health. *Nature* **2005**, *438*, 310–317. [CrossRef]
10. Bouchama, A. The 2003 European heat wave. *Intensive Care Med.* **2004**, *30*, 1–3. [CrossRef]
11. Argüeso, D.; Evans, J.P.; Fita, L.; Bormann, K.J. Temperature response to future urbanization and climate change. *Clim. Dyn.* **2014**, *42*, 2183–2199. [CrossRef]
12. Cai, M.; Kalnay, E. Impact of urbanization and land-use change on climate. *Nature* **2003**, *423*, 528–531. [CrossRef]
13. Georgescu, M.; Moustaoui, M.; Mahalov, A.; Dudhia, J. Summer-time climate impacts of projected megapolitan expansion in Arizona. *Nat. Clim. Chang.* **2013**, *3*, 37–41. [CrossRef]
14. Chen, X.; Zhao, H.-M.; Li, P.-X.; Yin, Z.-Y. Remote sensing image-based analysis of the relationship between urban heat island and land use/cover changes. *Remote Sens. Environ.* **2006**, *104*, 133–146. [CrossRef]
15. Heusinkveld, B.G.; Steeneveld, G.J.; Hove, V.L.W.A.; Jacobs, C.M.J.; Holtslag, A.A.M. Spatial variability of the Rotterdam urban heat island as influenced by urban land use. *J. Geophys. Res. Atmos.* **2014**, *119*, 677–692. [CrossRef]
16. Kikon, N.; Singh, P.; Singh, S.K.; Vyas, A. Assessment of urban heat islands (UHI) of Noida City, India using multi-temporal satellite data. *Sustain. Cities Soc.* **2016**, *22*, 19–28. [CrossRef]
17. Dominguez, F.; Villegas, J.C.; Breshears, D.D. Spatial extent of the North American Monsoon: Increased cross-regional linkages via atmospheric pathways. *Geophys. Res. Lett.* **2009**, *36*. [CrossRef]
18. Krayenhoff, E.S.; Moustaoui, M.; Broadbent, A.M.; Gupta, V.; Georgescu, M. Diurnal interaction between urban expansion, climate change and adaptation in US cities. *Nat. Clim. Chang.* **2018**, *8*, 1097–1103. [CrossRef]
19. Chapman, S.; Watson, J.E.M.; Salazar, A.; Thatcher, M.; McAlpine, C.A. The impact of urbanization and climate change on urban temperatures: A systematic review. *Landsc. Ecol.* **2017**, *32*, 1921–1935. [CrossRef]
20. Njoh, A.J. The experience and legacy of French colonial urban planning in sub-Saharan Africa. *Plan. Perspect.* **2004**, *19*, 435–454. [CrossRef]
21. Anas, A.; Rhee, H.-J. Curbing excess sprawl with congestion tolls and urban boundaries. *Reg. Sci. Urban Econ.* **2006**, *36*, 510–541. [CrossRef]
22. Cobbinah, P.B.; Aboagye, H.N. A Ghanaian twist to urban sprawl. *Land Use Policy* **2017**, *61*, 231–241. [CrossRef]
23. Chapman, S.; Thatcher, M.; Salazar, A.; James, E.M.W.; McAlpine, C.A. The effect of urban density and vegetation cover on the heat island of a subtropical city. *J. Appl. Meteorol. Climatol.* **2018**, *57*, 2531–2550. [CrossRef]
24. Seto, K.C.; Fragkias, M.; Güneralp, B.; Reilly, M.K. A meta-analysis of global urban land expansion. *PLoS ONE* **2011**, *6*, e23777. [CrossRef]

25. Fataar, R. Densification and the Ambition for a Democratic City. Available online: https://ethz.ch/content/dam/ethz/special-interest/conference-websites-dam/no-cost-housing-dam/documents/Fataar_Densificationandtheambitionforademocraticcity_RashiqFataar.pdf (accessed on 3 December 2020).

26. Todes, A.; Weakley, D.; Harrison, P. Densifying Johannesburg: Context, policy and diversity. *J. Hous. Built Environ.* **2018**, *33*, 281–299. [CrossRef]

27. Samarrai, F. Study: 'Unbearable' Heat Stress to Affect East Africans by Late 21St Century. Available online: https://news.virginia.edu/content/study-unbearable-heat-stress-affect-east-africans-late-21st-century-0 (accessed on 15 December 2020).

28. Gebrechorkos, S.H.; Hülsmann, S.; Bernhofer, C. Regional climate projections for impact assessment studies in East Africa. *Environ. Res. Lett.* **2019**, *14*, 044031. [CrossRef]

29. Anyah, R.O.; Qiu, W. Characteristic 20th and 21st century precipitation and temperature patterns and changes over the Greater Horn of Africa. *Int. J. Climatol.* **2012**, *32*, 347–363. [CrossRef]

30. Brown, M.E.; Brickley, E.B. Evaluating the use of remote sensing data in the US agency for international development famine early warning systems network. *J. Appl. Remote Sens.* **2012**, *6*, 063511. [CrossRef]

31. Conway, D.; Schipper, E.L.F. Adaptation to climate change in Africa: Challenges and opportunities identified from Ethiopia. *Glob. Environ. Chang.* **2011**, *21*, 227–237. [CrossRef]

32. Sun, R.; Lü, Y.; Yang, X.; Chen, L. Understanding the variability of urban heat islands from local background climate and urbanization. *J. Clean. Prod.* **2019**, *208*, 743–752. [CrossRef]

33. Seto, K.C.; Güneralp, B.; Hutyra, L.R. Global forecasts of urban expansion to 2030 and direct impacts on biodiversity and carbon pools. *Proc. Natl. Acad. Sci. USA* **2012**, *109*, 16083. [CrossRef]

34. Beck, H.E.; Zimmermann, N.E.; McVicar, T.R.; Vergopolan, N.; Berg, A.; Wood, E.F. Present and future Köppen-Geiger climate classification maps at 1-km resolution. *Sci. Data* **2018**, *5*, 1. [CrossRef]

35. Osman, M.M.; Sevinc, H. Adaptation of climate-responsive building design strategies and resilience to climate change in the hot/arid region of Khartoum, Sudan. *Sustain. Cities Soc.* **2019**, *47*, 101429. [CrossRef]

36. Zerboni, A.; Brandolini, F.; Mariani, G.S.; Perego, A.; Salvatori, S.; Usai, D.; Pelfini, M.; Williams, M.A. The Khartoum-Omdurman conurbation: A growing megacity at the confluence of the Blue and White Nile Rivers. *J. Maps* **2020**, 1–14. [CrossRef]

37. Ndetto, E.L.; Matzarakis, A. Basic analysis of climate and urban bioclimate of Dar es Salaam, Tanzania. *Theor. Appl. Climatol.* **2013**, *114*, 213–226. [CrossRef]

38. Egondi, T.; Kyobutungi, C.; Rocklöv, J. Temperature variation and heat wave and cold spell impacts on years of life lost among the urban poor population of Nairobi, Kenya. *Int. J. Environ. Res. Public Health* **2015**, *12*, 2735–2748. [CrossRef]

39. Oyugi, M.O.; Odenyo, V.A.; Karanja, F.N. The implications of land use and land cover dynamics on the environmental quality of Nairobi City, Kenya. *Am. J. Geogr. Inf. Syst.* **2017**, *6*, 111–127. [CrossRef]

40. (KNBS), K.N.B.o.S. *Republic of Kenya Population and Census Survey 2009*; Ministry of Planning: Nairobi, Kenya, 2010.

41. Angassa, K.; Leta, S.; Mulat, W.; Kloos, H. Seasonal characterization of municipal wastewater and performance evaluation of a constructed wetland system in Addis Ababa, Ethiopia. *Int. J. Energy Water Resour.* **2020**, *4*, 127–138. [CrossRef]

42. Feyisa, G.L.; Dons, K.; Meilby, H. Efficiency of parks in mitigating urban heat island effect: An example from Addis Ababa. *Landsc. Urban Plan.* **2014**, *123*, 87–95. [CrossRef]

43. Kabano, P.; Lindley, S.; Harris, A. Evidence of urban heat island impacts on the vegetation growing season length in a tropical city. *Landsc. Urban Plan.* **2021**, *206*, 103989. [CrossRef]

44. Richmond, A.; Myers, I.; Namuli, H. Urban Informality and Vulnerability: A Case Study in Kampala, Uganda. *Urban Sci.* **2018**, *2*, 22. [CrossRef]

45. Chakraborty, T.; Lee, X. A simplified urban-extent algorithm to characterize surface urban heat islands on a global scale and examine vegetation control on their spatiotemporal variability. *Int. J. Appl. Earth Obs. Geoinf.* **2019**, *74*, 269–280. [CrossRef]

46. Moser, B.K.; Stevens, G.R. Homogeneity of variance in the two-sample means test. *Am. Stat.* **1992**, *46*, 19–21. [CrossRef]

47. McGraw, K.O.; Wong, S.P. A common language effect size statistic. *Psychol. Bull.* **1992**, *111*, 361–365. [CrossRef]

48. Cohen, J. *Statistical Power Analysis for the Behavioral Sciences*; Lawrence Erlbaum Associates: New York, NY, USA, 1988; ISBN 0805802835.

49. Lenhard, W.; Lenhard, A. *Calculation of Effect Sizes*. Available online: https://www.psychometrica.de/effect_size.html (accessed on 4 June 2020).

50. Ellis, P.D. *The Essential Guide to Effect Sizes: Statistical Power, Meta-Analysis, and the Interpretation of Research Results*; Cambridge University Press: Cambridge, UK, 2010.

51. Team, R.; RStudio: Integrated Development for R. RStudio. Available online: http://www.rstudio.com/ (accessed on 6 June 2020).

52. Tran, H.; Uchihama, D.; Ochi, S.; Yasuoka, Y. Assessment with satellite data of the urban heat island effects in Asian mega cities. *Int. J. Appl. Earth Obs. Geoinf.* **2006**, *8*, 34–48. [CrossRef]

53. Chow, W.T.; Roth, M. Temporal dynamics of the urban heat island of Singapore. *Int. J. Climatol. A J. R. Meteorol. Soc.* **2006**, *26*, 2243–2260. [CrossRef]

54. Santamouris, M.; Papanikolaou, N.; Livada, I.; Koronakis, I.; Georgakis, C.; Argiriou, A.; Assimakopoulos, D. On the impact of urban climate on the energy consumption of buildings. *Sol. Energy* **2001**, *70*, 201–216. [CrossRef]

55. Rindfuss, R.R.; Myint, S.; Brazel, A.; Fan, C.; Zheng, B. Urban heat islands: Global variation and determinants in hot desert and hot semi-arid desert cities. In Proceedings of the Population Association of America 2016 Annual Meeting, Washington, DC, USA, 31 Mar–2 April 2016.
56. Imhoff, M.L.; Zhang, P.; Wolfe, R.E.; Bounoua, L. Remote sensing of the urban heat island effect across biomes in the continental USA. *Remote Sens. Environ.* **2010**, *114*, 504–513. [CrossRef]
57. Lazzarini, M.; Marpu, P.R.; Ghedira, H. Temperature-land cover interactions: The inversion of urban heat island phenomenon in desert city areas. *Remote Sens. Environ.* **2013**, *130*, 136–152. [CrossRef]
58. Peng, S.; Piao, S.; Ciais, P.; Friedlingstein, P.; Ottle, C.; Bréon, F.-M.; Nan, H.; Zhou, L.; Myneni, R.B. Surface urban heat island across 419 global big cities. *Environ. Sci. Technol.* **2012**, *46*, 696–703. [CrossRef] [PubMed]
59. Wu, X.; Wang, G.; Yao, R.; Wang, L.; Yu, D.; Gui, X. Investigating surface urban heat islands in south america based on MODIS data from 2003–2016. *Remote Sens.* **2019**, *11*, 1212. [CrossRef]
60. Yang, Q.; Huang, X.; Li, J. Assessing the relationship between surface urban heat islands and landscape patterns across climatic zones in China. *Sci. Rep.* **2017**, *7*, 1–11. [CrossRef]
61. Chieppa, J.; Bush, A.; Mitra, C. Using "Local Climate Zones" to detect urban heat island on two small cities in alabama. *Earth Interact.* **2018**, *22*, 1–22. [CrossRef]
62. Bhanjee, S.; Zhang, C.H. Mapping latest patterns of urban sprawl in dar es salaam, tanzania. *Pap. Appl. Geogr.* **2018**, *4*, 292–304. [CrossRef]
63. Kassa, F. Conurbation and urban sprawl in Africa: The case of the City of Addis Ababa. *Ghana J. Geogr.* **2013**, *5*, 73–89.
64. Abo-El-Wafa, H.; Yeshitela, K.; Pauleit, S. Exploring the future of rural-urban connections in sub-Saharan Africa: Modelling urban expansion and its impact on food production in the Addis Ababa region. *Geogr. Tidsskr.* **2017**, *117*, 68–81. [CrossRef]
65. Hou, H.; Estoque, R.C.; Murayama, Y. Spatiotemporal analysis of urban growth in three African capital cities: A grid-cell-based analysis using remote sensing data. *J. Afr. Earth Sci.* **2016**, *123*, 381–391. [CrossRef]
66. Du Toit, M.J.; Cilliers, S.S.; Dallimer, M.; Goddard, M.; Guenat, S.; Cornelius, S.F. Urban green infrastructure and ecosystem services in sub-Saharan Africa. *Landsc. Urban Plan.* **2018**, *180*, 249–261. [CrossRef]
67. Hope, S.K.R. Climate change and urban development in Africa. *Int. J. Environ. Stud.* **2009**, *66*, 643–658. [CrossRef]
68. Yuen, B.; Kumssa, A. *Climate Change and Sustainable Urban Development in Africa and Asia*; Springer: Dordrecht, The Netherlands, 2010.
69. Niang, I.; Ruppel, O.C.; Abdrabo, M.A.; Essel, A.; Lennard, C.; Padgham, J.; Urquhart, P. Africa. In *Climate Change 2014—Impacts, Adaptation and Vulnerability: Part B: Regional Aspects Working Group II Contribution to the IPCC Fifth Assessment Report*; Barros, V.R., Field, C.B., Estrada, Y.O., Genova, R.C., Girma, B., Kissel, E.S., Levy, A.N., MacCracken, S., Mastrandrea, P.R., White, L.L., Eds.; Cambridge University Press: Cambridge, UK, 2014; pp. 1199–1265.
70. Kombe, W.J.; Kreibich, V. *Informal Land Management in Tanzania. SPRING Centre, Faculty of Spatial Plannning*; University of Dortmund: Dortmund, Germany, 2000.
71. Augustijn-Beckers, E.-W.; Flacke, J.; Retsios, B. Simulating informal settlement growth in Dar es Salaam, Tanzania: An agent-based housing model. *Comput. Environ. Urban Syst.* **2011**, *35*, 93–103. [CrossRef]
72. Chakraborty, T.; Hsu, A.; Manya, D.; Sheriff, G. Disproportionately higher exposure to urban heat in lower-income neighborhoods: A multi-city perspective. *Environ. Res. Lett.* **2019**, *14*, 105003. [CrossRef]
73. Scott, A.A.; Misiani, H.; Okoth, J.; Jordan, A.; Gohlke, J.; Ouma, G.; Arrighi, J.; Zaitchik, B.F.; Jjemba, E.; Verjee, S. Temperature and heat in informal settlements in Nairobi. *PLoS ONE* **2017**, *12*, e0187300. [CrossRef]
74. Vermeiren, K.; Van Rompaey, A.; Loopmans, M.; Serwajja, E.; Mukwaya, P. Urban growth of Kampala, Uganda: Pattern analysis and scenario development. *Landsc. Urban Plan.* **2012**, *106*, 199–206. [CrossRef]
75. World Bank Group. Urban Development. Available online: https://www.worldbank.org/en/topic/urbandevelopment/overview#1 (accessed on 11 December 2020).
76. United Nation. 2018 Revision of World Urbanization Prospects. Available online: https://population.un.org/wup/Publications/Files/WUP2018-Report.pdf (accessed on 29 December 2020).
77. UN-Habitat. Addressing the Most Vulnerable First: Pro-Poor Climate Action in Informal Settlements. Available online: https://www.preventionweb.net/publications/view/62059 (accessed on 22 December 2020).
78. Awumbila, M. World Migration Report 2015: Linkages between Urbanization, Rural–Urban Migration and Poverty Outcomes in Africa. Available online: https://publications.iom.int/system/files/wmr2015_en.pdf (accessed on 21 December 2020).
79. Zhao, L.; Lee, X.; Schultz, N. A wedge strategy for mitigation of urban warming in future climate scenarios. *Atmos. Chem. Phys. Discuss.* **2017**, 1–28. [CrossRef]
80. Zhou, X.; Chen, H. Impact of urbanization-related land use land cover changes and urban morphology changes on the urban heat island phenomenon. *Sci. Total Environ.* **2018**, *635*, 1467–1476. [CrossRef] [PubMed]
81. Cavan, G.; Lindley, S.; Jalayer, F.; Yeshitela, K.; Pauleit, S.; Renner, F.; Gill, S.; Capuano, P.; Nebebe, A.; Woldegerima, T.; et al. Urban morphological determinants of temperature regulating ecosystem services in two African cities. *Ecol. Indic.* **2014**, *42*, 43–57. [CrossRef]
82. Feyisa, G.L.; Meilby, H.; Jenerette, G.D.; Pauliet, S. Locally optimized separability enhancement indices for urban land cover mapping: Exploring thermal environmental consequences of rapid urbanization in Addis Ababa, Ethiopia. *Remote Sens. Environ.* **2016**, *175*, 14–31. [CrossRef]

83. Scovronick, N.; Armstrong, B. The impact of housing type on temperature-related mortality in South Africa, 1996–2015. *Environ. Res.* **2012**, *113*, 46–51. [CrossRef]

84. He, B.-J.; Wang, J.; Liu, H.; Ulpiani, G. Localized synergies between heat waves and urban heat islands: Implications on human thermal comfort and urban heat management. *Environ. Res.* **2021**, *193*, 110584. [CrossRef] [PubMed]

85. Tewari, M.; Yang, J.; Kusaka, H.; Salamanca, F.; Watson, C.; Treinish, L. Interaction of urban heat islands and heat waves under current and future climate conditions and their mitigation using green and cool roofs in New York City and Phoenix, Arizona. *Environ. Res. Lett.* **2019**, *14*, 034002. [CrossRef]

86. White, R.; Turpie, J.; Letley, G.L. *Greening Africa's Cities: Enhancing the Relationship between Urbanization, Environmental Assets, and Ecosystem Services*; World Bank: Washington, DC, USA, 2017.

87. Larsen, L. Urban climate and adaptation strategies. *Front. Ecol. Environ.* **2015**, *13*, 486–492. [CrossRef]

88. UN-Habitat. UN-Habitat and Sudan sign Country Programme Document. Available online: https://www.developmentaid.org/#!/news-stream/post/25698/un-habitat-and-sudan-sign-country-programme-document (accessed on 13 December 2020).

89. Stringer, L.C.; Dougill, A.J.; Dyer, J.C.; Vincent, K.; Fritzsche, F.; Leventon, J.; Falcao, M.P.; Manyakaidze, P.; Syampungani, S.; Powell, P. Advancing climate compatible development: Lessons from southern Africa. *Reg. Environ. Chang.* **2014**, *14*, 713–725. [CrossRef]

90. Colenbrander, S.; Gouldson, A.; Roy, J.; Kerr, N.; Sarkar, S.; Hall, S.; Sudmant, A.; Ghatak, A.; Chakravarty, D.; Ganguly, D. Can low-carbon urban development be pro-poor? The case of Kolkata, India. *Environ. Urban.* **2017**, *29*, 139–158. [CrossRef]

91. Wanjiru, C.; Nunan, F. Enabling climate compatible development in the coastal region of Kenya. In *Making Climate Compatible Development Happen*; Taylor & Francis, Routledge: London, UK, 2017; pp. 223–244.

92. Ficklin, L.; Wood, B.T.; Dougill, A.J.; Sallu, S.M.; Stringer, L.C. Reconsidering climate compatible development as a new development landscape in Southern Africa. In *Making Climate Compatible Development Happen*; Taylor & Francis, Routledge: London, UK, 2017; pp. 44–65.

93. Alonso, D.; Bouma, M.J.; Pascual, M. Epidemic malaria and warmer temperatures in recent decades in an East African highland. *Proc. R. Soc. B Biol. Sci.* **2011**, *278*, 1661–1669. [CrossRef] [PubMed]

94. Baba, M.; Masiga, D.K.; Sang, R.; Villinger, J. Has Rift Valley fever virus evolved with increasing severity in human populations in East Africa? *Emerg. Microbes Infect.* **2016**, *5*, 1–10. [CrossRef] [PubMed]

95. UN-EMG. 2010–2020: UN Decade for Deserts and the Fight against Desertification. Available online: https://www.un.org/en/events/desertification_decade/value.shtml (accessed on 13 December 2020).

96. Hamada, S.; Tanaka, T.; Ohta, T. Impacts of land use and topography on the cooling effect of green areas on surrounding urban areas. *Urban For. Urban Green.* **2013**, *12*, 426–434. [CrossRef]

Article

The Role of Individual and Small-Area Social and Environmental Factors on Heat Vulnerability to Mortality Within and Outside of the Home in Boston, MA

Augusta A. Williams [1,*] , Joseph G. Allen [1], Paul J. Catalano [2,3] and John D. Spengler [1]

[1] Department of Environmental Health, Harvard T.H. Chan School of Public Health, Boston, MA 02115, USA; jgallen@hsph.harvard.edu (J.G.A.); spengler@hsph.harvard.edu (J.D.S.)
[2] Department of Biostatistics, Harvard T.H. Chan School of Public Health, Boston, MA 02115, USA; pcata@jimmy.harvard.edu
[3] Department of Data Sciences, Dana-Farber Cancer Institute, Boston, MA 02215, USA
* Correspondence: auw882@mail.harvard.edu

Received: 15 December 2019; Accepted: 4 February 2020; Published: 7 February 2020

Abstract: Climate change is resulting in heatwaves that are more frequent, severe, and longer lasting, which is projected to double-to-triple the heat-related mortality in Boston, MA if adequate climate change mitigation and adaptation strategies are not implemented. A case-only analysis was used to examine subject and small-area neighborhood characteristics that modified the association between hot days and mortality. Deaths of Boston, Massachusetts residents that occurred from 2000–2015 were analyzed in relation to the daily temperature and heat index during the warm season as part of the case-only analysis. The modification by small-area (census tract, CT) social, and environmental (natural and built) factors was assessed. At-home mortality on hot days was driven by both social and environmental factors, differentially across the City of Boston census tracts, with a greater proportion of low-to-no income individuals or those with limited English proficiency being more highly represented among those who died during the study period; but small-area built environment features, like street trees and enhanced energy efficiency, were able to reduce the relative odds of death within and outside the home. At temperatures below current local thresholds used for heat warnings and advisories, there was increased relative odds of death from substance abuse and assault-related altercations. Geographic weighted regression analyses were used to examine these relationships spatially within a subset of at-home deaths with high-resolution temperature and humidity data. This revealed spatially heterogeneous associations between at-home mortality and social and environmental vulnerability factors.

Keywords: heat-related mortality; built environment; urban resilience; extreme heat; climate change; urban heat island

1. Introduction

Global climate change is increasing the frequency, duration, and severity of extreme heat events around the world [1,2]. Extreme heat is one of the most severe public health impacts of climate change, resulting in increased mortality in many geographic locations [3–6], including in Boston, Massachusetts (MA), USA, the focus of this study [7–11]. In Boston, MA, of the annual number of hot days above 90 °F is expected to increase from 11 (1971–2000) to 40 days by 2030, and potentially 90 days by 2070 [12]. Projections have shown that that heat-related mortality has the potential to triple in the Northeast United States [13] without adequate climate change mitigation and adaptation strategies being implemented, with the effects being disproportionally experienced by the most vulnerable individuals.

There has been significant research done to further the concept of heat vulnerability, to determine and understand which individuals and communities are most vulnerable to the impacts of extreme heat. Much of this literature has focused on social factors, such as income, race, sex, education, and age, to determine vulnerability. For example, reviews of heat and mortality epidemiologic studies have found that non-Hispanic black individuals, women, those of a lower socioeconomic status, those with preexisting medical conditions (e.g., diabetes), elderly individuals over 65, and infants and children under 5 have been found to be most vulnerable to poor health outcomes, including death, during extreme heat events [4–6,14–16].

One group of factors that determines much of our exposure to extreme heat are the natural and built environments. The urban heat island (UHI) effect occurs when land surface temperatures in an urban area are higher than the surrounding rural or suburban areas, and an urban design with enhanced urban canyons, greater use of materials of high thermal absorption (e.g., asphalt, brick), and densely built areas can create temperatures much higher than surrounding areas and prevent urban areas from cooling at night. Furthermore, buildings modify our exposure to ambient temperature conditions, either exacerbating or mitigating indoor heat exposures given the same ambient conditions [17].

While buildings have become more energy efficient, a higher thermal insulation and greater airtightness have increased the risk of potential overheating during periods of power outages [17]. These increases in energy efficiency in heat-dominated climates, like that of Boston, MA, may potentially be adversely impacting the resiliency of buildings to extreme heat during power outages [18]. Overheating is a great risk in times of power loss when mechanical cooling may not be available. The risk of major and widespread power outages that affect a large number of customers and that would result in "heat disasters" appears to be growing [19]. Research has shown that the summertime overheating of residential buildings may increase by upwards of 25% by the middle of the 21st century [20].

As this body of work has progressed, some attention has been paid to other factors in the surrounding urban environment that may exacerbate or mitigate the underlying social vulnerability to extreme heat. Some of the risk factors associated with building and urban design have been incorporated into previous heat vulnerability research. For example, living in a multi-family apartment building has been found to increase the incidence of poor health outcomes during extreme heat events [21,22]. Building materials with a high thermal mass prevent a building from passively cooling during the overnight hours, increasing the need for cooling during subsequent days [23,24]. Having urban tree canopies can keep surrounding buildings cool, reducing the electricity consumption from cooling demands usage, while also keeping communities cooler and mitigating heat stress-related health conditions [25]; additionally, surrounding greenspace can reduce vulnerability to heat morbidity and mortality [16].

Using air conditioning (AC) has been the primary defense to extreme heat exposure indoors. However, solely relying on AC is currently not an equitable or sustainable adaptation strategy. A recent study in New York City (NYC) found that the odds of not having AC were found to be greatest for non-Hispanic black people and those with low-to-no income [26], and only 50% of NYC public housing residents have access to AC [27]. Approximately two-thirds of Massachusetts residents lack central AC at home, while 20% of households lack any type of AC [28]. Furthermore, the electricity used to power AC produces health-harmful pollutants, which are increased on hot days from enhanced power production [29]. Lastly, the refrigerants found within AC systems contain hydrofluorocarbons, a greenhouse gas more potent than carbon dioxide.

Despite spending up to 90% of time within buildings and built environments that largely determine our exposure to ambient conditions, few heat vulnerability and adaptation assessments include widespread information on buildings and the built environment. Although a framework for assessing heat vulnerability has been created [15], the past application of this at the local level found that only about half of all census tracts were correctly characterized as being vulnerable [30]. This demonstrates the need for the tailoring and implementation of heat vulnerability assessments at the local level. Furthermore, many existing heat vulnerability assessment frameworks and associated studies have used a single ambient meteorological measurement to determine exposure for an entire city. Since buildings and the urban environment modify our exposure to extreme temperatures, the use of ambient temperatures to define heat exposure has resulted in temperature exposure misclassification as this does not adequately characterize the temperatures people are experiencing indoors.

In 2015, a Boston-based heat vulnerability and urban heat island assessment, prepared for by the Trust for Public Land, was conducted by students at Tufts University [31]. This plan outlined several solutions that would reduce Boston's heat vulnerability, mainly through community engagement and adaptations to the built environment and was largely based on helpful and informative case studies of strategies implemented here and elsewhere in the US. While this assessment included some built environment factors like land use and central AC penetration, addition local research is needed on the role of buildings and the built environment in driving heat vulnerability in Boston in order to most effectively enact these interventions.

In this study, we aim to further local heat vulnerability research by incorporating neighborhood built and natural environmental characteristics in combination with social factors to assess vulnerability to mortality on hot days. A case-only analysis was used to examine whether the mortality on hot days in Boston, MA varied based on individual and small-area environmental and social characteristics at a variety of temperature exposure definitions. These relationships were then further explored spatially across Boston with a gridded climate dataset to enhance temperature exposure definitions. This comprehensive assessment of the underlying vulnerabilities and their social and environmental drivers will be vital when implementing heat action planning in the future.

2. Materials and Methods

Hourly meteorological data was accessed for the Boston Logan International Airport via the National Centers for Environmental Information. The daily maximum ambient temperatures (T_{MAX}) $\geq$ 90 °F were used as a binary measure of it being a "hot day", a common temperature threshold used to enact local cooling strategies in Boston. Recent evidence in Boston, MA has found that health impacts, including mortality, increase at daily maximum temperatures below 90 °F [32], so $T_{MAX} \geq$ 85 °F was used to define "warm days". The heat index (HI)—a measure of the real feel temperature after the relative humidity is considered with the actual air temperature [33]—was also considered by using days where $HI_{MAX} \geq$ 86 °F, which was the 95th percentile of the HI_{MAX} during the warm season in Boston, from May to September.

These assessments use single day exposures, and are consistent with findings that the mortality risk or effect modification of heat-mortality relationships increase on individual days of extreme heat [5,34–36]. These temperature metrics are also at values that are below the current heat advisory (HI of 95–99 °F for at least 2 h over 2 consecutive days), or heat wave (at least 3 days where $T_{MAX} \geq$ 90 °F) criteria, which trigger local messaging and communications, as well as city-wide interventions (e.g., opening of cooling centers). All analyses were conducted using only the warm season (May to September) data, to remove any seasonal confounding that exists between the temperature and mortality.

Small-area neighborhood data was assigned at the census tract (CT) level for each at-home mortality record, corresponding to the 2010 Census. CTs are small, relatively permanent subdivisions, usually include around 4000 people, and are designed to be "homogeneous with respect to population characteristics, economic status, and living conditions" [37]. CT-based assessments of the socioeconomic status (SES) have been found to be adequate measurements for individual estimates in MA [38]. There are 178 CTs in the City of Boston, but 14 of these CTs have little or no population, leaving 164 CTs for analysis. We obtained data on the CT social characteristics including the population density, inclusion of utilities in rent, the Gini coefficient for income inequality, and unemployment from the 2010 Census [39]. The proportion of those in each CT with a disability, older adults $\geq$ 65 years, children ≤ 5 years, non-Caucasian (non-Hispanic whites), low income, limited English proficiency, and with medical illnesses was accessed from the 2016 City of Boston's Climate Ready Boston Social Vulnerability Data, which draws upon the 2008–2012 American Community Survey 5-year Estimates [40].

We obtained data on the CT environmental characteristics, including the availability of street trees in 2011, from Boston Open Data [41], the 2005 impervious surface fraction and mean albedo from MassGIS, and as summarized by the Boston Area Research Initiative (BARI) [42]. 2017 building assessments from the City of Boston Assessing Department and as summarized by BARI provided CT-level summaries of residential buildings, including the decade/year built or last renovated. BARI also provides a mean residential energy efficiency score, which is an aggregated variable based on the age of the building, the heating system, and the cooling type, with higher values indicating a more energy efficient residence. We also used the mean value of the residential building/land and per area [43]. When evaluating the local real estate prices in Boston, triple decker homes and luxury apartment buildings, which are 2 common residential building archetypes in Boston, were found to have the highest value per area.

The data on all deaths occurring in Massachusetts for the period January 2000-December 2015 were obtained from The Commonwealth of Massachusetts Executive Office of Health and Human Services Department of Public Health. The data included the primary causes of death, classified using the International Classification of Disease, 10th Revision (ICD-10) codes of age, sex, race, place of death, education, occupation, and industry of work. All deaths, regardless of the primary causes, were included in this analysis given the wide range of health outcomes that can be negatively impacted or exacerbated by extreme heat [35,44–51]. All deaths from 2000–2015 that occurred at home or outside of the home (excluding those that occurred an inpatient or nursing facility) were included in this analysis. Deaths that occurred at home were used in assessing the at-home heat vulnerability based the surrounding neighborhood's social and environmental characteristics.

A case-only analysis, which was originally proposed to examine gene-environment interactions [52], can also be used to study how slow-varying characteristics modify the effects of a time-varying environmental exposure on a specified outcome [53]. This methodology is applied here to analyze whether the mortality on hot days is modified by individual and small-area (CT) social and environmental characteristics throughout the entire 2000–2015 study period. Three levels of modifiers were assessed here, including personal factors, primary cause of death, and area-level characteristics, similar to Zanobetti et al. (2013) [54]. To conduct the case-only analyses, a logistic regression was used to examine whether modifiers of interest were associated with increased relative odds of death within or outside of the home on days where $T_{MAX} \geq 90$ °F, $T_{MAX} \geq 85$ °F, or $HI_{MAX} \geq 86$ °F during the warm season. Individuals were considered to either have a personal modifier or not, or were dichotomized based on being either below the 25th percentile or above the 75th percentile of the values for neighborhood modifiers across Boston or dying from that specified primary cause of death, following previous studies using the same methodology [54]. Effect estimates and 95% confidence intervals are reported.

In addition to the temperature thresholds defined for the entire city using the airport weather station, we wanted to reduce a temperature exposure misclassification by assigning a local temperature estimate to each of the mortality records used in the geographically weighted regression. The Parameter-Elevation Regressions on Independent Slopes Model (PRISM) was used to provide higher spatial resolution (800 m^2) modeled temperature values, derived from a combination of weather station data, elevation models, and other spatial datasets to generate gridded estimates of climatic parameters [55,56]. It relies on the assumption that elevation is one of the most important determinants of temperature and precipitation, but also factors in the horizontal distance between weather stations, the spatial closeness between weather stations, the vertical layer of the atmosphere, topography, proximity to a coast, and the ability of terrain to affect precipitation. The distribution of these factors in relation to Boston, MA, results in the coastal proximity and elevation being the 2 factors that are most influential to the local temperature and precipitation. More details on PRISM and its development can be found in Daly et al., 2002 and Daly et al., 2007 [55,57]. The PRISM dataset has been found to be suitable for epidemiological studies looking at extreme heat and health relationships [58,59]. The daily maximum temperature at an 800 m^2 resolution along a uniform square grid was used to assign the maximum daily temperature for hot days where $T_{MAX} \geq 90\,^\circ F$ at Logan International Airport. This was also done for hot and humid days where $HI_{MAX} \geq 86\,^\circ F$ at the airport, using the daily maximum temperature and vapor pressure to assign an estimate of HI_{MAX} at the residence of each mortality record included in this subset of analyses. This daily data was available from 2003–2015, allowing us to capture all but 3 years of mortality data.

A subset of at-home deaths that occurred from 2003–2015 (i.e., those that had corresponding PRISM data) were spatially analyzed at the residential address using a geocoded latitude and longitude, and the outcome in the model was summarized as the daily tract-specific death rate per 100,000 people. The original objective of this research question was to pair building specific details with the death records, but too few death records with complete building information existed to do so, so area-level summaries were used for the building characteristics, along with the social and environmental parameters. A generalized linear model (GLM) with a Poisson distribution was used to fit the model to assess the impact of the temperature, and neighborhood social and environmental parameters on the tract- and day-specific mean at-home death rate on hot days. The linear model was fit using backwards stepwise fitting, first considering all covariates of interest in the model, eliminating those with $p < 0.20$. An ANOVA goodness of fit test determined that the nested model was an adequate fit compared to the full model. This same model was then used in the generalized geographically weighted regression (GWR) with a Poisson distribution with an appropriate covariance structure.

The use of mortality data was approved by both the Harvard TH Chan School of Public Health and The Commonwealth of Massachusetts Institutional Review Boards.

3. Results

During this 16-year study period, there was a total of 14,200 deaths in Boston that occurred at home. Death records missing a correct death date were excluded, and only records that occurred during the warm season (May–September) were kept in the analysis, leaving 6102 deaths that occurred at home. Excluding those deaths that happened within an inpatient or nursing facility, 34,404 deaths that had complete information during this time period occurred outside of the home (Table 1). Of the deaths that happened at home, 197 (8.5%) occurred on days with $T_{MAX} \geq 90\,^\circ F$, 186 (8.0%) on days with $HI_{MAX} \geq 86\,^\circ F$, and 475 (20.4%) on days with $T_{MAX} \geq 85\,^\circ F$. The descriptive statistics of the CTs in Boston are provided in Table 1. A map of Boston neighborhoods is available in Figure S1, and maps of the distribution of each small-area social and environmental factor are provided in Figure S2.

When evaluating the PRISM data of the spatial distribution of daily high temperatures, there was an average 2.97 $^\circ F$ temperature difference (ΔT_{MAX}) throughout the City of Boston on days when the airport temperature data indicated that daily $T_{MAX} \geq 90\,^\circ F$. The daily modeled vapor pressure and T_{MAX} were used to calculate the daily heat index on days when $HI_{MAX} \geq 86\,^\circ F$ at Logan International Airport. On average, the ΔHI_{MAX} was 8.1 $^\circ F$ across Boston, with the lowest HI_{MAX} and highest HI_{MAX} across Boston being 83.6 $^\circ F$ and 91.8 $^\circ F$, respectively.

Table 1. Descriptive statistics for the Boston, MA warm season climate, analyzed deaths from all causes from 2000–2015, and small-area social and environmental parameters at the census tract (CT) level. The frequency (n) represents the number of days meeting that meteorological threshold or the number of deaths that met that characteristic.

Characteristic	
Warm Season Maximum Temperature [°F] (mean (range))	75.7 (46–103)
Warm Season Mean Temperature [°F] (mean (range))	67.9 (44–92)
Warm Season Minimum Temperature [°F] (mean (range))	60.1 (37–81)
Frequency of $T_{MAX} \geq 90$ °F (n)	201
Frequency of $HI_{MAX} \geq 86$ °F (n)	165
Frequency of $T_{MAX} \geq 85$ °F (n)	502
At-Home Deaths (*n*)	6102
Male (*n* (%))	3197 (52.4)
Race, Non-Caucasian (*n* (%))	2276 (37.3)
≥65 years old (*n* (%))	3865 (63.3)
Outside of the Home Deaths (*n*)	34,404
Male (*n* (%))	20,744 (60.3)
Race, Non-Caucasian (*n* (%))	4322 (12.6)
≥65 years old (*n* (%))	19,732 (57.4)
Assessed value of residential building/area ($)(mean (SD))	5910 (28,000)
Assessed value of residential land/area ($)(mean (SD))	408 (463)
Energy efficiency score of residential buildings (mean (SD))	6.25 (0.352)
Year residential buildings were built/renovated (mean (SD))	1970 (15.9)
Decade residential buildings were built/renovated (mean (SD))	1980 (33.9)
Albedo (mean (SD))	0.124 (0.0114)
Impervious Surface Fraction (mean (SD)) [%]	0.787 (0.141)
Number of street trees/CT area (mean (SD))	0.00017 (0.00008)
Proportion of population with at least one disability (mean (SD)) [%]	11.4 (7.4)
Proportion of population that is ≤5 years old (mean (SD)) [%]	16.6 (10.4)
Proportion of population with at least 1 medical illness (mean (SD)) [%]	38.6 (3.19)
Proportion of population that is ≥65 years old (mean (SD)) [%]	10.5 (6.79)
Proportion of population that receives low-to-no income (mean (SD)) [%]	28.0 (17.1)
Proportion of population with limited English proficiency (mean (SD)) [%]	38.5 (17.8)
Proportion of population that is not Caucasian (mean (SD)) [%]	51.6 (30.3)
Proportion of population with utilities excluded from rent (mean (SD)) [%]	82.2 (17.6)
Proportion of population that is unemployed	0.0905 (0.0662)
Gross Rent ($/month)	1180 (471)
Ratio of Females-to-Males	1.06 (0.324)
Gini Coefficient	0.416 (0.0829)
Population Density (mean (SD)) [people/mile^2]	23,800 (18,300)

3.1. Case-Only Analysis

3.1.1. Deaths at Home

There were not significantly higher relative odds of dying at home than outside of the home for all temperature exposure definitions: $T_{MAX} \geq 90$ °F odds ratio (OR) = 1.01 (95% confidence interval [CI]: 0.91, 1.11), $HI_{MAX} \geq 86$ °F OR = 1.06 (95% CI: 0.95, 1.17), and $T_{MAX} \geq 85$ °F OR = 1.07 (95% CI: 1.00, 1.14). Compared with at-home deaths occurring on other warm season days, individuals living in CTs with a higher proportion of low-to-no income persons (OR = 1.30, 95% CI: 1.06, 1.59) or with limited English proficiency (OR = 1.29, 95% CI: 1.05, 1.57) had higher relative odds of dying at home on days with $T_{MAX} \geq 90$ °F than those in communities without these traits. Individuals who died at home living in CTs with a higher assessed building value per area were more highly represented among all at-home deaths than on days with $T_{MAX} \geq 90$ °F with an OR = 1.26 (95% CI: 1.04, 1.53) (Table 2).

Table 2. The relative odds ratio (OR) of dying at home during extreme heat, during the warm season for those who had the following characteristic (Personal, Primary Cause of Death) or lived in a census tract with the characteristic (Social, Environmental), compared with those who did not have that characteristic, Boston, 2000–2015. Individuals were considered to either have a personal modifier or not, or were dichotomized based on being either below the 25th percentile or above the 75th75th percentile of the values for neighborhood modifiers across Boston or dying from that specified primary cause of death. Bold indicates the OR is significant at an $p < 0.05$.

Personal	$T_{MAX} \geq 90\,°F$		$HI_{MAX} \geq 86\,°F$		$T_{MAX} \geq 85\,°F$	
	OR	95% CI	OR	95% CI	OR	95% CI
Female	1.12	(0.94, 1.34)	1.01	(0.83, 1.22)	1.05	(0.93, 1.19)
Not Married	1.14	(0.94, 1.39)	1.13	(0.91, 1.40)	1.08	(0.95, 1.24)
Race, Non-Caucasian	0.90	(0.75, 1.08)	0.99	(0.81, 1.21)	0.99	(0.87, 1.12)
Age < 57	1.02	(0.83, 1.25)	1.13	(0.91, 1.41)	1.03	(0.89, 1.19)
Age > 81	0.88	(0.72, 1.08)	1.06	(0.85, 1.32)	0.94	(0.82, 1.08)
Social						
Low Income	**1.30**	**(1.06, 1.59)**	0.92	(0.73, 1.16)	1.14	(0.99, 1.32)
Unemployment	1.03	(0.84, 1.25)	1.07	(0.86, 1.32)	0.98	(0.85, 1.12)
GINI Index	1.00	(0.81, 1.22)	1.01	(0.81, 1.27)	0.98	(0.86, 1.13)
Population Density	1.14	(0.93, 1.39)	1.08	(0.86, 1.35)	1.10	(0.95, 1.27)
Sex Ratio F:M	1.02	(0.84, 1.25)	0.98	(0.79, 1.22)	1.04	(0.91, 1.19)
Utilities not Included	1.13	(0.92, 1.39)	0.89	(0.71, 1.13)	1.08	(0.93, 1.25)
Disability	1.19	(0.97, 1.46)	1.09	(0.87, 1.37)	1.10	(0.95, 1.28)
Children	1.17	(0.95, 1.44)	1.03	(0.82, 1.30)	1.11	(0.96, 1.28)
Elderly	1.01	(0.83, 1.24)	1.01	(0.81, 1.26)	0.99	(0.86, 1.14)
Limited English Prof.	**1.29**	**(1.05, 1.57)**	0.88	(0.69, 1.12)	1.14	(0.99, 1.32)
Race, Non-Caucasian	1.19	(0.97, 1.46)	1.18	(0.94, 1.48)	1.14	(0.98, 1.32)
Medical Illness	0.97	(0.79, 1.20)	0.89	(0.70, 1.12)	1.01	(0.87, 1.17)
Environmental						
Energy Efficiency	0.96	(0.78, 1.18)	0.95	(0.76, 1.20)	0.94	(0.82, 1.09)
Assessed Value of Res Land/Total Property Area	0.91	(0.74, 1.12)	0.87	(0.69, 1.10)	1.05	(0.91, 1.20)
Assessed Value of Res Building/Gross Floor Area	**1.26**	**(1.04, 1.53)**	1.04	(0.83, 1.29)	1.10	(0.96, 1.27)
Year Built/Renovated	1.13	(0.91, 1.38)	**0.77**	**(0.62, 0.95)**	1.12	(0.97, 1.29)
Trees/CT Area	0.92	(0.75, 1.13)	**0.76**	**(0.60, 0.97)**	0.96	(0.83, 1.10)
Albedo	0.93	(0.78, 1.11)	0.93	(0.76, 1.13)	0.93	(0.82, 1.05)
Impervious Surface Fraction	0.96	(0.78, 1.17)	0.96	(0.77, 1.20)	1.05	(0.91, 1.20)
Primary Cause of Death						
Infection	1.46	(0.72, 2.96)	1.72	(0.81, 3.65)	**1.69**	**(1.02, 2.80)**
Cancer	0.84	(0.69, 1.01)	0.85	(0.69, 1.04)	0.93	(0.82, 1.06)
Inflammatory Disease	0.87	(0.35, 2.18)	1.11	(0.44, 2.79)	0.66	(0.33, 1.30)
Endocrine/Nutritional/Metabolic Disease	1.05	(0.71, 1.54)	1.12	(0.75, 1.69)	0.94	(0.72, 1.24)
Mental/Behavioral/Neurodevelopmental Disease	1.26	(0.86, 1.85)	0.76	(0.46, 1.26)	1.02	(0.76, 1.36)
Nervous System Disease	1.24	(0.76, 2.04)	1.06	(0.59, 1.88)	1.07	(0.74, 1.54)
Heart Disease	**1.21**	**(1.01, 1.46)**	1.11	(0.9, 1.37)	1.11	(0.97, 1.26)
Respiratory Disease	1.12	(0.76, 1.67)	1.12	(0.73, 1.74)	1.08	(0.82, 1.43)
Injury/Accident/Event	0.75	(0.49, 1.16)	0.96	(0.62, 1.49)	0.97	(0.74, 1.29)
Self-Harm	0.85	(0.39, 1.86)	0.71	(0.29, 1.78)	0.72	(0.41, 1.24)
Assault-Related Altercation	0.79	(0.4, 1.56)	1.47	(0.82, 2.63)	1.01	(0.66, 1.54)

When considering days where $HI_{MAX} \geq 86\,°F$, none of the personal or neighborhood social factors were significant modifiers between days where $HI_{MAX} \geq 86\,°F$ and at-home mortality. However, individuals living in CTs with newer or more recently renovated residential buildings or a greater density of street trees in the CT had a lower relative risk of dying at home with OR = 0.77 (95% CI:

0.62, 0.95) and OR = 0.76 (95% CI: 0.60, 0.97), respectively (Table 2). There were no primary causes of death that were significant modifiers of this association on hot and humid days. There were also no significant individual, community social, or community environmental modifiers on days where $T_{MAX} \geq 85$ °F for at-home deaths, but those whose primary cause of death was infection-related were more highly represented with OR = 1.69 (95% CI: 1.02, 2.80)

3.1.2. Deaths Outside of the Home

When considering deaths that occurred outside of the home but not within an inpatient or nursing facility, we first examined all deaths that occurred within the city of Boston (including those that did not reside in the city of Boston).. On warm days, where $T_{MAX} \geq 85$ °F, we saw that deaths from circulatory/heart-related disease were less represented in deaths outside of the home, with OR = 0.94 (95% CI: 0.89, 0.99). On these warm days, individual characteristics significantly modified the association between heat and deaths such that individuals over the age of 82 (the 75th percentile of ages represented in deaths occurring in Boston outside of the home) (OR = 0.93; 95% CI: 0.87, 0.98) had lower relative odds of death, but individuals younger than 52 (the 25th percentile) had higher relative odds of death (OR = 1.09; 95% CI: 1.02, 1.16). There was no significant effect modification for any factors on hot or hot and humid days (Table 3).

Table 3. The relative odds ratio (OR) of dying outside of the home during extreme heat, during the warm season for those who had the following characteristic (Personal, Primary Cause of Death) or who lived in a census tract with the characteristic (Social, Environmental), compared with those who did not have that characteristic, Boston, 2000–2015. Individuals were considered to either have a personal modifier or be dying from that specified primary cause of death, or not. Bold indicates that the OR is significant at $p < 0.05$.

Personal	$T_{MAX} \geq 90$ °F		$HI_{MAX} \geq 86$ °F		$T_{MAX} \geq 85$ °F	
	OR	95% CI	OR	95% CI	OR	95% CI
Sex	1.03	(0.95, 1.12)	0.97	(0.88, 1.05)	1.03	(0.98, 1.09)
Not Married	1.04	(0.96, 1.12)	1.05	(0.96, 1.14)	1.02	(0.97, 1.07)
Race, Non-Caucasian	0.94	(0.84, 1.06)	1.03	(0.90, 1.17)	0.92	(0.85, 1.00)
Age > 82	0.96	(0.88, 1.04)	0.99	(0.90, 1.09)	**0.93**	**(0.87, 0.98)**
Age < 52	1.07	(0.98, 1.17)	1.03	(0.93, 1.14)	**1.09**	**(1.02, 1.16)**
Primary Cause of Death						
Infection	0.81	(0.55, 1.21)	1.11	(0.76, 1.62)	0.99	(0.77, 1.28)
Cancer	0.99	(0.89, 1.10)	1.01	(0.90, 1.14)	1.03	(0.96, 1.11)
Blood/Immune	1.06	(0.62, 1.81)	0.95	(0.51, 1.75)	0.87	(0.59, 1.29)
Endocrine/Nutritional/Metabolic	1.12	(0.94, 1.34)	1.09	(0.90, 1.33)	1.12	(0.99, 1.26)
Mental/Behavioral/Neurodevelopmental	0.84	(0.63, 1.14)	1.01	(0.74, 1.37)	0.9	(0.73, 1.10)
Nervous System Disease	0.96	(0.70, 1.33)	0.83	(0.57, 1.22)	0.88	(0.70, 1.11)
Circulatory Disease	0.99	(0.91, 1.07)	0.92	(0.84, 1.01)	**0.94**	**(0.89, 0.99)**
Respiratory Disease	0.86	(0.71, 1.04)	1.04	(0.85, 1.28)	0.92	(0.80, 1.05)
Congenital Disease	1.54	(0.70, 3.41)	0.24	(0.03, 1.74)	1.4	(0.77, 2.53)
Injury/Accident/Event	1.12	(0.96, 1.31)	1.02	(0.86, 1.22)	1.09	(0.98, 1.22)
Self-Harm	1.03	(0.79, 1.34)	0.87	(0.63, 1.19)	1.12	(0.94, 1.34)
Assault-Related Altercation	1.05	(0.85, 1.30)	1.23	(0.98, 1.53)	1.09	(0.94, 1.26)

We then restricted these deaths outside of the home to just those that involved Boston residents. On hot days, where $T_{MAX} \geq 90$ °F, deaths from substance abuse (OR = 2.88; 95% CI: 1.42, 5.86) and unknown causes (OR = 2.38; 95% CI: 1.13, 5.02) were more highly represented, as were those in CTs with a greater proportion of elderly individuals or with a greater CT energy efficiency score. On days were $HI_{MAX} \geq 86$ °F, individuals in the youngest quartile of mortality records, below age 58, had greater relative odds of death, with OR = 1.53 (95% CI: 1.07, 2.18). On warm days, where $T_{MAX} \geq 85$ °F, individuals under 58 were still more highly represented (OR = 1.31; 95% CI: 1.05, 1.62), and deaths from assault-related altercations had 1.79 times the relative odds of occurring on warm days (95% CI: 1.24, 2.58) (Table 4).

Table 4. The relative odds ratio (OR) of Boston residents dying outside of the home during extreme heat, during the warm season, for those who had the following characteristic (Personal, Primary Cause of Death) or who lived in a census tract with the characteristic (Social, Environmental), compared with those who did not have that characteristic, Boston, 2000–2015. Individuals were considered to either have a personal modifier or not, or were dichotomized based on being either below the 25th percentile or above the 75th percentile of the values for neighborhood modifiers across Boston or dying from that specified primary cause of death. Bold indicates that the OR is significant at $p < 0.05$.

Personal	$T_{MAX} \geq 90\,°F$		$HI_{MAX} \geq 86\,°F$		$T_{MAX} \geq 85\,°F$	
	OR	95% CI	OR	95% CI	OR	95% CI
Sex	0.81	(0.54, 1.21)	0.74	(0.47, 1.16)	0.77	(0.59, 1.00)
Not Married	1.20	(0.78, 1.83)	1.38	(0.85, 2.24)	1.12	(0.85, 1.47)
Race, Non-Caucasian	1.05	(0.72, 1.54)	1.2	(0.78, 1.84)	1.18	(0.92, 1.52)
Age > 82	0.98	(0.69, 1.39)	1.07	(0.73, 1.56)	0.81	(0.64, 1.02)
Age < 58	1.15	(0.82, 1.60)	**1.53**	**(1.07, 2.18)**	**1.31**	**(1.05, 1.62)**
Social						
Low Income	0.75	(0.52, 1.07)	0.81	(0.55, 1.19)	0.99	(0.79, 1.24)
Unemployment	1.03	(0.74, 1.45)	1.16	(0.81, 1.67)	1.06	(0.85, 1.32)
GINI Index	1.34	(0.96, 1.88)	0.91	(0.61, 1.35)	0.92	(0.73, 1.16)
Population Density	0.94	(0.62, 1.43)	0.9	(0.56, 1.43)	0.8	(0.61, 1.06)
Sex Ratio F:M	0.93	(0.66, 1.30)	1.01	(0.70, 1.46)	0.85	(0.68, 1.06)
Utilities not Included	1.07	(0.76, 1.51)	0.67	(0.44, 1.02)	1.09	(0.87, 1.36)
Children	0.81	(0.57, 1.15)	0.96	(0.67, 1.39)	1.07	(0.87, 1.33)
Elderly	**1.49**	**(1.07, 2.07)**	0.92	(0.62, 1.37)	**1.33**	**(1.06, 1.66)**
Limited English Prof.	0.8	(0.56, 1.13)	0.92	(0.63, 1.35)	0.99	(0.79, 1.24)
Race, Non-Caucasian	0.97	(0.70, 1.36)	0.91	(0.62, 1.31)	1.16	(0.94, 1.43)
Medical Illness	1.36	(0.97, 1.91)	1.24	(0.85, 1.80)	1.14	(0.91, 1.43)
Low Income	0.75	(0.52, 1.07)	0.81	(0.55, 1.19)	0.99	(0.79, 1.24)
Environmental						
Energy Efficiency	**1.44**	**(1.03, 2.02)**	1.08	(0.73, 1.59)	1.11	(0.88, 1.40)
Assessed Value of Res Land/Total Property Area	0.86	(0.56, 1.32)	0.97	(0.62, 1.52)	0.78	(0.59, 1.03)
Assessed Value of Res Building/Gross Floor Area	0.74	(0.51, 1.07)	0.85	(0.57, 1.27)	0.98	(0.78, 1.23)
Year Built/Renovated	0.83	(0.54, 1.28)	0.85	(0.53, 1.38)	0.81	(0.61, 1.07)
Trees/CT Area	0.86	(0.56, 1.34)	1.12	(0.72, 1.76)	0.95	(0.72, 1.25)
Albedo	1.15	(0.85, 1.56)	0.90	(0.64, 1.25)	1.1	(0.90, 1.34)
Impervious Surface Fraction	0.76	(0.53, 1.10)	0.90	(0.61, 1.33)	0.83	(0.66, 1.05)
Primary Cause of Death						
Infection	0.54	(0.07, 4.02)	1.40	(0.60, 1.14)	**0.77**	**(0.62, 0.96)**
Liver Disease	0.46	(0.06, 3.45)	3.58	(0.59, 2.39)	1.14	(0.70, 1.86)
Cancer	0.88	(0.59, 1.31)	0.85	(0.77, 1.48)	0.96	(0.77, 1.21)
Diabetes	1.16	(0.55, 2.44)	1.67	(0.30, 1.93)	1.26	(0.75, 2.10)
Heart Disease	0.9	(0.64, 1.24)	0.65	(0.12, 2.10)	1.26	(0.68, 2.35)
Nervous System Disease	1.11	(0.49, 2.50)	0.78	(0.46, 3.91)	0.8	(0.33, 1.93)
Substance Abuse	**2.88**	**(1.42, 5.86)**	1.16	(0.60, 3.98)	1.75	(0.90, 3.39)
Inflammatory Disease	0.88	(0.21, 3.77)	1.07	(0.69, 8.62)	1.6	(0.56, 4.58)
Cerebrovascular Disease	0.27	(0.04, 1.95)	0.33	(0.06, 3.48)	1.14	(0.42, 3.08)
Digestive System Related Disease	0.66	(0.09, 4.97)	**1.74**	**(1.17, 6.73)**	1.75	(0.89, 3.47)
Unknown Causes	**2.38**	**(1.13, 5.02)**	1.98	(0.82, 4.76)	1.66	(0.92, 2,99)
Assault-Related Altercation	1.49	(0.87, 2.54)	0.92	(0.39, 38.5)	**1.79**	**(1.24, 2.58)**
Injury/Accident/Event	0.8	(0.42, 1.49)	1.42	(0.75, 2.68)	1.11	(0.77, 1.60)
Self-Harm	1.00	(0.36, 2.80)	0.92	(0.28, 2.98)	0.99	(0.50, 1.94)

3.2. Geographic Weighted Regression

Geographic weighted regression analyses were used to evaluate the role of environmental and social factors in driving the spatial distribution of at-home mortality from 2003–2015 on hot days (n = 389) with the gridded climate data defining the local T_{MAX} at the residence of each death record. The nested GLM that was found to be an appropriate fit using the model-building criteria previously described was compared to the full model using an ANOVA goodness of fit test, as shown in Equation (1).

$$
\begin{aligned}
Total\ Daily\ Death\ Rate\ \sim\ &\beta_0 + \beta_1(Albedo) + \beta_2\left(\tfrac{Street\ Trees}{CT\,Area}\right) + \beta_3\left(\tfrac{Value}{Area}\right) + \\
&\beta_4(Impervious\ Surface\ Fraction) + \beta_5(Population\ Density) + \\
&\beta_6(Prop.\ Disability) + \beta_7(Prop.\ Older) + \beta_8(Prop.\ Low\ Income) + \\
&\beta_9(Prop.People\ of\ Color) + \beta_{10}(T_{MAX-LOCAL}),
\end{aligned}
\tag{1}
$$

The covariates from the GLM showed that the density of street trees in a CT and the proportion of those in the CT who were older adults were significant predictors of the daily at-home death rate (Table S1). Applying the same GLM model to the GWR, the GWR model (Equation (2)) was found to have a pseudo R^2 = 0.89.

$$
\begin{aligned}
Total\ Daily\ Death\ Rate\ \sim\ &\beta_0 + \beta_1(Albedo) + \beta_2\tfrac{(Street\ Trees)}{CT\ Area} + \beta_3\left(\tfrac{Building\ Value}{Building\ Area}\right) + \\
&\beta_4(Impervious\ Surface\ Fraction) + \beta_5(Population\ Density) + \\
&\beta_6(Prop.\ Disability) + \beta_7(Prop.\ Older) + \beta_8(Prop.\ Low\ Income) + \\
&\beta_9(Prop.People\ of\ Color) + \beta_{10}\left(T_{MAX_{local}}\right) + offset(\log(Population)),
\end{aligned}
\tag{2}
$$

There were wide ranges in the value of the coefficients when examining them spatially, indicating relationships with at-home mortality on hot days that vary based on the local context. The mean CT albedo, population density, proportion of those with a disability, and the proportion of those with low-to-no income or who are non-Caucasian had local regression coefficients that were negative, indicating a lower daily mean death rate (Table S2). The median values of the local coefficients for the daily T_{MAX}, trees/CT area, impervious surface fraction, and the proportion of those who are older adults aged 65 or older were positive, indicating a higher daily mean death rate.

Figure 1 provides a map of the local coefficients for each of the covariates, included in the GWR model, that was restricted to just the deaths occurring at home on hot days. The strongest positive local coefficients (which would contribute to a higher at-home mortality) were seen with the proportion of non-Caucasian individuals in East Boston, Roxbury, and Dorchester, the population density in parts of Mission Hill and into Jamaica Plain, the proportion of trees per the CT area in Dorchester, the value of building per area in East Boston, and the surface albedo from Roslindale through Hyde Park. The strongest negative local coefficients were seen with the proportion of those with low-to-no income from Mission Hill south through Roslindale and Hyde Park, and the surface albedo from Charlestown through Downtown to Dorchester.

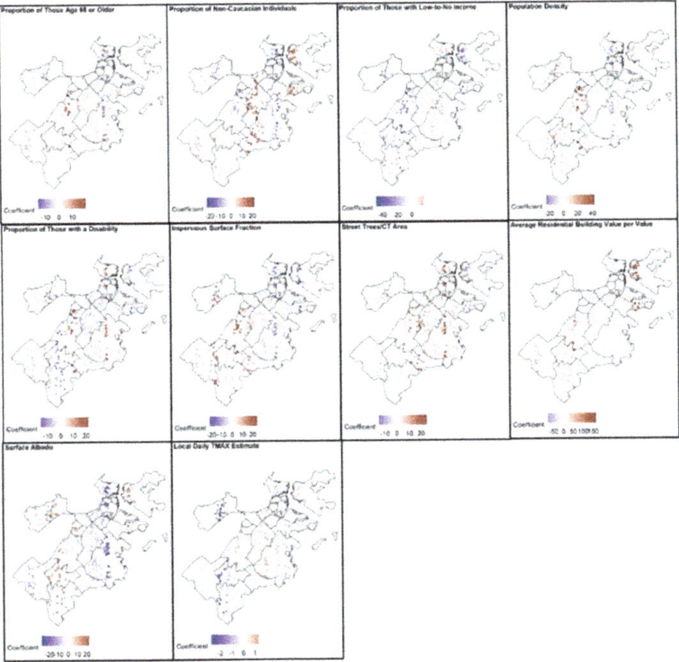

Figure 1. The local GWR coefficients, with blue colors representing negative coefficients (lower daily at-home mortality rate) and red colors representing positive coefficients (higher daily at-home mortality rate), in relation to daily at-home death rates on days where $T_{MAX} \geq 90\,°F$ at Logan International Airport.

The same GWR analyses were used to evaluate the role of the environmental and social factors in driving the spatial distribution of the at-home mortality on hot and humid days, where $HI_{MAX} \geq 86\,°F$ at Logan International Airport (n = 358) with the gridded climate data defining the HI_{MAX} at the residence of each death record. A nested GLM that included the proportion of non-Caucasian individuals but excluded the population density and building value/area was found to be an appropriate fit using the model-building criteria previously described, compared to the full model using an ANOVA goodness of fit test, and is found in Equation (3).

$$\begin{aligned}
\text{Total Daily Death Rate} \ \sim \ &\beta_0 + \beta_1(Albedo) + \beta_2\left(\tfrac{Street\ Trees}{CTArea}\right) + \\
&\beta_3(Impervious\ Surface\ Fraction) + \beta_4(Prop.\ Disability) + \beta_5(PProp.\ Older) + \\
&\beta_6(Prop.\ Low\ Income) + \beta_7(Prop.People\ of\ Color) + \beta_8(HI_{MAX-LOCAL}),
\end{aligned} \tag{3}$$

The covariates from the GLM for the days when $HI_{MAX} \geq 86\,°\,F$ showed that the impervious surface fraction and the proportion of elderly individuals were significantly associated with the daily at-home death rate (Table S3). Applying the same GLM model to the GWR, the GWR model (Equation (4)) was found to have a pseudo $R^2 = 0.75$.

$$\begin{aligned}
\text{Total Daily Death Rate} \ \sim \ &\beta_0 + \beta_1(Albedo) + \beta_2\tfrac{(Street\ Trees)}{CT\ Area} + \\
&\beta_3(Impervious\ Surface\ Fraction) + \beta_4(Prop.\ Disability) + \\
&\beta_5(Prop.Older) + \beta_6(Prop.Low\ Income) + \\
&\beta_7(Prop.\ People\ of\ Color) + \beta_8\left(T_{MAX_{local}}\right) + \\
&\beta_9(offset(\log(Population)))
\end{aligned} \tag{4}$$

Again, there is a spatial heterogeneity in the value of the coefficients (Table S4). Similar to those days when $T_{MAX} \geq 90\,°F$ at the airport, the mean CT albedo, proportion of those with a disability, and the proportion of those with low-to-no income or who are non-Caucasian had local regression coefficients that were negative, indicating a negative association with the daily mean death rate on days when $HI_{MAX} \geq 86\,°F$ at the airport. The median values of all the other local coefficients, including the local HI_{MAX}, were positive, indicating a higher daily mean death rate. Figure 2 provides a map of the local coefficients for each of the covariates included in the GWR model that was restricted to just the deaths occurring at home on hot and humid days. Similar trends were seen in these analyses, with many of the strongest positive local coefficients seen with social and environmental vulnerability factors in East Boston and Dorchester.

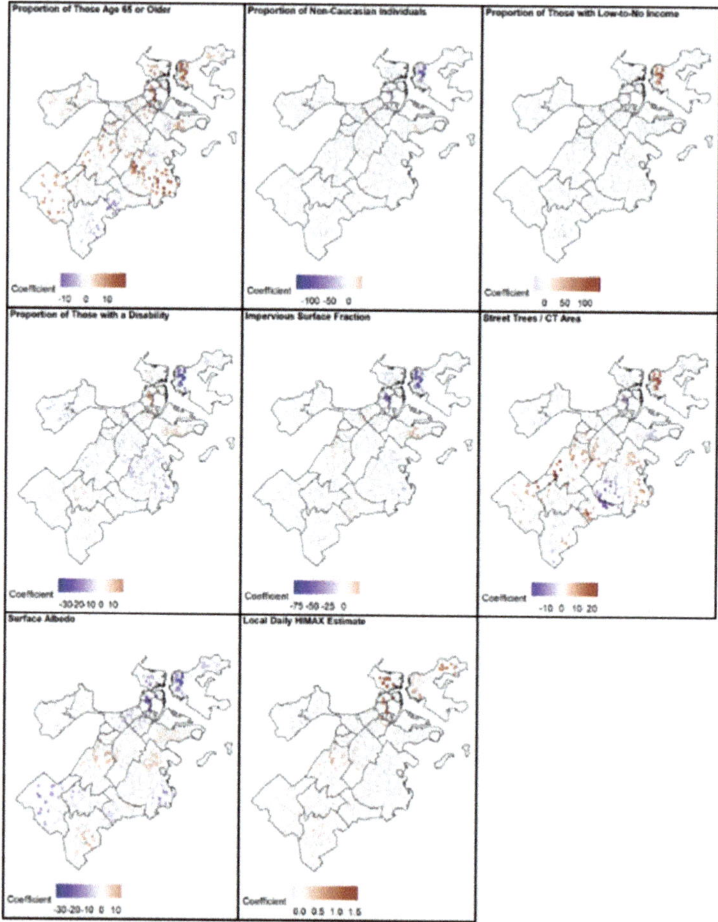

Figure 2. The local GWR coefficients, with blue colors representing negative coefficients and red colors representing positive coefficients, in relation to the daily at-home death rate on days where $HI_{MAX} \geq 86\,°F$ at Logan International Airport. CT denotes census tract.

4. Discussion

Heat vulnerability assessments have been found to provide valuable evidence for the design and strategic implementation of adaptation solutions that most effectively protect vulnerable populations.

The results from this study provide valuable local heat vulnerability knowledge to Boston, MA, while also furthering the evidence of heat-mortality relationships in the Northeast US. This study examined several small area-level and individual social and environmental factors that are associated with an increased likelihood of dying at home during extreme heat events. In the case-only analyses, those living in CTs with a greater prevalence of well-known social vulnerability factors, like those with low-to-no income or with a limited English proficiency had increased relative odds of dying at home on a hot day during the warm season in Boston, MA. This supports much of the existing research on heat vulnerability [15,26]. However, some small-area environmental factors, like density of street trees and the more recently built/renovated residential buildings, were less represented among at-home deaths on hot days, which can inform adaptation strategies throughout the city.

There was not a significantly higher relative risk for elderly individuals or CTs with a greater proportion of elderly individuals, dying at home at any temperature definition, which is similar to past research on heat vulnerability in NYC [26], despite older adults' enhanced susceptibility to extreme heat and previous findings documenting an increased vulnerability [4,5]. We hypothesize that these individuals, who are extremely vulnerable to heat-related mortality, may be more frequently dying within inpatient or nursing facilities, and were therefore not captured in this analysis. Furthermore, those who are oldest may be themselves aware of their enhanced susceptibility to heat-related medical complications, so may take precautions on these days (or have a support system that aids them in doing so).

Some environmental parameters, like more recently renovated/built residential buildings or a greater density of street trees, were able to reduce the relative odds of death for those dying at home on hot and humid days. However, other environmental factors, like the value of the residential building per its area (at-home deaths) and the CT residential energy efficiency (deaths of Boston residents outside of the home), were associated with higher relative odds of death. In Boston, an evaluation of local real estate prices revealed that triple decker homes had one of the highest values per area. These homes were historically used to house immigrant workers, and when their popularity surged in the 1980s it was common for them to be bought by absentee landlords and be poorly maintained [60]. Furthermore, apartment buildings are more energy efficient than single-family detached homes, and living in multi-family homes has previously found to be a risk factor for heat-related mortality in past severe heat events [21]. The specific context of an individual building should be taken into consideration when designing or retrofitting a building to better elucidate these relationships.

The small-area social and environmental factors surrounding the homes of those who died were not as important for deaths that occurred outside of the home as compared to those that happened at home. This makes sense, as the home neighborhood characteristics will likely not play as much of a role for deaths that occur elsewhere as they would for deaths that occur at home. However, Boston residents who were in the youngest quartile of deaths outside of the home were more highly represented within those that died on warm, as well as on hot and humid days. Future studies could investigate the breakdown of this lowest quartile, determining the relative risk of death for vulnerable populations that are young, such as children or occupational populations.

Interestingly, the temperature thresholds that were used did yield modifications by different primary causes of death. Deaths at home were more highly represented by those who died of heart disease or infection, which is similar to previous findings [4,61]. On hot and humid days, deaths from digestive-related issues were more highly represented among Boston residents who died outside of the home, which is also supported by the past literature [51]. Boston residents who died outside of the home during this time period had 2.88 and 2.38 times the relative odds of dying from substance abuse or unknown causes when $T_{MAX} \geq 90\,°F$ days and 1.79 times the relative odds of dying from an assault-related altercation on days when $T_{MAX} \geq 85°F$. Given that past research has shown that the heat-related mortality is frequently misclassified, and that sometimes proper causes are not listed at all, we believe that the high relative odds of dying from unknown causes presents a clear signal of heat-related mortality effects. Furthermore, research has demonstrated that aggressive and violent behaviors increase on hot days [62–65]. In Boston, there are significant increases in police, fire, and medical services across the city at temperatures of around 83–85 °F [66], which provides evidence that societal services are impacted by heat at these lower temperatures [67]. Our findings of increased relative odds of dying from substance abuse follow patterns seen in past studies in other geographic locations [68,69]. To our knowledge, these causes of deaths are not captured in local assessments of the public health impacts of extreme heat and mortality risk factors, despite some anecdotal information provided to us on these associations at the start of this study (personal communications with the City of Boston Public Health Commission).

The relationships between social and environmental modifiers of at-home mortality during extreme heat events varied spatially. When comparing this information to previous heat vulnerability assessments in Boston, MA, a few interesting trends emerge. A 2015 assessment of a heat vulnerability index found that 11 CTs in Boston had the greatest heat vulnerability despite access to cooling resources (e.g., cool centers, spray pads, etc.), including in East Boston, Chinatown, Fenway, the South End, Mattapan, Roslindale, and 5 CTs in Roxbury, all primarily driven by social and environmental vulnerability [31]. The area of East Boston that was found to be highly vulnerable also corresponded to areas of East Boston that were found to have an increased social and environmental vulnerability to extreme heat. These analyses did not find that Chinatown, Fenway, or the South End were particularly vulnerable to dying at home on a hot day. However, mortality is the most extreme outcome that is possible with extreme heat exposure, with deaths at home comprising a small fraction of all deaths, so these other neighborhoods may still be vulnerable to other negative health outcomes, if not to at-home mortality. Alternatively, with Boston's recent adaptation of heat mitigation strategies, it would be worth investigating if any social or environmental interventions had taken place during this study period that could attenuate the at-home death rate on hot days in certain neighborhoods.

The 2016 Climate Ready Boston report highlights the CTs where the greater proportion of socially vulnerable individuals live in the city, including those who are non-Caucasian, of older age, younger age, with a disability or a medical illness, with limited English proficiency, or of low-to-no income. The neighborhoods of Dorchester, Roxbury, East Boston, Brighton, Downtown, South End, and Mission Hill appear to be the most vulnerable from these factors. While these neighborhoods have some overlap with the findings of this study—especially Dorchester, Roxbury, and East Boston—the others do not have as much overlap. Again, the results of this study are not saying that residents of these other neighborhoods are not vulnerable, but instead that they are not as socially or environmentally vulnerable as those Dorchester, Roxbury, and East Boston to dying at home on hot days.

The results of this analysis highlight the spatial variability of heat vulnerability factors, which would have been masked by traditional regression analyses that assumed spatially consistent relationships between all of these factors and at-home mortality. From this, interesting trends in some of the environmental vulnerability factors emerged, where in some areas features considered to be beneficial, like street trees, demonstrate a negative relationship with daily at-home mortality. We hypothesize that this may be related to the differential quality of environmental features that are present. One example of this is present with a discussion of the quality over quantity of street trees. In many neighborhoods, the GWR analyses suggest a positive association between the density of street trees and at-home mortality on hot days, insofar as the density of street trees was associated with a higher at-home mortality rate, which is counterintuitive to what one may expect. The quality of these trees is important in determining how much shading effect may be available, but we did not have quantitative information on the quality of the trees in each neighborhood. Figure S3 demonstrates some of the features of trees that may be influencing their role in shading residential buildings, including if they are in an ill state of health, if they are young trees, or if they are not equally distributed within neighborhoods (i.e., only located on one side of a street), all of which would limit the shading potential the trees had to offer. We are eager to explore the role of this building archetype and design and this street tree quality on indoor temperature exposures in future studies.

One of the most unique aspects of this research is the application of a gridded climate dataset to refine the temperature exposure throughout Boston. Many of the strongest coefficients, in either direction, align with Interstate-93 in the easternmost part of Boston. This boundary may also be indicative of the extent of the sea breeze on hot days. On these hot days, the mean wind direction at Logan International Airport during the day was 89.5°, indicating easterly winds off the Atlantic Ocean. Past analyses of Boston's sea breeze have found that it is one of the most developed in mid-latitudes with peaks in July and August. Within the city of Boston, there is both a smaller bay breeze effect, as well as a larger sea breeze effect, which often does not penetrate much further inland than the coastline due to abrupt topographic shifts in western Boston [70]. We hypothesize that this boundary, seen stretching north-to-south from Downtown to Dorchester, may be due to the proximity to a major roadway, which has documented negative health effects on surrounding residents [71], or to stark changes in temperature resulting from a microclimatic effect, like a bay breeze or the marine layer, or perhaps to a combination of these elements and requires further research. By using a refined exposure assessment and capturing temperature variations across the city, we can distinguish this local climate effect, which influences heat exposure and thus heat vulnerability.

The results of this study highlight the differential social and environmental drivers of heat vulnerability in Boston, MA. The goal of this study was to demonstrate how our surrounding built and natural environment exists concurrently with our social environment and how it can provide an important point of intervention, relying less on changing behaviors, that can be used in tandem with the existing communication of heat-harm-reduction strategies. Locally, there have been active efforts to mitigate poor health outcomes during extreme heat events for those who have been found to be socially vulnerable. However, it is important to remember that the natural and built environment play a large role in determining an individual's exposure to extreme heat and interact with the social factors to create unique vulnerabilities for each neighborhood. As was seen in East Boston, the proportion of street trees/CT area increased, had negative location coefficients in relationship to at-home mortality rate. A recent study found that nearly 80% of those living in almost 100 different US cities were in neighborhoods with less than 20% tree cover [25]. Most CTs in Boston are also below this percentage of tree cover [31]. Researchers then simulated in these cities an improved urban tree canopy and found that 245–346 deaths per year could be avoided, while also reducing building-related heat stress so that the cooling demand and attendant electricity consumption could decrease [25]. Lessons learned from the effective deployment of street trees in neighborhoods where trees have been shown to be protective of poor health outcomes may inform future adaptation strategies that increase the urban tree canopy most effectively.

Adaptation strategies like improved urban materials, cool roofs, evaporative roofs, and the increased density of street trees have the potential to reduce the range of summertime temperatures in the Northeast US and reduce the impact of the increased number of hot days that we are expected to experience by the end of the century [50]. The median age of residential buildings in Boston, MA is 54 years, and some of the most common residential building archetypes in Boston were predominantly built in the late 19th and early 20th centuries. Older buildings, which tend to use more traditional materials, have been found to have higher thermal masses [72], making them prone to overheating during times of power outages. Low-income individuals are more likely to live in homes that are overcrowded and/or that are of poor environmental and physical quality [73]. While these relationships between housing and public health are not new, demonstrating the quantitative effect that these built environment factors have on a climate change-relevant public health outcome provides local actionable evidence.

Currently, Boston has primarily relied on mechanical cooling to mitigate hot indoor temperatures, but without that cooling many of our residential buildings are not passively habitable and cannot provide safe and survivable indoor conditions without mechanical cooling in times of extreme heat [24]. In order to be the most resilient to the future extreme heat events of the next century, our residential buildings and surrounding built environments need to consider climate adaptive architecture and design that reduce the thermal loads of our urban areas. Planning for the health impacts of extreme heat in the design and retrofitting of buildings is vital for protecting public health in a changing climate. This will be important in urban areas, but also in those suburban areas and satellite cities around major urban centers that have been found to show a rapidly increasing vulnerability to extreme heat [74].

It is important to consider the limitations of these analyses. A case-only analysis is limited in that it only assesses the relative risk of mortality with individual and neighborhood characteristics, and not the greatest absolute risk. Additionally, in a case-only analysis you can only evaluate one modifier at a time, and we have evaluated many personal, social, and environmental factors. Given the large number of factors evaluated, there could be increases in a type I error that could potentially lead to the spurious identification of factors that modify the association between extreme heat and mortality, and we are unable to assess various combinations of environmental and social vulnerability. Although we have removed the effect of season, residual confounding may still be present if there is an interaction between season and any of the effect modifiers being examined. This study relies heavily on CT-level boundaries and information. There were likely changes in the social/environmental parameters over this study period, but only cross-sectional estimates that were available at different points in this study period were utilized to assign these parameters across an entire 15-year time period. Finally, one well known heat-mitigating environmental factor—the proximity to greenspace—was not included in this study, as Boston is only one of 2 cities in the US where 100% of the population is within 10 min from a park [75], so there was no significant variation amongst those deaths included in this study in relation to this environmental factor.

Furthermore, there are many ways to measure temperature and assess extreme heat. The temperature thresholds used within this study are supported by recent research, including findings by Guo et al. (2017), according to which daily maximum temperatures for defining a heat wave are better at predicting mortality than the minimum temperatures are, [76] as well as findings by Kingsley et al. (2016), which state that there are significant increases in mortality at temperatures from 75–85 °F in Rhode Island [32]. The thresholds used in these analyses also align with local heat interventions that are enacted in Boston during extreme heat events, including the opening of cooling centers and transportation for vulnerable persons, and they are therefore important to local heat responses. The use of modeled daily temperatures at the location of each of the residences examined in this study further reduces the temperature misclassification that is commonly present in heat-health research. However, we also acknowledge that metrics like the duration of extreme heat and whether or not an extreme heat event is the first of a season or not are important to consider [34], and that buildings influence heat exposure; consequently, further work is needed to fully remove any bias from temperature exposure misclassification.

Despite these limitations, some of which also exist in much of the preexisting heat vulnerability research, there are many strengths to the analyses within this paper. A case-only analysis that assesses the change in the heat-related mortality risk by individual and neighborhood characteristics can be used to create composite heat-related vulnerability indices to prioritize the most vulnerable neighborhoods based on social and environmental factors. This information can be used to compare how non-time varying characteristics—or characteristics that change very slowly over time—modify the effect of a time-varying environmental exposure like extreme heat on excess mortality. A case-only analysis that is focused on the warm season only also removes the effect of season, which reduces potential seasonal confounding, reduces potential confounding by variables typically associated with mortality (e.g., smoking), simplifies modeling, and reduces the model's sensitivity to misspecification bias. By including some health-promotive covariates, like street trees and albedo, we can assess protective neighborhood features instead of just assessing harmful features. The inclusion of both positive and adverse covariates allows us to both focus on strategies that create more health-protective environments and to reduce harmful environments, respectively, for all. This study also took a vast body of heat vulnerability research and metrics that have been demonstrated to be important for heat-mortality relationships, and assessed them at a local level, where the context and scale that were considered were tailored to Boston, MA. Finally, the use of the gridded climate dataset allows us to reduce temperature exposure misclassification and better characterize local temperature variability throughout Boston.

Future studies on the specific local building archetypes and how these modify indoor temperature exposures will fill a gap that remains from this study and will provide the most actionable evidence when constructing or retrofitting buildings to be more heat resilient. As Boston is currently exploring altering building codes to enhance resilience to flooding and sea-level rising, it will be critical to evaluate these policy measures for extreme heat as well. Integrating this work with a simulated adaptation strategy implementation can elucidate which building-focused adaptation strategies will be the most effective at reducing poor health outcomes at home during extreme heat events.

5. Conclusions

Future climate change adaptation will be implemented at the local level, so decision makers need data and evidence at that scale to best inform policy and infrastructure decisions. This study examined mortality at home and outside of the home on hot days in Boston, MA. We evaluated how individual and small-area social, built environment, and natural environment characteristics modify the association between heat and mortality at three temperature thresholds. While some neighborhoods had a greater social vulnerability, some of the environmental factors examined were able to reduce the relative odds of death within and outside the home. While many neighborhoods have been found to be socially vulnerable in past studies, there are some neighborhoods—like Roxbury, Dorchester, and East Boston—where both social and environmental vulnerability exists, highlighting priority locations for the implementation of a wide range of adaptation solutions. Furthermore, even at temperatures below current local thresholds used for warnings, advisories, and local interventions, there were significantly higher relative odds of death from unknown causes, substance abuse, injury/accidents, and assault, most of which are not currently being incorporated into heat vulnerability assessments and adaptation planning.

Supplementary Materials: The following are available online at http://www.mdpi.com/2225-1154/8/2/29/s1, Figure S1. Neighborhoods of Boston, MA; Figure S2. Distribution of the all small-area social and environmental vulnerability factors, Boston, 2000–2015; Table S1. Summary of results from GLM analyses limited to only hot days; Table S2. Summary of results from semiparametric GWR model analyses; Figure S3. Image of triple-decker homes that are common in Boston, MA; Figure S4. Images of street trees of varying quality.

Author Contributions: Conceptualization, A.A.W., and J.D.S..; methodology, A.A.A., J.G.A., P.J.C.,, and J.D.S.; software, A.A.W.; validation, A.A.W.; formal analysis, A.A.W. and P.J.C.; investigation, A.A.W.; resources, A.A.A.; data curation, A.A.W.; writing—original draft preparation, A.A.W.; writing—review and editing, A.A.A., J.G.A., P.J.C., and J.D.S.; visualization, A.A.W.; supervision, J.D.S., J.G.A., P.J.C.; project administration, J.D.S.; funding acquisition, A.A.W. and J.G.A. All authors have read and agreed to the published version of the manuscript.

Funding: This research was funded by the Harvard University President's Climate Change Solutions Fund. The acquisition of spatial meteorological data was funded by NIH Grant #P30ES000002.

Acknowledgments: The authors wish to acknowledge Joel Schwartz for providing mortality data, and input on the study design, as well as Tracy Sachs and Anna Kosheleva for administrative support through this study.

Conflicts of Interest: The authors declare no conflict of interest.

References

1. IPCC. 2018: Summary for Policymakers. In *Global Warming of 1.5 °C*; An IPCC Special Report on the Impacts of Global Warming of 1.5 °C above Pre-Industrial Levels and Related Global Greenhouse Gas Emission Pathways, in the Context of Strengthening the Global Response to the Threat of Climate Change, Sustainable Development, and Efforts to Eradicate Poverty; World Meteorological Organization: Geneva, Switzerland, 2018; pp. 1–32.

2. Reidmiller, D.R.; Avery, C.W.; Easterling, D.R.; Kunkel, K.E.; Lewis, K.L.M.; Maycock, T.K.; Stewart, B.C. *Fourth National Climate Assessment. Volume II: Impacts, Risks, and Adaptation in the United States, Report-in-Brief*; U.S. Global Change Research Program: Washington, DC, USA, 2018.

3. Zhang, K.; Chen, Y.-H.; Schwartz, J.D.; Rood, R.B.; O'Neill, M.S. Using Forecast and Observed Weather Data to Assess Performance of Forecast Products in Identifying Heat Waves and Estimating Heat Wave Effects on Mortality. *Environ. Health Perspect.* **2014**, *122*, 912–918. [CrossRef] [PubMed]

4. Basu, R. High ambient temperature and mortality: A review of epidemiologic studies from 2001 to 2008. *Environ. Health* **2009**, *8*, 13. [CrossRef] [PubMed]

5. Medina-Ramón, M.; Zanobetti, A.; Cavanagh, D.P.; Schwartz, J. Extreme Temperatures and Mortality: Assessing Effect Modification by Personal Characteristics and Specific Cause of Death in a Multi-City Case-Only Analysis. *Environ. Health Perspect.* **2006**, *114*, 1331–1336. [CrossRef]

6. Son, J.-Y.; Liu, J.C.; Bell, M.L. Temperature-related mortality: A systematic review and investigation of effect modifiers. *Environ. Res. Lett.* **2019**, *14*, 073004. [CrossRef]

7. Zanobetti, A.; Schwartz, J. Temperature and Mortality in Nine US Cities. *Epidemiology* **2008**, *19*, 563–570. [CrossRef] [PubMed]

8. Saha, M.V.; Davis, R.E.; Hondula, D.M. Mortality Displacement as a Function of Heat Event Strength in 7 US Cities. *Am. J. Epidemiol.* **2014**, *179*, 467–474. [CrossRef]

9. Gosling, S.N.; McGregor, G.R.; Páldy, A. Climate change and heat-related mortality in six cities Part 1: Model construction and validation. *Int. J. Biometeorol.* **2007**, *51*, 525–540. [CrossRef]

10. Gosling, S.N.; McGregor, G.R.; Lowe, J.A. Climate change and heat-related mortality in six cities Part 2: Climate model evaluation and projected impacts from changes in the mean and variability of temperature with climate change. *Int. J. Biometeorol.* **2009**, *53*, 31–51. [CrossRef]

11. Curriero, F.C. Temperature and Mortality in 11 Cities of the Eastern United States. *Am. J. Epidemiol.* **2002**, *155*, 80–87. [CrossRef]

12. The City of Boston and the Green Ribbon Commission. *Climate Ready Boston*; The City of Boston and the Green Ribbon Commission: Boston, MA, USA, 2016; pp. 1–199.

13. Petkova, E.; Horton, R.; Bader, D.; Kinney, P. Projected Heat-Related Mortality in the U.S. Urban Northeast. *Int. J. Environ. Res. Public Health* **2013**, *10*, 6734–6747. [CrossRef]

14. Reid, C.E.; Mann, J.K.; Alfasso, R.; English, P.B.; King, G.C.; Lincoln, R.A.; Margolis, H.G.; Rubado, D.J.; Sabato, J.E.; West, N.L.; et al. Evaluation of a Heat Vulnerability Index on Abnormally Hot Days: An Environmental Public Health Tracking Study. *Environ. Health Perspect.* **2012**, *120*, 715–720. [CrossRef] [PubMed]

15. Reid, C.E.; O'Neill, M.S.; Gronlund, C.J.; Brines, S.J.; Brown, D.G.; Diez-Roux, A.V.; Schwartz, J. Mapping Community Determinants of Heat Vulnerability. *Environ. Health Perspect.* **2009**, *117*, 1730–1736. [CrossRef] [PubMed]

16. Kim, E.-J.; Kim, H. Effect modification of individual- and regional-scale characteristics on heat wave-related mortality rates between 2009 and 2012 in Seoul, South Korea. *Sci. Total Environ.* **2017**, *595*, 141–148. [CrossRef]

17. Anderson, M.; Carmichael, C.; Murray, V.; Dengel, A.; Swainson, M. Defining indoor heat thresholds for health in the UK. *Perspect. Public Health* **2013**, *133*, 158–164. [CrossRef]

18. Baniassadi, A.; Heusinger, J.; Sailor, D.J. Energy efficiency vs resiliency to extreme heat and power outages: The role of evolving building energy codes. *Build. Environ.* **2018**, *139*, 86–94. [CrossRef]

19. Sailor, D.J.; Baniassadi, A.; O'Lenick, C.R.; Wilhelmi, O.V. The growing threat of heat disasters. *Environ. Res. Lett.* **2019**, *14*, 054006. [CrossRef]

20. Baniassadi, A.; Sailor, D.J.; Krayenhoff, E.S.; Broadbent, A.M.; Georgescu, M. Passive survivability of buildings under changing urban climates across eight US cities. *Environ. Res. Lett.* **2019**, *14*, 074028. [CrossRef]

21. Semenza, J.C.; Rubin, C.H.; Falter, K.H.; Selanikio, J.D.; Flanders, W.D.; Howe, H.L.; Wilhelm, J.L. Heat-Related Deaths during the July 1995 Heat Wave in Chicago. *N. Engl. J. Med.* **1996**, *335*, 84–90. [CrossRef]

22. Smargiassi, A.; Fournier, M.; Griot, C.; Baudouin, Y.; Kosatsky, T. Prediction of the indoor temperatures of an urban area with an in-time regression mapping approach. *J. Expo. Sci. Environ. Epidemiol.* **2008**, *18*, 282–288. [CrossRef]

23. Quinn, A.; Tamerius, J.D.; Perzanowski, M.; Jacobson, J.S.; Goldstein, I.; Acosta, L.; Shaman, J. Predicting indoor heat exposure risk during extreme heat events. *Sci. Total Environ.* **2014**, *490*, 686–693. [CrossRef]

24. Holmes, S.H.; Phillips, T.; Wilson, A. Overheating and passive habitability: Indoor health and heat indices. *Build. Res. Inf.* **2016**, *44*, 1–19. [CrossRef]

25. McDonald, R.I.; Kroeger, T.; Zhang, P.; Hamel, P. The Value of US Urban Tree Cover for Reducing Heat-Related Health Impacts and Electricity Consumption. *Ecosystems* **2019**, *23*, 137–150. [CrossRef]

26. Madrigano, J.; Ito, K.; Johnson, S.; Kinney, P.L.; Matte, T. A Case-Only Study of Vulnerability to Heat Wave–RelatedMortality in New York City (2000–2011). *Environ. Health Perspect.* **2015**, *123*, 672–678. [CrossRef] [PubMed]

27. Gonzalez, S. Without AC, Public Housing Residents Swelter Through the Summer. *WNYC News*, 2016.

28. Energy Information Administration. *Household Energy Use in Massachusetts: A Closer Look at Residential Energy Consumption*; Energy Information Administration: Washington, DC, USA, 2009; p. 2. Available online: www.eia.gov/consumption/residential/reports/2009/state_briefs/pdf/ma.pdf (accessed on 6 February 2020).

29. Abel, D.; Holloway, T.; Kladar, R.M.; Meier, P.; Ahl, D.; Harkey, M.; Patz, J. Response of Power Plant Emissions to Ambient Temperature in the Eastern United States. *Environ. Sci. Technol.* **2017**, *51*, 5838–5846. [CrossRef] [PubMed]

30. Chuang, W.-C.; Gober, P. Predicting Hospitalization for Heat-Related Illness at the Census-Tract Level: Accuracy of a Generic Heat Vulnerability Index in Phoenix, Arizona (USA). *Environ. Health Perspect.* **2015**, *123*, 606–612. [CrossRef] [PubMed]

31. Coutts, E.; Ito, K.; Nardi, C.; Vuong, T. *Planning Urban Heat Island Mitigation in Boston*; Trust for Public Land: Boston, MA, USA, 2015; p. 3. Available online: https://www.tpl.org/sites/default/files/UHI_Tufts_Executive%20Summary.pdf (accessed on 6 February 2020).

32. Kingsley, S.L.; Eliot, M.N.; Gold, J.; Vanderslice, R.R.; Wellenius, G.A. Current and Projected Heat-Related Morbidity and Mortality in Rhode Island. *Environ. Health Perspect.* **2016**, *124*, 460–467. [CrossRef]

33. National Weather Service. NWS Heat Index. Available online: https://www.weather.gov/safety/heat-index (accessed on 6 February 2020).

34. Anderson, G.B.; Bell, M.L. Heat Waves in the United States: Mortality Risk during Heat Waves and Effect Modification by Heat Wave Characteristics in 43 U.S. Communities. *Environ. Health Perspect.* **2010**, *119*, 210–218. [CrossRef]

35. Braga, A.L.F.; Zanobetti, A.; Schwartz, J. The effect of weather on respiratory and cardiovascular deaths in 12 U.S. cities. *Environ. Health Perspect.* **2002**, *110*, 859–863. [CrossRef]

36. Wellenius, G.A.; Eliot, M.N.; Bush, K.F.; Holt, D.; Lincoln, R.A.; Smith, A.E.; Gold, J. Heat-related morbidity and mortality in New England: Evidence for local policy. *Environ. Res.* **2017**, *156*, 845–853. [CrossRef]

37. Iceland, J.; Steinmetz, E. *The Effects of Using Census Block Groups Instead of Census Tracts When Examining Residential Housing Patterns*; Bureau of the Census: Suitland-Silver Hill, MD, USA, 2003.

38. Subramanian, S.V.; Chen, J.T.; Rehkopf, D.H.; Waterman, P.D.; Krieger, N. Comparing Individual- and Area-based Socioeconomic Measures for the Surveillance of Health Disparities: A Multilevel Analysis of Massachusetts Births, 1989–1991. *Am. J. Epidemiol.* **2006**, *164*, 823–834. [CrossRef]

39. United States Census. 2010. Available online: https://www.factfinder.census.gov/faces/nav/jsf/pages/index.xhtml (accessed on 6 February 2020).

40. Climate Ready Boston Social Vulnerability. 2017. Available online: https://www.data.boston.gov/dataset/climate-ready-boston-social-vulnerability (accessed on 6 February 2020).

41. Trees. 2019. Available online: https://www.bostonopendata-boston.opendata.arcgis.com/datasets/ce863d38db284efe83555caf8a832e2a_1 (accessed on 6 February 2020).

42. Impervious Surface Fraction—Census Tracts. 2013. Available online: https://www.worldmap.harvard.edu/data/geonode:impervious_surface_fraction_census_tract_5yd (accessed on 6 February 2020).

43. Shields, M.; O'Brien, D.; de Benedictis-Kessner, J. Property Assessment 2018.

44. Song, X.; Wang, S.; Hu, Y.; Yue, M.; Zhang, T.; Liu, Y.; Tian, J.; Shang, K. Impact of ambient temperature on morbidity and mortality: An overview of reviews. *Sci. Total Environ.* **2017**, *586*, 241–254. [CrossRef] [PubMed]

45. Lin, S.; Luo, M.; Walker, R.J.; Liu, X.; Hwang, S.-A.; Chinery, R. Extreme High Temperatures and Hospital Admissions for Respiratory and Cardiovascular Diseases. *Epidemiology* **2009**, *20*, 738–746. [CrossRef] [PubMed]

46. Mastrangelo, G.; Fedeli, U.; Visentin, C.; Milan, G.; Fadda, E.; Spolaore, P. Pattern and determinants of hospitalization during heat waves: An ecologic study. *BMC Public Health* **2007**, *7*, 200. [CrossRef] [PubMed]

47. Gronlund, C.J.; Zanobetti, A.; Schwartz, J.D.; Wellenius, G.A.; O'Neill, M.S. Heat, Heat Waves, and Hospital Admissions among the Elderly in the United States, 1992–2006. *Environ. Health Perspect.* **2014**, *122*, 1187–1192. [CrossRef] [PubMed]

48. Wang, Y.-C.; Lin, Y.-K. Association between Temperature and Emergency Room Visits for Cardiorespiratory Diseases, Metabolic Syndrome-Related Diseases, and Accidents in Metropolitan Taipei. *PLoS ONE* **2014**, *9*, e99599. [CrossRef] [PubMed]

49. Hajat, S.; Haines, A.; Sarran, C.; Sharma, A.; Bates, C.; Fleming, L.E. The effect of ambient temperature on type-2-diabetes: Case-crossover analysis of 4+ million GP consultations across England. *Environ. Health* **2017**, *16*, 73. [CrossRef] [PubMed]

50. Sun, S.; Weinberger, K.R.; Spangler, K.R.; Eliot, M.N.; Braun, J.M.; Wellenius, G.A. Ambient temperature and preterm birth: A retrospective study of 32 million US singleton births. *Environ. Int.* **2019**, *126*, 7–13. [CrossRef]

51. Lin, S.; Sun, M.; Fitzgerald, E.; Hwang, S.-A. Did summer weather factors affect gastrointestinal infection hospitalizations in New York State? *Sci. Total Environ.* **2016**, *550*, 38–44. [CrossRef]

52. Khoury, M.J.; Flanders, W.D. Nontraditional Epidemiologic Approaches in the Analysis of Gene Environment Interaction: Case-Control Studies with No Controls! *Am. J. Epidemiol.* **1996**, *144*, 207–213. [CrossRef]

53. Armstrong, B.G. Fixed Factors that Modify the Effects of Time-Varying Factors: Applying the Case-Only Approach. *Epidemiology* **2003**, *14*, 467–472. [CrossRef]

54. Zanobetti, A.; O'Neill, M.S.; Gronlund, C.J.; Schwartz, J.D. Susceptibility to Mortality in Weather Extremes: Effect Modification by Personal and Small-Area Characteristics. *Epidemiology* **2013**, *24*, 809–819. [CrossRef]

55. Daly, C.; Gibson, W.; Taylor, G.; Johnson, G.; Pasteris, P. A knowledge-based approach to the statistical mapping of climate. *Clim. Res.* **2002**, *22*, 99–113. [CrossRef]

56. PRISM Climate Group, Oregon State University. *Parameter-Elevation Regressions on Independent Slopes Model (PRISM)*; PRISM Climate Group, Oregon State University: Corvallis, OR, USA, 2018.

57. Daly, C.; Smith, J.W.; Smith, J.I.; McKane, R.B. High-Resolution Spatial Modeling of Daily Weather Elements for a Catchment in the Oregon Cascade Mountains, United States. *J. Appl. Meteorol. Climatol.* **2007**, *46*, 1565–1586. [CrossRef]

58. Spangler, K.R.; Weinberger, K.R.; Wellenius, G.A. Suitability of gridded climate datasets for use in environmental epidemiology. *J. Expo. Sci. Environ. Epidemiol.* **2019**, *29*, 777–789. [CrossRef] [PubMed]

59. Weinberger, K.R.; Spangler, K.R.; Zanobetti, A.; Schwartz, J.D.; Wellenius, G.A. Comparison of temperature-mortality associations estimated with different exposure metrics. *Environ. Epidemiol.* **2019**, *3*, e072. [CrossRef]

60. *Triple Decker Trends*; Department of Neighborhood Development Research & Development Unit: Boston, MA, USA, 1999; p. 4.

61. MacFadden, D.R.; McGough, S.F.; Fisman, D.; Santillana, M.; Brownstein, J.S. Antibiotic resistance increases with local temperature. *Nat. Clim. Chang.* **2018**, *8*, 510–514. [CrossRef]

62. Anderson, C.A. Temperature and Aggression: Ubiquitous Effects of Heat on Occurrence of Human Violence. *Psychol. Bull.* **1989**, *106*, 74. [CrossRef]

63. Butke, P.; Sheridan, S.C. An Analysis of the Relationship between Weather and Aggressive Crime in Cleveland, Ohio. *Weather Clim. Soc.* **2010**, *2*, 127–139. [CrossRef]

64. Anderson, C.A. Heat and Violence. *Curr. Dir. Psychol. Sci.* **2001**, *10*, 33–38. [CrossRef]

65. Michel, S.J.; Wang, H.; Selvarajah, S.; Canner, J.K.; Murrill, M.; Chi, A.; Efron, D.T.; Schneider, E.B. Investigating the relationship between weather and violence in Baltimore, Maryland, USA. *Injury* **2016**, *47*, 272–276. [CrossRef]

66. Williams, A.A.; Allen, J.G.; Catalano, P.J.; Buonocore, J.J.; Spengler, J.D. The influence of heat on emergency services in Boston, MA: Relative risk and time-series analyses of police, medical, and fire dispatches. *Am. J. Public Health* **2020**. In Press.

67. Obradovich, N.; Tingley, D.; Rahwan, I. Effects of environmental stressors on daily governance. *Proc. Natl. Acad. Sci. USA* **2018**, *115*, 8710–8715. [CrossRef]

68. Page, L.A.; Hajat, S.; Kovats, R.S.; Howard, L.M. Temperature-related deaths in people with psychosis, dementia and substance misuse. *Br. J. Psychiatry* **2012**, *200*, 485–490. [CrossRef] [PubMed]

69. Simning, A.; van Wijngaarden, E.; Conwell, Y. Anxiety, mood, and substance use disorders in United States African-American public housing residents. *Soc. Psychiatry Psychiatr. Epidemiol.* **2011**, *46*, 983–992. [CrossRef] [PubMed]

70. Barbato, J. Areal parameters of the Sea Breeze and its vertical structure in the Boston Basin. *BAMS* **1978**, *59*, 1420–1431. [CrossRef]

71. Lue, S.-H.; Wellenius, G.A.; Wilker, E.H.; Mostofsky, E.; Mittleman, M.A. Residential proximity to major roadways and renal function. *J. Epidemiol. Community Health* **2008**, *67*, 629–634.

72. Tillson, A.-A.; Oreszczyn, T.; Palmer, J. Assessing impacts of summertime overheating: Some adaptation strategies. *Build. Res. Inf.* **2013**, *41*, 652–661. [CrossRef]

73. Krieger, J.; Higgins, D.L. Housing and Health: Time Again for Public Health Action. *Am. J. Public Health* **2002**, *92*, 758–768. [CrossRef]

74. Ho, H.C.; Knudby, A.; Chi, G.; Aminipouri, M.; Lai, D.Y.-F. Spatiotemporal analysis of regional socio-economic vulnerability change associated with heat risks in Canada. *Appl. Geogr.* **2018**, *95*, 61–70. [CrossRef]

75. ParkScore. 2019. Available online: https://www.tpl.org/city/boston-massachusetts (accessed on 6 February 2020).

76. Guo, Y.; Gasparrini, A.; Armstrong, B.G.; Tawatsupa, B.; Tobias, A.; Lavigne, E.; Coelho, M.D.S.Z.S.; Pan, X.; Kim, H.; Hashizume, M.; et al. Heat Wave and Mortality: A Multicountry, Multicommunity Study. *Environ. Health Perspect.* **2017**, *125*, 087006. [CrossRef]

climate

MDPI

Improving the Indoor Air Quality of Residential Buildings during Bushfire Smoke Events

Priyadarsini Rajagopalan * and Nigel Goodman

Sustainable Building Innovation Lab, School of Property, Construction and Project Management, RMIT University, Melbourne, Victoria 3000, Australia; nigel.goodman@rmit.edu.au
* Correspondence: priyadarsini.rajagopalan@rmit.edu.au; Tel.: +61399252293

Abstract: Exposure to bushfire smoke is associated with acute and chronic health effects such as respiratory and cardiovascular disease. Residential buildings are important places of refuge from bushfire smoke, however the air quality within these locations can become heavily polluted by smoke infiltration. Consequently, some residential buildings may offer limited protection from exposure to poor air quality, especially during extended smoke events. This paper evaluates the impact of bushfire smoke on indoor air quality within residential buildings and proposes strategies and guidance to reduce indoor levels of particulates and other pollutants. The paper explores the different monitoring techniques used to measure air pollutants and assesses the influence of the building envelope, filtration technologies, and portable air cleaners used to improve indoor air quality. The evaluation found that bushfire smoke can substantially increase the levels of pollutants within residential buildings. Notably, some studies reported indoor levels of $PM_{2.5}$ of approximately $500\mu g/m^3$ during bushfire smoke events. Many Australian homes are very leaky (i.e., >15 ACH) compared to those in countries such as the USA. Strategies such as improving the building envelope will help reduce smoke infiltration, however even in airtight homes pollutant levels will eventually increase over time. Therefore, the appropriate design, selection, and operation of household ventilation systems that include particle filtration will be critical to reduce indoor exposures during prolonged smoke events. Future studies of bushfire smoke intrusion in residences could also focus on filtration technologies that can remove gaseous pollutants.

Keywords: bushfire smoke; indoor air quality; filtration; building envelope; energy

check for **updates**

Citation: Rajagopalan, P.; Goodman, N. Improving the Indoor Air Quality of Residential Buildings during Bushfire Smoke Events. *Climate* **2021**, 9, 32. https://doi.org/10.3390/cli9020032

Received: 10 January 2021
Accepted: 11 February 2021
Published: 15 February 2021

Publisher's Note: MDPI stays neutral with regard to jurisdictional claims in published maps and institutional affiliations.

1. Introduction

Indoor air quality (IAQ) may be defined as the air quality within buildings that can impact occupant comfort, health and wellbeing [1]. In most developed countries people spend the majority of their time indoors, and in particular, within residential buildings [2]. Residential indoor environments can be important places of exposure to air pollution, including hazardous air pollutants [3]. Common sources of indoor pollutants include emissions from building and furnishing materials, fragranced consumer products, and occupant activities such as cooking and cleaning [3–5]. In addition, when ambient air becomes heavily polluted, such during bushfires, concentrations of indoor pollutants can increase substantially, resulting in poor IAQ. During bushfire smoke events, the levels of indoor pollutants may initially be lower than outdoors, however, as smoke persists for days (or weeks), the levels indoors and outdoors can be similar. Due to the long periods of time spent at home and the relative scarcity of residences equipped with air filtration technologies capable of removing pollutants, exposure to bushfire smoke in residences may be considerable. One important recommendation given by authorities to residents affected by bushfire smoke is to stay indoors. Doing so can provide a level of protection from exposure to smoke, however the degree of protection will depend on factors such as the duration of the smoke event [6], the design of the building envelope, occupant activities, and type of ventilation system (if any). The effectiveness of homes in protecting occupants from poor air quality is not very well understood.

This paper evaluates the impact of bushfire smoke on the air quality within residential buildings. It focuses on (i) the indoor levels of pollutants (e.g., particles, gases) within residences during bushfire smoke events, (ii) recent advancements in the methods of monitoring pollutants (e.g., low-cost sensors), and (iii) strategies to improve IAQ, such as building envelope design, ventilation air filtration, and use of portable air cleaners. There are many different terms used to describe bushfires. These include wildfires, forest fires, vegetation fires, and landscape fires [7,8]. Barn et al. [9] proposed a definition of uncontrolled fires as "wildfires" and controlled fires as "landscape fires," although this terminology will differ between countries. In this paper we will report the terminology used in the cited article, including "bushfire," "wildfire," "forest fire," and "landscape fire".

2. Impacts of Bushfire Smoke

Smoke from bushfires can be a significant risk to health. Exposure to bushfire smoke has been associated with increased morbidity [10] and mortality [11]. Smoke from landscape fires has been attributed to an estimated 340,000 deaths every year [8]. Vulnerable members of the population are particularly at risk from health complications associated with exposure to bushfire smoke. For instance, children, the elderly, and individuals with respiratory and heart disease are at a greater risk of harm from exposure to bushfire and wildfire smoke [12,13]. Also, the association between maternal exposure to wildfire smoke and reduced birthweight is evident [14].

Under most climate change scenarios, the frequency and severity of bushfires are predicted to increase [15,16]. The main drivers include increased occurrences and durations of droughts as well as higher ambient temperatures. Modelling estimates of wildfire activity in the western United States suggest an increase of up to 54% in the areas burned by 2050 (compared to 2009), coupled with an approximate doubling of carbonaceous aerosol emissions [17]. Recent fires in California and other parts of the world have reinforced these concerns. In Australia, bushfires have always been a feature of the natural environment, however their impact has increased over the years with fire seasons extending for a longer time and extreme weather becoming more severe [18]. Climate change, smoke and other emissions from bushfires may also influence urban microclimates and exacerbate phenomena such as the urban heat island effect. A recent study conducted in Sydney, Australia [19] compared urban heat island intensity during the 2019/2020 bushfire season to historic meteorological data from the previous 20 years and found an exacerbation of urban heat island events compared to median levels. Analysis of the combined effects of extreme pollution, heat waves, and droughts demonstrated dependencies between environmental factors such as air temperature, relative humidity, particle concentration, wind speed, and rain, and anomalies in the intensity of the urban heat island in comparison to historical trends. The 2019–20 bushfires in Australia caused significant damage to natural and built environments in many states and exposed millions of people to extreme levels of air pollution. During these "Black Summer" fires, more than 15,300 bushfires burned an area of 18,983,588 hectares, destroyed 3113 houses, and took 33 lives. An estimated 1 billion vertebrate animals were lost, and the economic impact has been estimated to be in the order of AUD $40 billion [20]. Approximately 80% of the population was impacted by bushfire smoke for prolonged periods of time (i.e., weeks) [21]. An analysis of global data for the remotely sensed burned areas of all major global forest biomes over the past 20 years found that this massive bushfires in Australia burnt 21% of the total temperate broadleaf and mixed biome, pointing to the likelihood that the projected "flammable future" has arrived earlier than anticipated [22].

The predicted increases in the levels of outdoor pollutants from sources such as bushfires may result in higher indoor pollutant concentrations and increased exposures [23]. Fisk [24] reviewed the potential health consequences of climate change on indoor environments. Projected effects during wildfires include twice the number of heat-related deaths, increased hospitalizations due to asthma, pneumonia, and cardiovascular effects, and increased mortality and hospitalizations linked to ozone. The authors make the compelling point that a significant proportion of these adverse exposures are likely to occur indoors. Highlighting the challenges that the protracted occurrence of bushfire smoke creates, Vardoulakis et al. [25] called for "more nuanced health advice to protect populations and individuals from exposure to bushfire smoke." For example, the authors suggest the need for additional methods to communicate air quality information, and for an evaluation of the current health protection advice so that it can be adapted for longer periods of smoke exposure.

In summary, the frequency and severity of bushfires are likely to increase in the future. Smoke from bushfires can have a substantial impact on the levels of pollutants within residences and other indoor environments. Therefore, the extent to which people are exposed to pollutants is also likely to increase, and strategies to minimize exposure and health risks are needed. Some of these will be explored in the following sections of this article.

3. General Guidelines and Occupant Behaviour

Key factors that impact health-related symptoms are the concentrations of pollutants in the smoke, the duration of exposure, level of protection that can be utilised, and the underlying health status of those exposed [26]. Government agencies typically use an air quality index (AQI) to communicate the air pollution level to the public. Different countries have specific AQIs that correspond to relevant national air quality standards and use a combination of pollutants including $PM_{2.5}$, PM_{10}, ozone, sulphur dioxide and nitrogen dioxide. WHO [27] advises maximum levels of 10 $\mu g/m^3$ per year and 25 $\mu g/m^3$ per 24-h period for $PM_{2.5}$ levels. In Australia, the ambient $PM_{2.5}$ guideline is 8 $\mu g/m^3$ averaged over one year and 25 $\mu g/m^3$ averaged over one day [28]. As per the United States National ambient air quality standards [29], the one-year $PM_{2.5}$ standards for primary and secondary are 12 $\mu g/m^3$, and 15 $\mu g/m^3$, respectively, and the standard for 24 h is 35 $\mu g/m^3$. Primary standards are for protecting the health of sensitive populations and secondary standards provide protection against decreased visibility and damage to buildings, animals and vegetation. During recent bushfires in Australia, the concentrations of $PM_{2.5}$ measured in major cities were as high as 500 $\mu g/m^3$, more than 20 times the ambient air quality guidelines [30,31].

Advice for reducing exposure to bushfire smoke at home includes publicly available websites and factsheets (e.g., [32,33]) and academic literature [9]. The recommendations include staying indoors with windows and doors closed, reducing strenuous physical exercise (outdoors), going to an airconditioned facility or public building (e.g., shopping centre or library), attending a clean air shelter (if available), using a portable air cleaner, and using well fitted P2 facemasks (e.g., [32,33]). For vulnerable individuals such as asthmatics, the California Department of Public Health [32] suggest that residents consider temporarily evacuating their homes until air quality conditions improve. Some agencies also stress the importance of removing residual smoke that has adsorbed and deposited on surfaces [33]. However, cleaning activities such as vacuuming can increase particle levels in homes and should be avoided when a wildfire smoke is present [34].

Occupant behaviour during wildfires can strongly influence indoor levels of pollutants and occupant comfort. For instance, occupant movement has been associated with high indoor levels of $PM_{2.5}$ [35], therefore minimising the resuspension of indoor pollutants is an important strategy for preventing further exposure [36]. Using computational fluid dynamics to explore indoor exposure risks, Luo et al. [37] found that indoor airflow patterns and pollutant concentrations were significantly impacted by occupant behaviour. The use of air conditioning, operation of doors and windows, use of products, and movement were all important factors that impact indoor air quality during smoke events. Monitoring studies also found that indoor activities such as smoking, cooking, and burning incense or candles can significantly contribute to indoor levels of $PM_{2.5}$ [38]. Reducing or discontinuing the use of fragranced consumer products, scented candles, and air fresheners has been shown to improve indoor air quality, therefore the use of these products should be minimized, especially when wildfire smoke is present [32,39].

4. Monitoring Studies

Bushfire smoke contains a complex mix of particles and gases including PM_{10}, $PM_{2.5}$, carbon dioxide, carbon monoxide, sulphur dioxide, nitrogen dioxide, benzene, acetaldehyde, formaldehyde, polycyclic aromatic hydrocarbons, and ozone [40–42]. Experimental and monitoring studies of bushfire smoke in homes have predominantly focused on indoor $PM_{2.5}$ due to well documented health effects of exposure to fine particulate matter [7,8], and existence of health base exposure guidelines such as the National Environment Protection (Ambient Air Quality) Measure e.g., [32]. The relative availability of particle sensors to take measurements during a fire may also have contributed to the research focus on $PM_{2.5}$. The levels of $PM_{2.5}$ during bushfires can be many of times higher than guideline values. For instance, during the 2019–20 bushfires in Australia, average 24-h $PM_{2.5}$ concentrations in Sydney exceeded 100 $\mu g/m^3$ and peaked at approximately 500 $\mu g/m^3$ [30].

Pantelic et al. [43] monitored $PM_{2.5}$ generated during the Chico forest fire in California with the use of a combination of sensor networks outdoors and inside buildings. The study focused on two buildings with different modes of ventilation (i.e., mechanical, natural), in an urban area. The results showed that a mechanically ventilated building was more resilient to outdoor pollution with lower indoor-outdoor concentration ratios and lower indoor $PM_{2.5}$ levels. A study conducted in Denver (CO, USA) during the 2016/2017 wildfires season used laser-based optical particle counters (Dylos-1700) to evaluate levels of particulate matter within 28 low-income homes [44]. The study found median indoor levels of $PM_{2.5}$ were 4.6 times higher than outdoors. Homes that used mechanical ventilation systems with low efficiency filters had indoor-outdoor ratios that were 18% higher for $PM_{2.5}$ than those that did not [44], indicating that the outdoor pollutants are brought indoors through ventilation supply air even when filters were in place. This study also evaluated levels of black carbon (BC), CO, and NO2 and found that indoor levels of BC during wildfires were about twice the levels recorded when no wildfire smoke was present. In addition, the levels of BC, CO, and NO_2 were consistently higher in homes closer to roads compared to those further away (i.e., >200 m), reflecting the importance of traffic emissions on air quality. Furthermore, activities such as opening a window increased levels of BC, however decreased CO concentrations, suggesting an internal source of CO emissions (such as a gas pilot light) [44]. These studies highlight the variability of $PM_{2.5}$ levels in buildings with mechanical and natural ventilation systems. The studies also highlight the importance of occupant behaviour and proximity to background pollution sources in preventing indoor exposure to air pollutants.

Sapkota et al. [45] investigated transport of particles generated in forest fires using a range of measurement techniques (i.e., laser, time of flight aerosol spectrometers, oscillating microbalances). Three ambient sites and four indoor sites were monitored. $PM_{2.5}$ concentration averaged over 24 h reached 86 $\mu g/m^3$ during the event and the peak level $PM_{2.5}$ reached a maximum concentration of 199 $\mu g/m^3$, which is eight times above background levels. With median 0.91 indoor-outdoor ratio, indoor and outdoor concentration levels were found to be similar [45]. In Australia, Reisen et al. [46] evaluated levels of $PM_{2.5}$ within homes impacted by smoke from the prescribed burning of biomass. Four homes were monitored in the town of Ovens, located in the alpine region of north eastern Victoria. The study also monitored outdoor levels of $PM_{2.5}$ and ozone in the same region and at another rural location in Western Australia. For particle concentrations two different instrumental approaches were used: continuous $PM_{2.5}$ measurements were taken at 1-minute intervals using light scattering techniques (DustTrak, TSI Incorporated, Shoreview, MN, USA); and gravimetric mass measurements (MicroVol-1100, Ecotech Pty Ltd, Knoxfield, Australia) were made at weekly intervals to provide reference data and to calibrate the DustTraks. Ozone was measured using a photometric analyser (Model 49 UV, TECO). In two of the four residences, indoor $PM_{2.5}$ levels exceeded 25 $\mu g/m^3$, and maximum daily concentrations peaked at 89 $\mu g/m^3$. Air infiltration rates were also measured in this study using a carbon dioxide tracer gas method. The reported air exchange rates ranged between 0.29 and 0.9 air changes per hour, suggesting that these homes were well sealed. The indoor levels of $PM_{2.5}$ depended on the duration of the smoke event and the ventilation rate of the houses. Household activities were found to be key influencers of indoor pollutant levels when smoke events were short; and external conditions and household ventilation critical during more persistent smoke events.

Other studies have monitored pollutants including air toxics and polycyclic aromatic hydrocarbons (PAHs). A study of the indoor and outdoor levels of 63 PAHs (24 h) found that during wildfire events, the levels of indoor gas phase PAHs were consistently equal to or higher than outdoor levels, thus suggesting the importance of these compounds in evaluation of air pollution risk assessment [47]. Another study of gaseous emissions from bushfire and wildfire smoke compared the trace gas emissions factors of the smoke from Australian and North American research [48]. The study found that approximately 20% of the gases identified had similar emission factors including hydrogen cyanide, ethene, methanol, formaldehyde, and 1,3-butadiene, and others such as acetic acid, ethanol, monoterpenes, ammonia, acetonitrile, and pyrrole differed by a factor of two or more.

Ozone is an important secondary pollutant that can be detrimental to the health and the environment. It is formed by the interaction of nitrogen oxides (NOx) and non-methane organic carbon molecules (e.g., VOCs) in the presence of sunlight. Exposure to high levels of ozone has been linked to a range of adverse health effects including increased short-term mortality [49]. Wildfires produce approximately 170 Tg of ozone globally every year [40], and levels of ozone increase significantly in smoke plumes and can disperse over large regions [50]. For instance, wildfires in northern Quebec resulted in a 10ppbv increase in downwind ozone concentrations [50]. In addition, ozone can react with anthropogenic sources of volatile organic compounds (e.g., from consumer products) and generate hazardous air pollutants such as formaldehyde and ultrafine particles [51,52], further contributing to levels of hazardous air pollutants in urban areas.

Sensor Technology for Monitoring

Real time information relating to the severity and location of bushfire smoke can help reduce exposure and prevent adverse health effects. Air quality sensors that meet international standards are mostly located at fixed site locations. These sparsely distributed sensors are capable of fulfilling the regulatory needs and providing the public with access to air quality data. However, they have limitations as they can only provide data for a relatively local area. Also, due to the small number of sensors, detailed information about the spatial distribution of pollutants may not be available, making it difficult to identify hotspots [53]. In addition, fixed air quality sensors are characterized by relatively high purchase costs, regular maintenance, and the need for support infrastructure such as enclosures and a reliable power supply. A high spatial scale is important as the levels of air pollutants are extremely location dependent and the nearest fixed station may not be representative of the local pollution concentrations due to geographical and topographical differences.

Low-cost sensors (e.g., $PM_{2.5}$) are widely being used by researchers, private organisations and the general public to monitor air quality. Low-cost particle sensors typically use light scattering techniques to count and determine particle size. However, there is inadequate information about the accuracy of low-cost sensors, particularly on how they perform under conditions of heavy smoke [54]. Furthermore, limited performance information is provided by the manufacturers of the sensors. Field calibration under similar conditions to the actual measurement environment is crucial for obtaining accurate measurements from low cost sensors. Many particle sensors have high correlation (i.e., moderate to high R^2 values) with research-grade calibrated reference instruments, however sensor response has been found to decline at particle concentrations above 50–100 $\mu g/m^3$ [55]. In an evaluation of two types of low-cost particle sensors (i.e., PMS 1003s and PMS 5003s, Plantower, Shunyi

District, Beijing, China), Sayahi et al. [56] observed that $PM_{2.5}$ concentrations were typically 0.3–1.25 times the concentration of the reference sensors concentrations and over 1.5 times the reference values during wildfire smoke events. However, this study was limited as the hourly smoke concentrations only reached ~60 $\mu g/m^3$. Delp and singer [57] note that adjustment factors specific to bushfire smoke are needed for low cost monitors as the optical sensors used varied with aerosol properties. The four low-cost particle sensors assessed, i.e., AirVisual Pro (IQAir, Goldach, Switzerland), PurpleAir (PurpleAir, Draper, UT, USA), Air Quality Egg (Wicked Device LLC, Ithaca, NY, USA), and the Indoor Air Quality Pro Station (eLichens, Grenoble, France), showed accurate data with correction factors ranging between 0.48 to 0.60 compared to a reference instrument, a tapered element oscillating microbalance (i.e., TEOM-FDMS, Thermo Scientific, Waltham, MA, USA). This study was conducted during wildfire season when smoke concentrations were as high as 150 $\mu g/m^3$. The authors conclude that even though the adjustment factor can vary according to location and over time during a fire event, a global adjustment factor can reduce bias significantly. Holder et al. [54] collated three types of low-cost fine $PM_{2.5}$ sensors with reference instruments during a number of fire events and found moderate to strong correlation with reference instruments, however the sensors overpredicted $PM_{2.5}$ concentrations (with normalized root mean square errors = 80–167%). The authors developed different correction equations for each sensor. Field calibration, under similar environmental and measurement conditions is critical to obtain more accurate measurements from low-cost PM sensors [55,58]. Mehadi et al. [59] assessed seven instruments including two low cost sensors (Dylos, Riverside, CA, USA and PurpleAir, Draper, UT, USA) and found that the ratio of the median Purple Air $PM_{2.5}$ sensor concentrations to reference sensor concentrations varied from about 1.5 to 2 when the smoke intensity increased whereas the same for Dylos was around 25%. Many low-cost sensors have been found to perform better under stable laboratory conditions compared with field conditions [55]. Key factors that influence this variability are the changing particle compositions, sizes, and environmental factors. Therefore, site specific calibration rather than laboratory calibration may be necessary for low-cost sensors. It was also found that sensor performance was sensitive to the composition of smoke. Holder et al. [54] noted that further monitoring and comparison studies focusing on higher particle concentrations are required as optical-based sensors can saturate when the concentrations are very high.

Digital technologies such as smartphone apps that access real-time data are becoming more prevalent, and popular with users. For instance, a near real time smartphone app "AirRater" has been developed by Australian researchers to provide air quality information to vulnerable individuals such as asthmatics and the general public [21]. The app can track user symptoms, monitor air quality conditions, and help reduce personal exposure to air pollution, including smoke from bushfires [21]. An evaluation of the technology during smoke events revealed that the app helped users avoid exposure and effects of smoke by advising them on precautionary measures, such as (i) staying inside, (ii) rescheduling or planning outdoor activities, (iii) changing locations to less affected areas, and (iv) informing decisions on medication use [21].

5. Factors Influencing IAQ

5.1. Building Envelope

The integrity of the residential building envelope is critical for preventing smoke and other pollutants from infiltrating into the building. A tighter building envelope will be less leaky and will reduce the rate of infiltration through cracks, gaps and openings (provided windows and doors remain closed). Therefore, tighter buildings may perform better during smoke events. Attention to detail during design and construction is critical for creating tight building envelope. A more airtight envelope is achieved through the use of air barriers inside and outside and careful sealing of every construction joint in the building envelope such as around windows and doors, wall to roof and wall to floor junctions, and sealing of all service penetrations such as electrical fixtures, wiring plumbing and ducts. The leakiness of a building can be tested by measuring the number of air changes per hour (ACH). For instance, in the US, ASHRAE 62.1 recommends that to achieve acceptable indoor air quality in buildings, the ACH should be above 0.35 air changes per hour. Increases in ambient temperatures due climate change are expected to affect the ACH in buildings. In some US locations, increased infiltration has been predicted to increase by up to 30% in summer months. Consequently, patterns of exposures to both indoor and outdoor pollutants will likely occur [60]. In residential buildings that rely on fresh air for ventilation, the levels of particles from outdoor sources are often more pronounced, especially when outdoor air quality is poor.

A recent study of Australian homes found that the level of protection offered from peak outdoor $PM_{2.5}$ levels during controlled burning ranged from 12–76% [38]. The authors found that $PM_{2.5}$ infiltration was significantly influenced by the age of the house and ventilation behaviour (e.g., windows open/closed). Australian homes are generally considered leaky compared to homes in other countries, and older Australian homes are often much leakier than those built more recently [61]. In a study of air tightness in new Australian homes (i.e., <3 years old) the average ACH ranged between 7.9–28.5 @ 50 Pa, depending on state or territory. Among the leakiest homes were in Melbourne and Sydney, notably, these cities were severely impacted by smoke during the summer of 2019–2020 bushfires. By contrast, Canberra homes were found to have ACH slightly below the national average of 15.5 ACH @ 50 Pa [61].

In a more recent study in Melbourne during the 2019–20 bushfire season, Munro and Seagren [62] found that the $PM_{2.5}$ levels in a conventional leaky building were under 500 $\mu g/m^3$ when outdoor levels were close to 600 $\mu g/m^3$, while the levels in air tight homes reached peak of 320 to 380 $\mu g/m^3$ which is 30% lower. A centralised mechanical ventilation with an F7-grade filter was used in the airtight home that had air permeability of 0.93 m^3/m^2h at 50 Pa. The authors also monitored two identical airtight homes that use mechanical ventilation: one with a standard F7 filter and the other with a HEPA filter and found that the home with the HEPA filter achieved lower $PM_{2.5}$ concentrations, that are within recommended guidelines. In summary, even though an airtight building alone is unlikely to keep particles within threshold concentrations, more airtight construction provides an opportunity to better control indoor air quality.

5.2. Filter Technology and Portable Air Cleaners

Air filtration and purification technologies can improve indoor air quality during bushfire smoke. Fisk and Chan [63] found that using air filters during wildfires can substantially reduce hospitalisations and deaths from exposure to wildfire smoke. They also found that the greatest benefits will be realised by focusing on the homes of individuals who experience the most severe health effects such as the elderly. The authors state that the economic benefits of air filtration (particles) in buildings exceed the costs of installation and operation by more than a factor of ten. Common filters used in residential air conditioning such as G4 (~MERV 7–8) will only capture a small percentage of the smoke particles, say around 10%; and a finer F6 (~MERV 10–11) filter may capture around half of these particles. As discussed in the previous section, air filters may not be effective if houses are excessively leaky, therefore, the use of personal respirators independent of the mechanical ventilation system could be explored [6]. Studies have shown that portable air cleaners can reduce indoor levels of fine particles in office environments [64] and residential environments [65,66]. A Canadian study of 31 homes found that using portable air cleaners (Filtrete Ultra Clean Air Purifier model series FAP02-R, 3M, London, ON, Canada) equipped with a patented air filter can reduce indoor levels of $PM_{2.5}$ by a median of 52% [65]. An earlier Canadian study found that using portable air cleaners (with HEPA filters) can reduce $PM_{2.5}$ indoor levels of residential wood smoke in winter and forest fire smoke in summer [67]. In both Canadian studies, air filters were operated without filters ("placebo mode") and with filters during the smoke events to ascertain differences. In a randomised cross over intervention study conducted in Taipei, Chuang et al. [68]

examined the effect of long-term indoor air conditioner filtration on the link between cardiovascular health and air pollution. The results showed that increased levels of $PM_{2.5}$ was associated with cardiovascular health of adults. A review by Barn et al. [9] found that portable air filters can reduce indoor $PM_{2.5}$ concentrations by 32–88% with important factors of variation being study design and airflow into the room. The review found that high efficiency particulate air (HEPA) filters and electrostatic precipitators can reduce indoor $PM_{2.5}$ concentrations and respiratory and cardiovascular health effects.

Gaseous pollutants from wildfires are not generally the focus of studies as the main pollutant of concern is considered to be $PM_{2.5}$. However, air toxics such as benzene [69] and PAH's [47] are also present in smoke. The WHO [70] reports that there is no safe level of exposure to benzene as it is carcinogenic to humans. In Australia, the National Environmental Protection (Air Toxics) Measure specifies the national limits for outdoor air pollutants, including formaldehyde, benzene, and polycyclic aromatic hydrocarbons [71]. Therefore, strategies that minimise exposure during bushfires are also needed. Filtration technologies that remove other pollutants such as air toxics (in addition to $PM_{2.5}$) may provide further benefits in terms of acute and chronic health effects than technologies that exclusively remove particles. Laboratory studies have found activated carbon effective at reducing indoor concentrations of VOCs including benzene and formaldehyde [72], and toluene, cyclohexane, and ethyl acetate [73]. However, there is little or no research to demonstrate the effectiveness of gas-phase filtration systems in actual buildings [63, 72]. If a residential building has a heating ventilation and air conditioning (HVAC) system, it is likely to be less standardised by comparison to those installed in commercial buildings [74] and this can impact the effectiveness of the system.

Impact on Energy

The particle removal efficiency and energy consumption of an air filtration system is influenced by many factors including the system design and the type of installed filter. For instance, high efficiency particulate air filters (such as HEPA) can stop a large proportion of smoke particles. However, the air handling system must be capable of running the additional power load due to resistance to airflow created by the HEPA filter [75]. Also, switching to a higher rated filter has a consequence on energy consumption. Increased pressure drop across the filters will increase fan power consumption. Alavy and Seigel [74] note that the particle removal efficiency of filters can impact three energy related parameters: airflow rate, fan motor energy, and system energy. Prediction of long-term filtration is challenging because particle removal depends on many parameters related to buildings and associated systems such as airflow rate, runtime and particle loading in filters [74]. Using a filter with a higher minimum efficiency reporting value (MERV) causes a higher filter pressure drop and a reduced airflow rate and this will reduce fan energy use. However, the cooling energy will increase because the compressor will have to operate for longer periods of time. For a system with speed control, the fan speed will adjust to maintain the airflow and the total system energy use will remain approximately unchanged. For a system with no speed control, the energy will increase with increasing MERV. Alavy and Seigel [74] summarize that the major factor that determines the energy impact of higher efficiency filters is the increased system runtime due to reduced airflow. This is particularly important in systems without fan control. In summary, there are many factors affecting the relationship between energy use and filter efficiency and these are beyond the scope of the current paper.

A theoretical analysis by Stephens et al. [76] showed that the extent of energy impacts in relation to high-efficiency filters are likely to be small. They also measured results in a test system and confirmed that there was no difference in energy consumption with the use of high-efficiency filters in comparison to low-efficiency filters. Similarly, Walker et al. [77] measured energy usage and filter pressure drop in ten California homes for a year. When low MERV filters are replaced with MERV 10-13 filters, the effects on blower energy use are either negligible or moderate but MERV 16 filters introduced about 20% blower power increases. They found that climates with more cooling had bigger impacts with a change to higher MERV filtration. The authors suggested that requiring filter manufacturers to label filters with static pressure drop would allow contractors and consumers to make filter replacement decisions based on air flow resistance. However, others caution that modification of HVAC systems, such as by inclusion of additional or higher rated filters, needs be done carefully in order to avoid "unintended effects" such as reduced airflow, reduced heating/cooling capacity, filter bypass, life-cycle costs, and infiltration of the building envelope [75].

6. Discussion

Access to clean, fresh air is an enduring environmental and public health challenge. The need for fresh air is even more pronounced during bushfire smoke events when levels of air pollutants (e.g., $PM_{2.5}$) can be many times the levels specified in national ambient air quality guidelines. Bushfire smoke can persist for many weeks and can substantially impact air quality in local communities as well as those that are hundreds of kilometres away from the fire front [45], including densely populated urban areas. In a measurement In a measurement study that included ground-based air quality measurements, aircraft air quality measurements, and computer modelling, Wotawa and Trainer [78] found that wildfires in Canada contributed to elevated levels of carbon monoxide, VOCs, and ozone in the southeastern United States—approximately 3500 km away.

Therefore, strategies and guidance to reduce indoor levels of particulates and other pollutants from bushfires are needed. Vulnerable members of the population could be provided with timely advice on how to avoid smoke exposure. For example, schools and childcare centres could provide warnings to limit outside activities on these days [79]. Aged care homes could be equipped with a 'smoke plan' where appropriate preventative measures based on the concentration of smoke can be detailed. Awareness of, and access to, localised, reliable, real time, user-friendly data is important for residents so that they can minimise their personal exposure to air pollution. The inclusion of air quality, health and activity advice in smart phone apps such as "Air Rater" is an innovation that has been shown to help vulnerable members of the community such as asthmatics.

Very fine smoke particles can penetrate through filtered ventilation systems. The design of ventilation system is critical during persistent smoke events. Ventilation system using efficient filters such as HEPA filters can be effective only if the building construction is relatively airtight (i.e., low ACH). It may not be possibly to improve the efficiency of a filtration system simply by replacing a low efficiency filter with a high efficiency filter due to the pressure drop and limited capacity of the air handling system. Additionally, there is cost associated with the use of HEPA filters due to the pressure drop and increased cooling energy. In order to conserve energy, it may be necessary to only apply additional filtration (e.g., HEPA) in response to smoke events, provided filters can be easily retrofitted.

Low-cost particle sensors have been investigated by many researchers. With the advancements in sensor technology and with an increasing number of low-cost sensors available in the market, there is a need for additional studies of sensor performance focused on monitoring particle levels in bushfire smoke. The sensors tested in studies should be compared to reference instruments under the same conditions and in same locations and calibration factors applied that relate to the individual sensor tested as each sensor will have a unique response to smoke. Appropriate correction factors must be applied to ensure data reliability if low cost sensors are used.

Due to the potential adverse health effects of exposure to gaseous pollutants such as benzene and formaldehyde and the lack of studies investigating removal of these pollutants from bushfire smoke future studies should also focus on the benefits of filtration systems for removal of gaseous pollutants from indoor air in addition to particulate matter. However, monitoring individual organic pollutants (e.g., formaldehyde, benzene) at low levels can be difficult and expensive. The development of low-cost sensors capable of detecting individual organic compounds (e.g., formaldehyde) in real time at ppb levels is another area for exploration [80]. Figure 1 shows the summary of findings from this study highlighting areas of future studies.

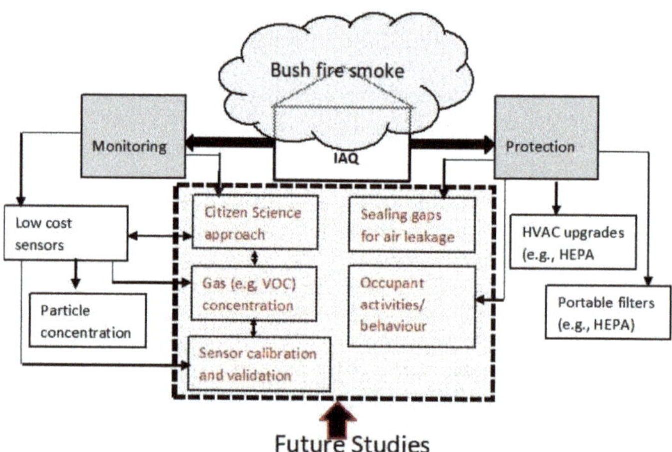

Figure 1. Summary of findings.

Even though it is advised that residential buildings should be constructed to provide an airtight envelope to reduce smoke infiltration, it can be expensive to renovate an entire leaky building to achieve an airtight envelope. Instead of renovating the whole house, the possibility of creating a dedicated indoor "clean air" space (e.g., bedroom) that is protected from smoke intrusion could be explored. This room could include a portable air filter with HEPA filters to remove particles and activated carbon filters for reduction of gaseous pollutants. Citizen science approaches can provide important insights into the effectiveness of low-cost interventions to reduce smoke infiltration into homes. It may be helpful for vulnerable members of the community, such as asthmatics, the elderly, and pregnant women to have access to portable respirators, clean air shelters, and opportunities for temporary relocation.

Studies also demonstrated the importance of reducing smoke pollution from other sources and highlight the need for technologies to improve IAQ even when bushfire smoke is not a threat. For example, Desservettaz et al. [81] compared the atmospheric composition of smoke from domestic wood heaters and hazard reduction burns. The study found no significant differences in the composition but concluded that despite the higher peak pollution levels of hazard reduction, the overall exposure of air toxins was greater from domestic wood smoke due to the greater frequency and total duration of use.

7. Conclusions

This study found that bushfire smoke can substantially degrade residential indoor air quality. While many studies have investigated the use of filtration systems in controlling particle levels, such studies concerning gaseous pollutants are limited. The study revealed that air filtration technologies can be effective at reducing indoor particle levels, although the effectiveness of these depends on many factors, and in particular the leakiness of the building envelope. Many residential buildings in Australia are very leaky (i.e., >15.5 ACH). This is of particular concern as many of these residences are located in densely populated urban areas that can be severely impacted by bushfire smoke. Strategies such as improving the building envelope (e.g., sealing gaps), modifying occupant behavior (e.g., minimizing indoor sources), and increasing awareness of resources (e.g., smartphone apps) can help protect occupants from exposure to bushfire smoke. Areas for future research include evaluations of air filtration technologies focused on the removal of gaseous pollutants within buildings, and calibration and validation of low-cost sensors for monitoring elevated levels of $PM_{2.5}$ and other pollutants. Also there are limited studies where lay people participate in monitoring and apply various intervention strategies in their home. The use of low-cost sensors by homeowners and participants of citizen science programs will help to deliver real time information to occupants.

Author Contributions: Conceptualization, P.R.; methodology, P.R. and N.G.; writing—original draft preparation, P.R. and N.G.; writing—review and editing, P.R. and N.G.; All authors have read and agreed to the published version of the manuscript.

Funding: This research received no external funding.

Conflicts of Interest: The authors declare no conflict of interest.

References

1. US EPA. United States Environmental Protection Agency. Introduction to Indoor Air Quality. Indoor Air Pollution and Health. 2020. Available online: https://www.epa.gov/indoor-air-quality-iaq/introduction-indoor-air-quality (accessed on 31 December 2020).
2. Klepeis, N.E.; Nelson, W.C.; Ott, W.R.; Robinson, J.P.; Tsang, A.M.; Switzer, P.; Behar, J.V.; Hern, S.C.; Engelmann, W.H. The national human activity pattern survey (Nhaps): A resource for assessing exposure to environmental pollutants. *J. Expo. Sci. Environ. Epidemiol.* **2001**, *11*, 231–252. [CrossRef]
3. Patel, S.; Sankhyan, S.; Vance, M.E.; Boedicker, E.K.; Decarlo, P.F.; Farmer, D.K.; Goldstein, A.H.; Katz, E.F.; Nazaroff, W.W.; Tian, Y.; et al. Indoor particulate matter during HOMEchem: Concentrations, size distributions, and exposures. *Environ. Sci. Technol.* **2020**, *54*, 7107–7116. [CrossRef]
4. Brown, S.K. Volatile organic pollutants in new and established buildings in Melbourne, Australia. *Indoor Air* **2002**, *12*, 55–63. [CrossRef]
5. Steinemann, A. Volatile emissions from common consumer products. *Air Qual. Atmosphere Heal.* **2015**, *8*, 273–281. [CrossRef]
6. Brown, S.K. Bushfires and indoor built environments. *Indoor Built Environ.* **2018**, *27*, 145–147. [CrossRef]
7. Dennekamp, M.; Abramson, M.J. The effects of bushfire smoke on respiratory health. *Respirology* **2011**, *16*, 198–209. [CrossRef] [PubMed]
8. Johnston, F.H.; Henderson, S.B.; Chen, Y.; Randerson, J.T.; Marlier, M.; DeFries, R.S.; Kinney, P.; Bowman, D.M.; Brauer, M. Estimated global mortality attributable to smoke from landscape fires. *Environ. Heal. Perspect.* **2012**, *120*, 695–701. [CrossRef] [PubMed]
9. Barn, P.K.; Elliott, C.T.; Allen, R.W.; Kosatsky, T.; Rideout, K.; Henderson, S.B. Portable air cleaners should be at the forefront of the public health response to landscape fire smoke. *Environ. Heal.* **2016**, *15*, 1–8. [CrossRef]
10. Reid, C.E.; Brauer, M.; Johnston, F.H.; Jerrett, M.; Balmes, J.R.; Elliott, C.T. Critical review of health impacts of wildfire smoke exposure. *Environ. Heal. Perspect.* **2016**, *124*, 1334–1343. [CrossRef]
11. Johnston, F.; Hanigan, I.; Henderson, S.; Morgan, G.; Bowman, D. Extreme air pollution events from bushfires and dust storms and their association with mortality in Sydney, Australia 1994–2007. *Environ. Res.* **2011**, *111*, 811–816. [CrossRef]
12. Liu, J.C.; Pereira, G.; Uhl, S.A.; Bravo, M.A.; Bell, M.L. A systematic review of the physical health impacts from non-occupational exposure to wildfire smoke. *Environ. Res.* **2015**, *136*, 120–132. [CrossRef] [PubMed]
13. Henderson, S.B.; Johnston, F.H. Measures of forest fire smoke exposure and their associations with respiratory health outcomes. *Curr. Opin. Allergy Clin. Immunol.* **2012**, *12*, 221–227. [CrossRef] [PubMed]
14. Holstius, D.M.; Reid, C.E.; Jesdale, B.M.; Morello-Frosch, R. Birth Weight following Pregnancy during the 2003 Southern California Wildfires. *Environ. Heal. Perspect.* **2012**, *120*, 1340–1345. [CrossRef] [PubMed]
15. Xu, R.; Yu, P.; Abramson, M.J.; Johnston, F.H.; Samet, J.M.; Bell, M.L.; Haines, A.; Ebi, K.L.; Li, S.; Guo, Y. Wildfires, global climate change, and human health. *N. Engl. J. Med.* **2020**, *383*, 2173–2181. [CrossRef]
16. Liu, Y.; Stanturf, J.; Goodrick, S. Trends in global wildfire potential in a changing climate. *For. Ecol. Manag.* **2010**, *259*, 685–697. [CrossRef]
17. Spracklen, D.V.; Mickley, L.J.; Logan, J.A.; Hudman, R.C.; Yevich, R.; Flannigan, M.D.; Westerling, A.L. Impacts of climate change from 2000 to 2050 on wildfire activity and carbonaceous aerosol concentrations in the western United States. *J. Geophys. Res. Space Phys.* **2009**, *114*. [CrossRef]
18. Sharples, J.J.; Cary, G.J.; Fox-Hughes, P.; Mooney, S.; Evans, J.P.; Fletcher, M.-S.; Fromm, M.; Grierson, P.F.; McRae, R.; Baker, P. Natural hazards in Australia: Extreme bushfire. *Clim. Chang.* **2016**, *139*, 85–99. [CrossRef]
19. Ulpiani, G.; Ranzi, G.; Santamouris, M. Experimental evidence of the multiple microclimatic impacts of bushfires in af-fected urban areas: The case of Sydney during the 2019/2020 Australian season. *Environ. Res. Comm.* **2020**, *2*, 065005.
20. Filkov, A.; Duff, T.; Penman, T. *Determining Threshold Conditions for eXtreme Fire Behaviour*; Annual Report 2019–2020, Report number 626; Bushfire and Natural Hazards CRC: Melbourne, VIC, Australia, November 2020.
21. Campbell, S.L.; Jones, P.J.; Williamson, G.J.; Wheeler, A.J.; Lucani, C.; Bowman, D.M.J.S.; Johnston, F.H. Using digital technology to protect health in prolonged poor air quality episodes: A case study of the airrater app during the Australian 2019–2020 fires. *Fire* **2020**, *3*, 40. [CrossRef]

22. Boer, M.M. Resco de Dios, V.; Bradstock, R.A. Unprecedented burn area of Australian mega forest fires. *Nat. Clim.* **2020**, *10*, 171–172. [CrossRef]
23. Nazaroff, W.W. Exploring the consequences of climate change for indoor air quality. *Environ. Res. Lett.* **2013**, *8*, 015022. [CrossRef]
24. Fisk, W.J. Review of some effects of climate change on indoor environmental quality and health and associated no-regrets mitigation measures. *Build. Environ.* **2015**, *86*, 70–80. [CrossRef]
25. Vardoulakis, S.; Jalaludin, B.B.; Morgan, G.G.; Hanigan, I.C.; Johnston, F.H. Bushfire smoke: Urgent need for a national health protection strategy. *Med, J. Aust.* **2020**, *212*, 349–353.e1. [CrossRef]
26. Williamson, G.J.; Bowman, D.M.S.; Price, O.F.; Henderson, S.B.; Johnston, F.H. A transdisciplinary approach to under-standing the health effects of wildfire and prescribed fire smoke regimes. *Environ. Res. Lett.* **2016**, *11*, 125009. [CrossRef]
27. WHO. *Air Quality Guidelines: Particulate Matter, Ozone, Nitrogen Dioxide and Sulfur Dioxide*; 2006. Available online: https://apps.who.int/iris/bitstream/handle/10665/69477/WHO_SDE_PHE_OEH_06.02_eng.pdf (accessed on 19 December 2020).
28. NEPM. *National Environment Protection (Ambient Air Quality) Measure*; 2016. Available online: https://soe.environment.gov.au/theme/ambient-air-quality/topic/2016/national-air-quality-standards (accessed on 20 December 2020).
29. USEPA. United States Environmental Protection Agency. *Reviewing National Ambient Air Quality Standards (NAAQS)*; 2020. Available online: https://www.epa.gov/naaqs/particulate-matter-pm-air-quality-standards (accessed on 5 January 2021).
30. Yu, P.; Xu, R.; Abramson, M.J.; Li, S.; Guo, Y. Bushfires in Australia: A serious health emergency under climate change. *Lancet Planet. Heal.* **2020**, *4*, e7–e8. [CrossRef]
31. Australian Government. National Environment Protection (Ambient Air Quality) Measure Review. National Environment Protection Council. 2016. Available online: https://www.legislation.gov.au/Details/F2016C00215 (accessed on 20 December 2020).
32. CDPH. California Department of Public Health. Wildfire Smoke FAQs. 2020. Available online: https://www.cdph.ca.gov/Programs/EPO/Pages/BP_Wildfire_FAQs.aspx (accessed on 4 January 2021).
33. EPA VIC. Environmental Protection Authority, Victoria, 2018, After a Fire: Cleaning up a Smoke-Affected Home. 2018. Available online: https://www.epa.vic.gov.au/about-epa/publications/1711 (accessed on 20 December 2020).
34. NSW Government. Protect Yourself from Bushfire Smoke. 2020. Available online: https://www.health.nsw.gov.au/environment/air/Documents/protect-yourself-from-bushfire-smoke.pdf (accessed on 25 November 2020).
35. Elbayoumi, M.; Ramli, N.A.; Yusof, N.F.F.M. Spatial and temporal variations in particulate matter concentrations in twelve schools environment in urban and overpopulated camps landscape. *Build. Environ.* **2015**, *90*, 157–167. [CrossRef]
36. Martins, N.R.; da Graça, G.C. Impact of PM2. 5 in indoor urban environments: A review. *Sustain. Cities Soc.* **2018**, *42*, 259–275. [CrossRef]
37. Luo, N.; Weng, W.; Xu, X.; Hong, T.; Fu, M.; Sun, K. Assessment of occupant-behavior-based indoor air quality and its impacts on human exposure risk: A case study based on the wildfires in Northern California. *Sci. Total Environ.* **2019**, *686*, 1251–1261. [CrossRef]
38. Reisen, F.; Powell, J.C.; Dennekamp, M.; Johnston, F.H.; Wheeler, A.J. Is remaining indoors an effective way of reducing exposure to fine particulate matter during biomass burning events? *J. Air Waste Manag. Assoc.* **2019**, *69*, 611–622. [CrossRef] [PubMed]
39. Steinemann, A. Ten questions concerning fragrance-free policies and indoor environments. *Build. Environ.* **2019**, *159*, 106054. [CrossRef]
40. Jaffe, D.A.; Wigder, N.L. Ozone production from wildfires: A critical review. *Atmospheric Environ.* **2012**, *51*, 1–10. [CrossRef]
41. Na, K.; Cocker, D.R. Fine organic particle, formaldehyde, acetaldehyde concentrations under and after the influence of fire activity in the atmosphere of Riverside, California. *Environ. Res.* **2008**, *108*, 7–14. [CrossRef]
42. Malilay, J. A Review of Factors Affecting the Human Health Impacts of Air Pollutants from Forest Fires. Health Guidelines For Vegetation Fire Events: Background Papers. 1999. Available online: https://www.who.int/docstore/peh/Vegetation_fires/Backgroundpapers/BackgrPap8.pdf (accessed on 20 December 2020).
43. Pantelic, J.; Dawe, M.; Licina, D. Use of IoT sensing and occupant surveys for determining the resilience of buildings to forest fire generated PM2. 5. *PLoS ONE* **2019**, *14*, 0223136.
44. Shrestha, P.M.; Humphrey, J.L.; Carlton, E.J.; Adgate, J.L.; Barton, K.E.; Root, E.D.; Miller, S.L. Impact of outdoor air pollution on indoor air quality in low-income homes during wildfire seasons. *Int. J. Environ. Res. Public Heal.* **2019**, *16*, 3535. [CrossRef]
45. Sapkota, A.; Symons, J.M.; Kleissl, J.; Wang, L.; Parlange, M.B.; Ondov, J.; Breysse, P.N.; Diette, G.B.; Eggleston, P.A.; Buckley, T.J. Impact of the 2002 Canadian forest fires on particulate matter air quality in Baltimore City. *Environ. Sci. Technol.* **2005**, *39*, 24–32. [CrossRef] [PubMed]
46. Reisen, F.; Meyer, C.M.; McCaw, L.; Powell, J.C.; Tolhurst, K.G.; Keywood, M.D.; Gras, J.L. Impact of smoke from biomass burning on air quality in rural communities in southern Australia. *Atmospheric Environ.* **2011**, *45*, 3944–3953. [CrossRef]
47. Messier, K.P.; Tidwell, L.G.; Ghetu, C.C.; Rohlman, D.; Scott, R.P.; Bramer, L.M.; Dixon, H.M.; Waters, K.M.; Anderson, K.A. Indoor versus outdoor air quality during wildfires. *Environ. Sci. Technol. Lett.* **2019**, *6*, 696–701. [CrossRef] [PubMed]
48. Guérette, E.-A.; Paton-Walsh, C.; Desservettaz, M.; Smith, T.E.L.; Volkova, L.; Weston, C.J.; Meyer, C.P. Emissions of trace gases from Australian temperate forest fires: Emission factors and dependence on modified combustion efficiency. *Atmospheric Chem. Phys. Discuss.* **2018**, *18*, 3717–3735. [CrossRef]
49. Koken, P.J.M.; Piver, W.T.; Ye, F.; Elixhauser, A.; Olsen, L.M.; Portier, C.J. Temperature, air pollution, and hospitalization for cardiovascular diseases among elderly people in Denver. *Environ. Heal. Perspect.* **2003**, *111*, 1312–1317. [CrossRef]

50. Kang, C.-M.; Gold, D.R.; Koutrakis, P. Downwind O3 and PM2.5 speciation during the wildfires in 2002 and 2010. *Atmospheric Environ.* **2014**, *95*, 511–519. [CrossRef]

51. Nazaroff, W.W.; Weschler, C.J. Cleaning products and air fresheners: Exposure to primary and secondary air pollutants. *Atmospheric Environ.* **2004**, *38*, 2841–2865. [CrossRef]

52. McDonald, B.C.; De Gouw, J.A.; Cui, Y.Y.; Kim, S.-W.; Gentner, D.R.; Isaacman-VanWertz, G.; Goldstein, A.H.; Harley, R.A.; Frost, G.J.; Roberts, J.M.; et al. Volatile chemical products emerging as largest petrochemical source of urban organic emissions. *Science* **2018**, *359*, 760–764. [CrossRef]

53. Liu, H.-Y.; Schneider, P.; Haugen, R.; Vogt, M. Performance Assessment of a Low-Cost PM2.5 Sensor for a near Four-Month Period in Oslo, Norway. *Atmosphere* **2019**, *10*, 41. [CrossRef]

54. Holder, A.L.; Mebust, A.K.; Maghran, L.A.; McGown, M.R.; Stewart, K.E.; Vallano, D.M.; Elleman, R.A.; Baker, K.R. Field evalua-tion of low-cost particulate matter sensors for measuring wildfire smoke. *Sensors* **2020**, *20*, 4796. [CrossRef]

55. Rai, A.C.; Kumar, P.; Pilla, F.; Skouloudis, A.N.; Di Sabatino, S.; Ratti, C.; Yasar, A.; Rickerby, D. End-user perspective of low-cost sensors for outdoor air pollution monitoring. *Sci. Total Environ.* **2017**, 691–705. [CrossRef]

56. Sayahi, T.; Butterfield, A.; Kelly, K. Long-term field evaluation of the Plantower PMS low-cost particulate matter sensors. *Environ. Pollut.* **2019**, *245*, 932–940. [CrossRef] [PubMed]

57. Delp, W.W.; Singer, B.C. Wildfire smoke adjustment factors for low-cost and professional PM2.5 Monitors with optical Sensors. *Sensors* **2020**, *20*, 3683. [CrossRef]

58. Morawska, L.; Thai, P.K.; Liu, X.; Asumadu-Sakyi, A.; Ayoko, G.; Bartonova, A.; Bedini, A.; Chai, F.; Christensen, B.; Dunbabin, M.; et al. Applications of low-cost sensing technologies for air quality monitoring and exposure assessment: How far have they gone? *Environ. Int.* **2018**, *116*, 286–299. [PubMed]

59. Mehadi, A.; Moosmüller, H.; Campbell, D.E.; Ham, W.; Schweizer, D.; Tarnay, L.; Hunter, J. Laboratory and field evaluation of real-time and near real-time PM2.5 smoke monitors. *J. Air Waste Manag. Assoc.* **2020**, *70*, 158–179. [CrossRef] [PubMed]

60. Ilacqua, V.; Dawson, J.; Breen, M.; Singer, S.; Berg, A. Effects of climate change on residential infiltration and air pollution exposure. *J. Expo. Sci. Environ. Epidemiol.* **2015**, *27*, 16–23. [CrossRef] [PubMed]

61. Ambrose, M.; Syme, M. Air tightness of New Australian residential buildings. *Procedia Eng.* **2017**, *180*, 33–40. [CrossRef]

62. Munro, C.; Seagren, J. The Smoke Infiltration Situation in the Aftermath of the Devastation Bushfire. Events. Ecolibrium, AIRAH. 2020. Available online: https://www.airah.org.au/Content_Files/EcoLibrium/2020/04-20-Eco-001.pdf (accessed on 20 December 2020).

63. Fisk, W.J.; Chan, W.R. Health benefits and costs of filtration interventions that reduce indoor exposure to PM2.5 during wildfires. *Indoor Air* **2016**, *27*, 191–204. [CrossRef]

64. Stauffer, D.A.; Autenrieth, D.A.; Hart, J.F.; Capoccia, S. Control of wildfire-sourced PM2.5 in an office setting using a commercially available portable air cleaner. *J. Occup. Environ. Hyg.* **2020**, *17*, 109–120. [CrossRef] [PubMed]

65. Wheeler, A.J.; Gibson, M.D.; MacNeill, M.; Ward, T.J.; Wallace, L.A.; Kuchta, J.; Seaboyer, M.; Dabek-Zlotorzynska, E.; Guernsey, J.R.; Stieb, D.M. Impacts of air cleaners on indoor air quality in residences impacted by wood smoke. *Environ. Sci. Technol.* **2014**, *48*, 12157–12163. [CrossRef] [PubMed]

66. Sublett, J.L. Effectiveness of air filters and air cleaners in allergic respiratory diseases: A review of the recent literature. *Curr. Allergy Asthma Rep.* **2011**, *11*, 395. [CrossRef] [PubMed]

67. Barn, P.; Larson, T.I.; Noullett, M.; Kennedy, S.; Copes, R.; Brauera, M. Infiltration of forest fire and residential wood smoke: An evaluation of air cleaner effectiveness. *J. Expo. Sci. Environ. Epidemiol.* **2007**, *18*, 503–511. [CrossRef]

68. Chuang, H.C.; Ho, K.F.; Lin, L.Y.; Chang, T.Y.; Hong, G.B.; Ma, C.M.; Liu, I.J.; Chuang, K.J. Long-term indoor air con-ditioner filtration and cardiovascular health: A randomized crossover intervention study. *Environ. Int.* **2017**, *106*, 91–96. [CrossRef]

69. Reisen, F.; Brown, S.K. Implications for community health from exposure to bushfire air toxics. *Environ. Chem.* **2006**, *3*, 235–243. [CrossRef]

70. World Health Organization. *Air Quality Guidelines for Europe*, 2nd ed.; WHO European Series No.91; WHO Regional Office for Europe: Copenhagen, Denmark, 2000. Available online: https://apps.who.int/iris/handle/10665/107335 (accessed on 4 January 2021).

71. Australian Government. *National Environment Protection (Air Toxics) Measure ACT, 2004*; Environment Protection and Heritage Council: Canberra, Australia, 2004.

72. Sidheswaran, M.A.; Destaillats, H.; Sullivan, D.P.; Cohn, S.; Fisk, W.J. Energy efficient indoor VOC air cleaning with activated carbon fiber (ACF) filters. *Build. Environ.* **2012**, *47*, 357–367. [CrossRef]

73. Haghighat, F.; Lee, C.-S.; Pant, B.; Bolourani, G.; Lakdawala, N.; Bastani, A. Evaluation of various activated carbons for air cleaning–Towards design of immune and sustainable buildings. *Atmospheric Environ.* **2008**, *42*, 8176–8184. [CrossRef]

74. Alavy, M.; Siegel, J.A. IAQ and energy implications of high efficiency filters in residential buildings: A review (RP-1649). *Sci. Technol. Built Environ.* **2019**, *25*, 261–271. [CrossRef]

75. Keefe, A. Evidence Review: Filtration in Institutional Settings during Wildfire Smoke Events. BC Centre for Disease Control, Vancouver BC. 2014. Available online: http://www.bccdc.ca/resource-gallery/Documents/Guidelines%20and%20Forms/Guidelines%20and%20Manuals/Health-Environment/WFSG_EvidenceReview_FiltrationinInstitutions_FINAL_v3_edstrs.pdf (accessed on 5 January 2021).

76. Stephens, B.; Novoselac, A.; Siegel, J. The Effects of Filtration on Pressure Drop and Energy Consumption in Res-idential HVAC Systems (RP-1299). *HVAC R Res.* **2010**, *16*, 273–294.
77. Walker, I.S.; Faulkner, D.; Dickerhoff, D.J.; Turner, W.J.N. Energy Implications of In Line Filtration in California Homes. *ASHRAE Trans.* **2013**, *119*, 399–417.
78. Wotawa, G.; Trainer, M. The influence of Canadian forest fires on pollutant concentrations in the United States. *Science* **2000**, *288*, 324–328. [CrossRef] [PubMed]
79. Walter, C.M.; Schneider-Futschik, E.K.; Knibbs, L.D.; Irving, L.B. Health impacts of bushfire smoke exposure in Australia. *Respirology* **2020**, *25*, 495–501. [CrossRef]
80. Spinelle, L.; Gerboles, M.; Kok, G.; Persijn, S.; Sauerwald, T. Review of portable and low-cost sensors for the ambient air monitoring of benzene and other volatile organic compounds. *Sensors* **2017**, *17*, 1520. [CrossRef] [PubMed]
81. Desservettaz, M.; Phillips, F.; Naylor, T.; Price, O.; Samson, S.; Kirkwood, J.; Paton-Walsh, C. Air quality impacts of smoke from hazard reduction burns and domestic wood heating in Western Sydney. *Atmosphere* **2019**, *10*, 557. [CrossRef]

Article

A Comparative Analysis of Different Future Weather Data for Building Energy Performance Simulation

Mamak P.Tootkaboni [1,*], Ilaria Ballarini [1], Michele Zinzi [2] and Vincenzo Corrado [1]

1   Department of Energy, Politecnico di Torino, 10129 Turin, Italy; ilaria.ballarini@polito.it (I.B.); vincenzo.corrado@polito.it (V.C.)
2   ENEA, Via Anguillarese 301, 00123 Rome, Italy; michele.zinzi@enea.it
*   Correspondence: mamak.ptootkaboni@polito.it

Abstract: The building energy performance pattern is predicted to be shifted in the future due to climate change. To analyze this phenomenon, there is an urgent need for reliable and robust future weather datasets. Several ways for estimating future climate projection and creating weather files exist. This paper attempts to comparatively analyze three tools for generating future weather datasets based on statistical downscaling (WeatherShift, Meteonorm, and CCWorldWeatherGen) with one based on dynamical downscaling (a future-typical meteorological year, created using a high-quality reginal climate model). Four weather datasets for the city of Rome are generated and applied to the energy simulation of a mono family house and an apartment block as representative building types of Italian residential building stock. The results show that morphed weather files have a relatively similar operation in predicting the future comfort and energy performance of the buildings. In addition, discrepancy between them and the dynamical downscaled weather file is revealed. The analysis shows that this comes not only from using different approaches for creating future weather datasets but also by the building type. Therefore, for finding climate resilient solutions for buildings, care should be taken in using different methods for developing future weather datasets, and regional and localized analysis becomes vital.

Keywords: climate change; future weather data; building energy performance; thermal comfort; statistical downscaling of climate models; dynamical downscaling of climate models

Citation: P.Tootkaboni, M.; Ballarini, I.; Zinzi, M.; Corrado, V. A Comparative Analysis of Different Future Weather Data for Building Energy Performance Simulation. *Climate* **2021**, *9*, 37. https://doi.org/10.3390/cli9020037

Academic Editor: Steven McNulty

Received: 2 February 2021
Accepted: 19 February 2021
Published: 23 February 2021

Publisher's Note: MDPI stays neutral with regard to jurisdictional claims in published maps and institutional affiliations.

1. Introduction

There is an urgent need to address climate change as a primary global problem. According to the World Meteorological Organization (WMO) report on global climate, recent years have seen a continued increase in greenhouse gas concentration, global mean temperature, global sea level, and melting cryosphere [1]. The fifth assessment report (AR5) of the Intergovernmental Panel on Climate Change (IPCC) points out that if the emissions continue to rise, the global average temperature will be 2.6-4.8 degrees Celsius (°C) higher than the present by the end of the 21st century. Even if the greenhouse gas emissions stop immediately, the temperature increase will persist for centuries due to the effect of already present greenhouse gases in the atmosphere [2].

In addition to the changes in average temperature trend, extreme events intensify in frequency and magnitude; as an example, over the period 1880 to 2005, the frequency of heat waves in Europe has doubled, and longer heatwaves are more than 90% definite as the climate pattern has been disrupted [3]. The August 2003 heatwave was responsible for around 45,000 excess deaths across 12 European countries [4]. From 2015 to 2019, heatwaves were the deadliest meteorological hazard in many countries, particularly in Europe and North America [1], with an increase of mortality [5] and morbidity, especially for the elderly [6]. Consequences of these extremes included higher energy uses in buildings [5–7], infrastructure failures [8,9], and negative economic impacts [10,11].

Future climate scenarios affirm these trends, projecting different warming rates built upon a number of possible scenarios of future anthropogenic greenhouse gas emissions. The first set of scenarios was Emissions Scenarios (SRES), which were introduced in the IPCC 4th Special Report in 1996 [12]. Later, in 2014, the IPCC adopted a new series of scenarios called "Representative Concentration Pathways (RCPs)" that are established using hypotheses about economic growth, choices of technology, and land use [2]. RCPs are identified by their associated warming effect (radiative forcing, which is measured in units of watts per meter squared) in the year 2100. Radiative forcing is a direct measure of the amount Earth's energy imbalance. The four RCPs (RCP2.6, RCP4.5, RCP6.0, and RCP8.5) span a range of assumptions about future controls on greenhouse gas (and other) emissions. The lowest RCP represents a very aggressive Green House Gases (GHG) mitigation scenario aimed at limiting global warming to about 2 °C, while the highest RCP corresponds to minimal effort to reduce GHG emissions this century. The RCP 8.5 assumes that atmospheric concentrations of CO_2 are three to four times higher than pre-industrial levels by 2100 [2].

Insights from these future scenarios demonstrate the urgent need for emission reduction in a wide range of relevant sectors. Among these sectors, the building stock is one of the main greenhouse gas emitters, which in the European Union accounts for 50% of the CO_2 emissions [13]. In addition, buildings will be also subjected to a warmer climate due to their long lifespan. The existing body of research on the effect of climate change on the future performance of buildings predicts that there will be a paradigm shift in building energy performance. Unsurprisingly, a drastic rise in cooling energy use and a moderate decrease in heating energy use is predicted [14–16]. Recently, in the study of Soutullo et al. [17], for two reference buildings in Madrid, the annual heating and cooling requirements were shown to be around 22% lower and higher, respectively, just by considering the effect of climate change in the last decade. In addition, many studies revealed that this trend may happen even in energy efficient dwellings. Da Guarda et al. [18] analyzed the vulnerability of a zero-energy building (ZEB) to the impact of climate change in 2020 (2011 to 2040), 2050 (2041 to 2070), and 2080 (2071 to 2100). Results showed that due to the increase in cooling energy consumption, there will be a power generation gap up to 40.2% for the 2080 period. In addition, the climate change impact magnitude is not equal for different scenarios, case studies, and regions. Zhai and Helman [19] studied a campus building stock energy prediction, using four future climate models, which are representative of 56 model scenarios for seven climate zones in US. The results demonstrated that cooling energy increases variously (5%, 28%, 20%, and 52%) for different scenarios. In another study, Chai et al. [20] analyzed the life cycle of a net zero energy building in typical climate regions of China. It was concluded that among the different climate regions, the impact of climate change on energy balance and thermal comfort varies significantly. So, in a changing climate, it becomes necessary to prepare buildings for the future at the regional scale, considering different scenarios, to avoid problems such as overheating and power outage, which brings health risks for the occupants.

To investigate the future performance of a building in the context of climate change, building energy simulation (BES) is a vital support tool. BES needs a robust weather dataset that defines the external boundary conditions the building will face during its lifetime. Typically, a representative year of hourly weather data is required to represent the typical regional climate condition and to define the dynamic energy behavior of the building. Several methodologies have been developed to create this one-year climate data from historical climate records [21]. The most commonly used methodology is the Typical meteorological year (TMY), which was introduced in 1978 [22]. TMY is a fictive year constructed of twelve representative typical months [23]. Representative months are selected by comparing the distribution of each month with the long-term distribution of that month for the available climate dataset (the Finkelstein–Schafer statistics) [24]. The analysis of the present climate is based on the observation of climate variables and the application of statistical methods for understanding the current trends. On the other hand, the analysis of future climate is based on future scenarios and the projections of climate models.

Future scenarios are the input data used to provide initial conditions for General Circulation Models or Global Climate Models (GCMs), which are models for forecasting climate change. GCMs provide climate information on the global scale with a typical spatial resolution of 150–600 km^2 [2]. Consequently, if they are used for building energy simulation, the climate change effect and related weather extremes at the local level will not be considered. In this case, the GCMs should be downscaled to applicable spatial (less than 100 km^2) and temporal resolution (less than monthly value). There are two main approaches to downscale GCMs: dynamical and statistical downscaling. Several studies compared different methodologies that use these approaches for the generation of future weather data. Jentsch et al. indicate that weather variability is not generated in the statistically downscaled weather dataset and this approach includes the effect of climate change independently between the variables [25]. On the other hand, Dias et al. point out that the statistical downscaling approach has the advantage of reducing the computational time so that various climate change scenarios can be applied [26], besides providing enough information to study the performance of the building [27]. In light of these studies, there is still a need to deeply analyze different methodologies for future weather data generation.

This study aims to contribute to evaluating the suitability and robustness of different future weather data for analyzing the future performance of reference buildings both in terms of thermal comfort and energy performance. It represents a comparative study of four future weather datasets. Three of them were produced using common weather generator tools available today (WeatherShift, Meteonorm, and CCWorldWeatherGen), which apply statistical downscaling. The other one is a TMY created using a high-quality regional climate models database (from Euro-Coordinated Regional Climate Downscaling Experiment (CORDEX)) that applies the dynamical downscaling. The study investigates the impact of each type of these future weather data in the building energy performance and thermal comfort predictions. It evaluates the heating and cooling demand, the overall energy performance in the presence of heating and cooling systems with continuous operation, and the overheating risk in a free-floating regime of two building types, which are representative of the existing residential building stock in Italy, using the EnergyPlus simulation engine [28]. The analysis was carried out for Rome, as it is one of the representative cities of Mediterranean hot summer climates according to Köppen classification [29]. Representative Concentration Pathways 8.5 (business as usual) [2] have been applied in this study for the mid-century period from 2040 to 2060. This period was used for the analysis, since GCMs uncertainties due to internal climate variability, climate model, and future scenarios increase significantly over time [30,31].

The next section provides a short background on downscaling of the global climate models for generating future weather files for BES. The following sections present the methods and case studies used in this study, the results and discussion, and the conclusion.

2. Review of GCMs Downscaling Methods

Global climate models are complicated numerical models that simulate the state and evolution of the atmosphere, including the atmospheric circulation and energy exchanges in terms of radiation, heat, and moisture. They simulate the processes related to cloud formation and precipitation and take into account the interaction with the ocean and the land [32]. To check if GCMs can simulate the evolution of the climate systems, they are validated against past climate conditions [33]. After verification and validation, GCMs are set to run by forcing greenhouse gas concentration scenarios as an initial condition. GCMs results have global or continental scale spatial resolution and long temporal resolution such as seasonal or annual periods. Due to these coarse resolutions, the direct use of GCMs outputs for building performance assessment is not possible. As previously mentioned, to reach local climate and applicable temporal resolution, downscaling of the GCMs is required. Statistical downscaling and dynamical downscaling are two main approaches; they are presented in Sections 2.1 and 2.2.

2.1. Statistical Downscaling

Statistical downscaling develops and applies statistical relationships between regional or local climate variables and large-scale climate data using deterministic or stochastic approaches [27]. This downscaling approach is a computationally less demanding alternative that facilitates achieving various sets of results. The simplicity of this method—in comparison with dynamical downscaling—persuades many researchers to favor it. This method is mostly applied to GCM projections, while it may also be applied to RCM output as being a better representative for the local climate [34]. In the two following sub-sections, major approaches for applying statistical downscaling are explained in more detail.

2.1.1. Stochastic Weather Generation

Stochastic weather generators are among statistical models, which fill in missing data and enable the production of long synthetic weather series indefinitely. This becomes possible through simulating major properties of observed meteorological records, including daily means, variances and covariances, frequencies, extremes, etc. [35]. These models rely on statistical analysis of recorded climate data in which a few independent weather variables—such as solar radiation—are adequate to derive all other relevant variables. The stochastic weather generation method has the advantage of enabling the integration of the distribution used for the climate change signal. In addition, it is accountable for potential changes in weather patterns and climate variability [32]. However, what appears to be a limitation of this method is the need for a large amount of data to train the model, since distributions for generating future data are based on the baseline data given to the model [35]. The well-known tool that uses this method is Meteonorm. More details about this software and the way it becomes applied in this study will be explained in Section 3.1.1.

2.1.2. Time Series Adjustment: Morphing

Morphing is the most common statistical downscaling method for the adjustment of time series toward the future. This method was firstly presented by Belcher et al. in 2005, assuming the current weather data as baseline [35]. In order to transform this baseline to a future time series, monthly climate change signals given by a GCM or Regional Climate Model (RCM) are used. There are three ways to morph data—shifting, scaling, or a combination of them—depending on the climate variable and expression of the climate change signal (absolute, relative):

- The Shift is applied when absolute monthly mean change (Δx_m) derived from a GCM or RCM is predicted for a given variable (x_0) such as atmospheric pressure, for the month m, according to Equation (1):

$$x_m = x_0 + \Delta x_m. \tag{1}$$

- The Stretch is applied when a relative monthly mean change (α_m) derived from a GCM or RCM is predicted for a given variable (x_0) such as wind speed, for the month m, according to Equation (2):

$$x_m = \alpha_m \cdot x_0. \tag{2}$$

- The combination of Shift and Stretch is applied when both absolute and relative monthly mean changes derived from a GCM or RCM are predicted for a given variable (x_0) such as dry-bulb temperature, for the month m, according to Equation (3):

$$x_m = x_0 + \Delta x_m + \alpha_m (x_0 - x_{0,m}) \tag{3}$$

where $x_{0,m}$ is the variable x_0 average over month m for all the considered averaging years of future data provided by the climate models.

CCWorldWeatherGen and WeatherShift are two available tools that use the morphing method to create future weather data. More details about these tools and their application in this study will be explained in Sections 3.1.2 and 3.1.3.

2.2. Dynamical Downscaling

Dynamical downscaling uses a nesting strategy to obtain climate information at a resolution of 2.5–100 km^2. To this aim, a Regional Climate Model (RCM) is used to derive local or regional climate information. This method simulates "atmospheric and land surface processes, while accounting for high resolution topographical data, land–sea contrasts, surface characteristics, and other components of the Earth-system" [36]. The climate information generated by RCMs has much finer spatial resolution compared to GCMs. This allows RCMs to better represent the spatial and temporal variability of local climate and guarantee physically consistent datasets [37]. However, a large amount of computational power and storage for data creation is one of the limitations of this method. Furthermore, the accuracy of the relevant GCM determines the overall quality of the output. In order to evaluate such uncertainties, different GCM–RCM pairings are combined, and a series of simulations are performed. ENSEMBLES [38] and EURO-CORDEX [39] projects are two of such efforts.

EURO-CORDEX—as the main reference framework for regional downscaling research—aims to facilitate the process of knowledge exchange and communication. Many sectors—e.g., building sector, agriculture, heat and fire risk, and air quality—utilize EURO-CORDEX, since it provides a consistent database of downscaled multi-year projections for various regions all over the world [40]. In addition, by providing a better understanding of the regional and local climate and its associated uncertainties, EURO-CORDEX evaluates and enhances different RCMs. CORDEX includes a large RCM database, and it is updated by new climate data from available domains all over the globe [41]. For European countries, the grid resolution provided by EURO-CORDEX projections equals 12.5 km. For Middle East and North Africa, this quantity is 25 km, while the rest of the world has the grid resolution of 50 km. The time scales—on which the data in the multi-layer format are available—include monthly, daily, every six hours, every three hours, and hourly during the historical period from 1976 to 2005 and for the future period, from 2006 to 2100. The data are available either for RCP 4.5 or RCP 8.5 scenarios, depending on the model [42]. Although most of the available data on the platform are not bias-adjusted, a number of bias-adjusted data are available for some specific models and climate variables. In this study, the hydrostatic version of the regional model REMO-2015 (from 0.11° resolution of the CORDEX European domain), developed by the Max Planck Institute for Meteorology in Hamburg, Germany and currently maintained at the Climate Service Center Germany (GERICS) in Hamburg is used [43,44]. The utilization of this model in creating future TMY for this study is discussed in Section 3.1.4.

3. Materials and Methods

3.1. Describing Future Weather Data Generation for Rome

Four future weather datasets to be analyzed in this work were generated for Rome, using Meteonorm, CCWorldWeatherGen, and WeatherShift weather generator tools, and one RCM (GERICS-REMO-2015) from the EURO-CORDEX project. The weather datasets were developed for the mid-century period from 2040 to 2060. In the study of Hawkins and Sutton, this period (around 2050) is indicated as the period in which temperature predictions will be best in comparison with other periods during the century. Uncertainties of GCMs due to the internal climate variability, climate model, and future scenarios increase significantly over time [30,31]. The following sub-sections describe the applied methodology in detail.

3.1.1. Meteonorm

By integrating the climate database with spatial interpolation of the principal weather variables and a stochastic weather generator, Meteonorm generates hourly weather data for any site in the world [45]. These data can be used as an input for building performance simulation. Weather variables such as global irradiance on a horizontal plane at the ground level, dry-bulb temperature, dew-point temperature, and wind speed are provided by Meteonorm. This tool can be also used for climate change studies. GCMs under the IPCC fourth assessment report (AR4) [46] are used in this tool to generate future weather data for different emission scenarios (B1, A1B, and A2), with 10-year intervals from 2010 until 2100 [47]. The Meteonorm version 7.2 was used in this study to generate a typical meteorological year of 2050 for the A2 emission scenario (pessimist scenarios) for the city of Rome.

3.1.2. CCWorldWeatherGen

The CCWorldWeatherGen is a Microsoft® Excel based tool developed by the Sustainable Energy Research Group of Southampton University [25]. It uses the Morphing methodology to create future weather datasets in Energy Plus Weather (EPW) format for different locations all over the world. The output data of UK Met-office, the Hadley Center Coupled Model 3 (HadCM3) [48] global climate model, forced with IPCC A2 emission scenarios is used in this tool. The HadCM3 climate model was chosen since by the time—in comparison with 29 other climate models—this model was the only one that had all necessary climate variables for the morphing procedure [49]. What HadCM3 provides as input for the Morphing procedure in CCWorldWeatherGen is the monthly value of relative changes regarding the period of 1961–1990. The Excel tool superimposes this input on the weather variables of the baseline weather data stored in an EPW file. The tool generates future weather data sets for 3 time slices: 2001–2040 (referred as '2020s'), 2041–2070 (referred as '2050s'), and 2071–2100 (referred as '2080s'). Being a free online tool is an advantage that makes it widely used. However, due to possible differences in the reference time frame between HadCM3 and the EPW data, inaccuracy in the outputs of the tool may occur [50]. In this study, the International Weather for Energy Calculation (IWEC) TMY file of Rome—downloaded from the Energy Plus database—was used to be morphed for the time slice of 2050s.

3.1.3. WeatherShift

The WeatherShift TM tool was developed upon morphing methodology by Arup and Argos Analytics for creating future weather data [51]. "The tool blends 14 of the more recently simulated GCMs (BCC-CSM1.1, BCC-CSM1.1(m), CanESM2, CSIRO- Mk3.6.0, GFDL-CM3, GFDL-ESM2G, GFDL-ESM2M, GISS-E2-H, GISS-E2-R, HadGEM2-ES, IPSL-CM5A-LR, IPSL-CM5A-MR, IPSL-CM5B-LR, NorESM1-M) into cumulative distribution functions (CDF) [52]. It is based on RCP 4.5 and 8.5 emission scenarios of the IPCC fifth assessment report. Creating CDFs allows a percentile distribution (called warming percentile factor) and "smooths out" the inter-modal uncertainty and stochastic climate behavior [53]. The tool produces future weather data for time periods of 20 years starting from 2011 and ending in 2100. The morphing method in this tool is applied in 8 climate variables of the reference TMY: the mean, maximum, and minimum daily temperature, relative humidity, daily total solar irradiance, wind speed, atmospheric pressure, and precipitation. The future projections are relative to the baseline period of 1976-2005. In this study, the 50th percentile and the RCP 8.5 emission scenarios were selected to set the tool for generating future weather datasets of Rome for the period of 2041–2060. The IWEC-TMY was the baseline for this procedure.

3.1.4. TMY out of GERICS-REMO-2015

As mentioned earlier, the GERICS-REMO-2015 regional climate model was used in this study to apply the dynamical downscaling method. The data for this model were downloaded from the EURO-CORDEX entry point through the Earth System Grid Federation (ESGF) for the Europe domain on a $0.11°$ grid in rotative coordinates (equivalent to a 12.5 km grid). These data are available in the NetCDF4 format, which is a file format for storing multidimensional scientific data. The extraction of the data for our case study (city of Rome) was performed through the Cordex Data Extractor software [54] that allows finding the closest data point on the grid to the desired latitude and longitude. The RCP 8.5 scenario was adapted to extract these data for the 2041–2060 period. The driving model considered in this study is MPI-M-MPI-ESM-LR, which is well supported according to the IPCC report on the evaluation of climate models [55].

In order to create a future typical meteorological year (F-TMY) for the city of Rome out of the 20 years of extracted data, the methodology of standard EN ISO 15927-4 [56] was used. This international standard covers the selection of appropriate meteorological data for the assessment of the long-term mean energy use for heating and cooling. TMY is constructed from 12 representative months (Best Months) from multi-year records. The selection of Best Months is done by comparing the cumulative distribution function of the single and reference years through the Finkelstein–Schafer (FS) statistics [24]. This method was used in this study since the criteria for selecting the Best Month is not merely limited to dry-bulb air temperature; it also takes the global solar irradiance, relative humidity, and wind speed into account.

3.2. Energy Performance and Thermal Comfort Assessment

The building energy performance was assessed by means of a detailed dynamic simulation model using EnergyPlus (version 9.0) with an hourly time-step. The results are discussed in terms of annual thermal energy need for space heating and space cooling ($EP_{H/C,nd}$) and electrical energy demand per unit of area (E_{el}/A_f). The latest indicator (E_{el}/A_f) was calculated according to Equation (4):

$$E_{el}/A_f = \frac{EP_{H,nd}}{\eta_{H,u} \cdot \eta_{H,g}} + \frac{EP_{C,nd}}{\eta_{C,u} \cdot \eta_{C,g}} \tag{4}$$

where $EP_{H/C,nd}$ is the annual thermal energy need for space heating/cooling, $\eta_{H/C,u}$ is the mean seasonal efficiency of the heating/cooling utilization (including emission, control, and distribution) subsystems, and $\eta_{H/C,g}$ is the mean seasonal efficiency of the heating/cooling generation subsystem.

The reference mean seasonal efficiency values of the utilization subsystems were assumed in compliance with the Italian Interministerial Decree of June 26th, 2015 [57]. As a reversible heat pump has been selected as a generation subsystem type to carry out the analysis, a future value of the mean seasonal generation efficiency was adopted to take into account the increase of the ambient temperature due to climate change. The generation efficiency was calculated assuming proportionality between the coefficient of performance (COP) and its maximum theoretical efficiency over different temperatures, as in Equation (5):

$$\eta_{H,g} \sim \frac{\sum_{\text{heating season}} \Phi_H}{\sum_{\text{heating season}} \left(\Phi_H \cdot \frac{T_{cond,out}}{T_{cond,out} - T_{evap,in}} \right)} \tag{5}$$

where Φ_H is the thermal energy load for heating, $T_{cond,out}$ is the condenser outlet temperature (hot water), and $T_{evap,in}$ is the evaporator inlet air temperature.

In the same way, proportionality between the energy-efficiency ratio (EER) and its maximum theoretical efficiency over different temperatures has been assumed, as in Equation (6):

$$\eta_{C,g} \sim \frac{\sum_{\text{cooling season}} \Phi_C}{\sum_{\text{cooling season}} \left(\Phi_C \cdot \frac{T_{evap,out}}{T_{cond,in} - T_{evap,out}} \right)} \tag{6}$$

where Φ_C is the thermal energy load for cooling, $T_{evap,out}$ is the evaporator outlet temperature (chilled water), and $T_{cond,in}$ is the condenser inlet air temperature.

The thermal comfort was assessed in accordance with the EN 16798-1 standard [58]. The adaptive comfort model was adopted to predict how the pattern of outside weather conditions affect the indoor thermal sensation of the user in free-floating condition. In this model, the optimal operative temperature ($\theta_{o,c}$, in °C) is calculated as in Equation (7):

$$\theta_{o,c} = 0.33 \cdot \theta_{r,m} + 18.8 \tag{7}$$

where $\theta_{r,m}$ is the outdoor running mean temperature, which is defined as an exponential running mean of the outdoor air temperature.

In this research, a medium level of occupant expectation (i.e., second category of indoor environmental quality, as defined in [58]) was applied, in which the range of comfort is between $\theta_{o,c} + 3$ °C (highest limit) and $\theta_{o,c} - 4$ °C (lowest limit). In addition, the hours of exceedance (*HE*) were calculated as an indicator to quantify indoor overheating. The *HE* indicator is equal to the number of hours during the cooling period in which the operative temperature of the zone is greater than the upper limit temperature of the thermal comfort range.

3.3. Definition of Case Studies

Representative buildings of the Italian residential building stock were used to carry out the comparison between the future weather datasets. The buildings have been selected from the Italian *Building Typology Matrix* developed in the Intelligent Energy Europe-*Typology Approach for Building Stock Energy Assessment* (IEE-TABULA) research project [59], which aimed at creating a harmonized definition of the residential building typology at the European level. The two selected building types (Figure 1) belong to the categories of single-family house (SFH) and apartment block (AB), respectively, and both were built in the construction period 1946–1960. The two building sizes present a significantly different shape factor—0.73 m^{-1} for the SFH and 0.46 m^{-1} for the AB—and window-to-wall ratio: 0.09 for the SFH and 0.23 for the AB.

(a)   (b)

Figure 1. Italian residential building-types of the construction period 1946–1960, according to the IEE-TABULA project: single-family house (**a**) and apartment block (**b**).

The buildings have uninsulated envelope components, as the construction period predates the first Italian law on energy savings issued in 1976. The opaque external wall is a solid brick masonry ($U = 1.48$ W·m^{-2} K^{-1} the SFH, and $U = 1.15$ W·m^{-2} K^{-1} the AB), while the horizontal envelope components are reinforced brick–concrete slabs ($U = 1.65$ W·m^{-2} K^{-1}). The transparent envelope components are single glazing ($g_{gl,n} = 0.85$) and wood-frame windows ($U = 4.90$ W·m^{-2} K^{-1}) with exterior wooden Venetian blinds ($g_{gl + sh} = 0.35$).

A reference technical building system was assumed for the case studies in the present work. It consists in an electrical reversible air-to-water heat pump coupled with fan coils.

The energy performance of the buildings was assessed assuming a standard user behavior for the quantification of the internal heat gains and the airflow rates by natural ventilation [60]. A continuous operation of the heating and cooling systems was set, considering 20 °C and 26 °C temperature set points, respectively. The heating season covers the period between October 15th and April 15th as fixed by the Italian energy regulations. The availability of the cooling system was assumed in the remaining part of the year (from April 16th to October 14th).

4. Results and Discussion

The aim of this research is to analyze different types of future weather datasets by comparing their relative impact on building energy performance predictions. In the first set of results, boxplots of the outdoor dry-bulb temperature (a) and the global horizontal solar irradiance during daily hours (b), which are the weather key variables in building energy simulation, are plotted (Figure 2). Boxplots show a pattern of increase in both variables due to climate change. All future weather files show almost similar mean values higher than the present weather file. Nevertheless, F-TMY—which is derived from a dynamical downscaling method—shows lower dispersion compared to other future files (statistically downscaling methods).

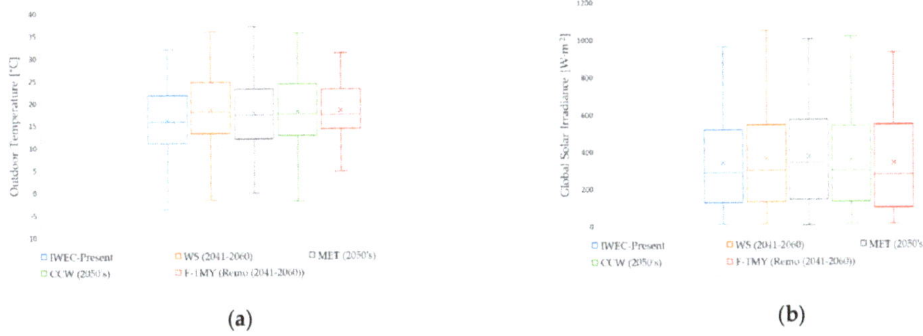

(a)   (b)

Figure 2. Boxplots of the outdoor dry-bulb temperature (**a**) and the global horizontal solar irradiance during daily hours (**b**) for IWEC (Present), Weathershift (WS), Meteonorm (MET), CCWorldWeatehr-Gen (CCW), and F-TMY. All future weather files are for 2050s considering Representative Concentration Pathway (RCP) 8.5.

Below, Figure 3 presents net thermal energy needs for heating and cooling normalized by the conditioned floor area for a mono family house (a) and apartment block (b) for present and different future weather data to assess the building energy performance of the case studies. The heating energy demand for the mono family house (MFH) dominates over cooling demand. In addition, MFH has also higher energy demand for heating compared to the apartment block (AB). This is due to the higher shape factor (S/V) ratio, which entails that heat transfer by transmission is the most relevant term of the energy balance, and outdoor temperature is the main driving force. Consequently, the decrease of $EP_{H,nd}$ in the future prevails over the increase of $EP_{C,nd}$. On the opposite, the AB shows closer values of $EP_{H,nd}$ and $EP_{C,nd}$. It appears that the heating need is slightly dominant in the present, but it will be overtaken by cooling in the future. For all future weather data except F-TMY, the relative change of $EP_{H,nd}$ is in the range of 30% to 34% for both buildings, while the relative change of $EP_{C,nd}$ is above 160% for MFH and above 100% for AB. This unevenness in relative variation is mainly related to the different magnitude of the present energy need. As regards F-TMY, lower values of $EP_{H,nd}$ and $EP_{C,nd}$ are shown compared to the other future weather data, meaning that $EP_{H,nd}$ will decrease more and $EP_{C,nd}$ will increase less. This trend is strictly dependent on the lower dispersion of temperature values for F-TMY compared to the other future weather datasets. Comparing the four sets of weather data, Weathershift (WS), Meteonorm (MET), and CCWorldWeatehrGen (CCW) show almost similar variations in $EP_{H,nd}$ and $EP_{C,nd}$, while the F-TMY presents a significantly different variations in the two indicated parameters. This comes from the fact that WS, MET, and CCW are all statistically downscaled weather datasets, and F-TMY is a dynamically downscaled weather dataset.

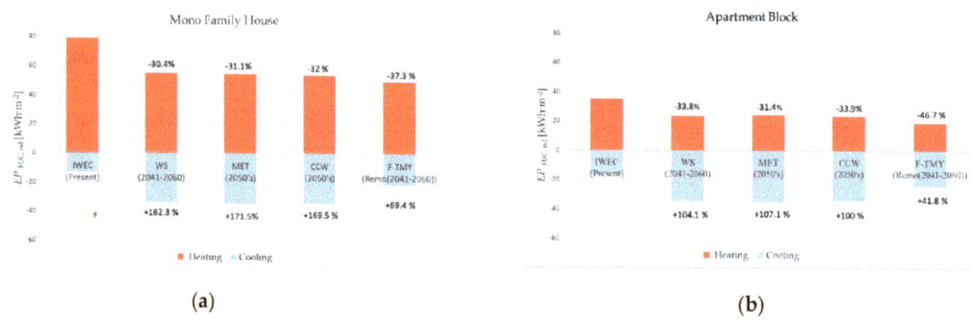

(a)

(b)

Figure 3. Net thermal energy needs for heating and cooling normalized by the conditioned floor area for a mono family house (**a**) and apartment block (**b**) for IWEC (Present), Weathershift (WS), Meteonorm (MET), CCWorldWeatehr-Gen (CCW), and F-TMY. All future weather files are for 2050s considering RCP 8.5.

In order to better present this trend, the box plots of thermal energy load for heating in the month of January and for cooling in the month of August are shown in Figures 4 and 5. The figures indicate that both for MFH and AB, the mean values for the month of January are almost the same for all future weather files, while the deviation of F-TMY is lower than the three other files. On the other hand, for the month of August, the mean values of WS, MET, and CCW are significantly higher than F-TMY. The reason lies in the fact that the dynamical downscaled weather data in comparison with the statistical ones better represents the temporal variability of climate, which leads to a more consistent dataset. As another outcome of the inconsistency in the statistical downscaled weather files, both figures demonstrate the overestimation of the data in the thermal energy load for cooling in August.

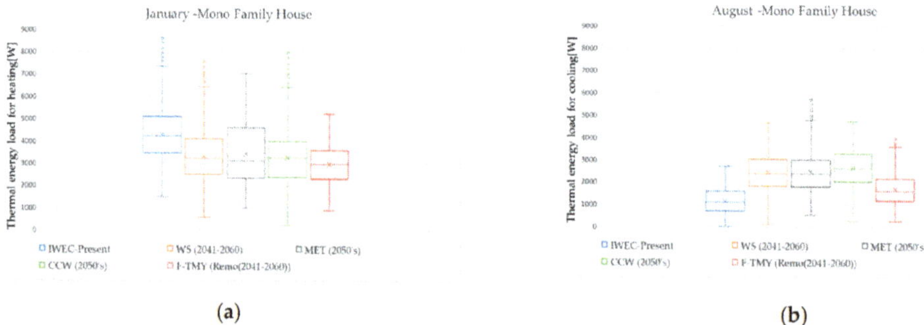

(a)

(b)

Figure 4. Boxplots of heating loads in January (**a**) and cooling loads in August (**b**) of the mono family house for IWEC (Present), Weathershift (WS), Meteonorm (MET), CCWorldWeatehr-Gen (CCW), and F-TMY. All future weather files are for 2050s considering RCP 8.5.

The adaptive comfort analysis in free floating condition, of MFH and AB for IWEC (Present), Weathershift (WS), Meteonorm (MET), CCWorldWeatehrGen (CCW), and F-TMY is presented in Figure 6. The graphs show the distribution of hours of the cooling period (April 16th until October 14th: 4368 hours) in three ranges: comfort, warm discomfort, and cold discomfort. The discrepancy between F-TMY and other future weather data is pointed out. The percentage of warm discomfort hours for the WS, MET, and CCW is almost the same and equals around 40% for MFH and 90% for AB. For the F-TMY, the percentage of warm discomfort hours is less for both cases (29% for MFH and 72% for AB). This discrepancy can be found in Figure 7 where boxplots of last floor operative temperature of MFH (a) and AB (b) in August for present and different future weather data are presented.

In this case, despite having similar dispersions, the mean values of last floor operative temperature of F-TMY is significantly lower than the mean value of the other three future weather datasets for both MFH and AB. This is strongly dependent on the lower dispersion of temperature values for F-TMY. If we now turn to the comparison of the two building types, occupants in AB will experience overheating much more often than the occupants in MFH because of a reduced potentiality of exploiting the heat transfer in AB through the envelope for ejecting heat produced by internal and solar sources. This is due to the lower S/V value and larger window-to-wall ratio (WWR) of the AB compared to MFH. In addition, hours of exceedance (HE) for all the weather datasets for MFH and AB are presented in Table 1. It is observed that the absolute change of the increase in the HE for statistically downscaled future weather datasets is almost the same, while for dynamically downscaled future weather, data are significantly lower.

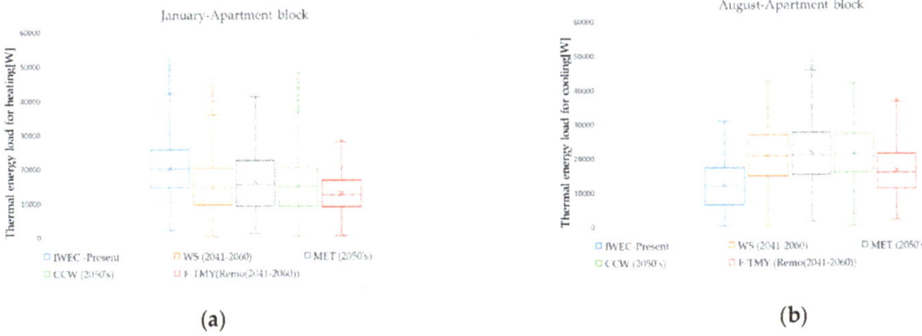

(a)   (b)

Figure 5. Boxplots of heating loads in January (**a**) and cooling load in August (**b**) of the apartment block for IWEC (Present), Weathershift (WS), Meteonorm (MET), CCWorldWeatehr-Gen (CCW), and F-TMY. All future weather files are for 2050s considering RCP 8.5.

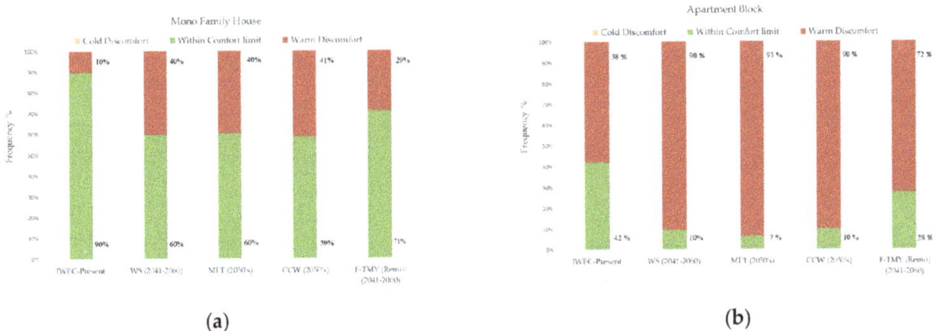

(a)   (b)

Figure 6. Adaptive comfort analysis for a mono family house (**a**) and apartment block (**b**) for IWEC (Present), Weathershift (WS), Meteonorm (MET), CCWorldWeatehr-Gen (CCW), and F-TMY. All future weather files are for 2050s considering RCP 8.5.

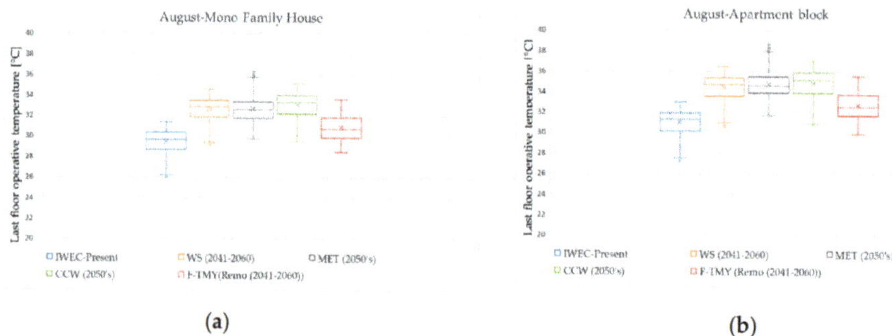

(a)

(b)

Figure 7. Boxplot of last floor operative temperature of a mono family house (**a**) and apartment block (**b**) in August, for IWEC (Present), Weathershift (WS), Meteonorm (MET), CCWorldWeatehr-Gen (CCW), and F-TMY. All future weather files are for 2050s considering RCP 8.5.

Table 1. Electrical energy demand per unit of area and hours of exceedance for a mono family house (MFH) and apartment block (AB), for IWEC (Present), Weathershift (WS), Meteonorm (MET), CCWorldWeatehr-Gen (CCW), and F-TMY. All future weather files are for 2050s considering RCP 8.5.

		IWEC	WS	Absolute Change	MET	Absolute Change	CCW	Absolute Change	TMY-R	Absolute change
SFH	E_{el}/A_f [kWh m^{-2}]	38.7	40.7	*2*	41.5	*2.8*	40.5	*1.8*	29.8	*−8.9*
	HE[h]	222	887	*665*	877	*655*	910	*688*	638	*416*
AB	E_{el}/A_f [kWh m^{-2}]	22.9	29	*6.1*	29.5	*6.6*	28.1	*5.2*	19.4	*−3.5*
	HE[h]	1273	1995	*722*	2060	*787*	1984	*711*	1596	*323*

In addition, Table 1 also summarizes the values of the electrical energy demand per unit of area (E_{el}/A_f). The E_{el}/A_f increases in MFH and AB similarly for WS, MET, and CCW, while the absolute change is not significantly high. On the other hand, E_{el}/A_f slightly decreases for F-TMY in both cases. The reason can be explained below: as mentioned before, a future value for the mean seasonal efficiency of the heating ($\eta_{H,g}$) and the cooling ($\eta_{C,g}$) generation subsystem was adopted to consider the increase of ambient temperature due climate change. The mean seasonal efficiency increases for the heating and decreases for the cooling for all the future weather datasets. However, due to the lower discrepancy of the temperature values for F-TMY compared to other future weather datasets, the increase for $\eta_{H,g}$ in the dynamical downscaled model is more, while the decreases in the $\eta_{C,g}$ is less. Consequently, according to Equation (4), the reduction in the energy for winter conditioning outweighs the cooling demand in the case of F-TMY. Finally, if the variation of E_{el}/A_f for MFH and AB are compared, the absolute changes are lower for MFH, which comes from its higher S/V value that skews the energy usage of it more toward the heating regime.

5. Conclusions

Statistical and dynamical are two main approaches to downscale global climate models for creating weather datasets to be used in building energy simulation. Considering there are different methodologies that use these approaches, evaluating their suitability and robustness is vital. This study set out to compare WeatherShift, Meteonorm, and CCWorldWeatherGen—which are common weather generator tools applying statistical downscaling—in addition to a TMY created using a high-quality regional climate models database (from Euro-CORDEX) that applies the dynamical downscaling. All future weather files are for the 2050s considering RCP 8.5. Two representative buildings of the Italian residential building stock, including a mono family house (MFH) and an apartment block (AB), were selected to perform the analysis.

The results of this investigation show that different statistical downscaled future weather datasets created by weather generators predict the future energy performance and comfort analysis of the buildings quite similarly, compared to the dynamical one. This is demonstrated by almost the same values in the mean outdoor dry-bulb temperature, relative changes of thermal energy need for heating and cooling normalized by the conditioned floor area, mean value of thermal energy load for heating and cooling, the hours of discomfort, and the absolute changes in the electrical energy demand per unit of area. However, when it comes to the dynamical downscaled weather data, the above-mentioned parameters follow a different pattern. As an example, while the discomfort hours percentage for the WS, MET, and CCW equals around 40% for MFH and 90% for AB, for the F-TMY, this percentage is 29% for MFH and 72% for AB. Consequently, it was verified that dynamical downscaling, by better representing the spatial and temporal variability of local climate, provides physically consistent datasets.

The other significant result of this study is reached by comparing different building types. In more detail, the observed discrepancy between the future predictions of statistical and dynamical downscaling is affected not only by using different approaches for creating future weather datasets but also by building type. As an example, the thermal energy need for cooling in MFH for statistical downscaled datasets increases around 170%, and for the dynamical one, it increases around 70%. On the other hand, in AB, this parameter increases 100% for statistical downscaled data and around 40% for the dynamical one. This inequality in relative variation comes from the different magnitude of the present energy need for different building types. For buildings with a higher shape factor (MFH), the heating energy demand dominates the cooling energy demand, which also make them more sensitive to the climate change.

Overall, this study has provided a deeper insight into analyzing the effect of climate change on the future energy performance of buildings by considering different future weather datasets and building types. Firstly, it was shown that the climate change impact magnitude is not equal for different case studies, so that in a changing climate, performing a regional and localized analysis becomes vital. In addition, the results demonstrated that morphing method—regardless of its way of application—can provide adequate information to perform comparative analysis on long-term changes in energy building performance. However, the existing inconsistency within this method may lead to high prediction errors. In this case, the dynamical downscaling method is found to be more reliable when the aim is to develop, assess, and communicate resilient solutions to withstand as well as prevent the future impacts of climate change on building energy performance. Further studies are suggested to be carried out to consider model uncertainties of RCMs by following an ensemble-based approach. In addition, it is important to bear in mind that RCMs have been run not only for future but also for historical period. So, they can be compared with the real data, and the biases associated with the climate model data can be adjusted to reduce uncertainties and increase their physical consistency. This possibility does not exist for statistical downscaling method tools, as they are based on transforming the actual real data; it is possible to say they are often "black-box" tools.

Author Contributions: Conceptualization, M.P. and V.C.; methodology, I.B., M.P, M.Z., and V.C.; software, M.P.; formal analysis, M.P.; investigation, M.P. and M.Z.; resources, M.P. and M.Z.; data curation, I.B. and M.P.; writing—original draft preparation, M.P.; writing—review and editing, I.B., M.Z., and V.C.; visualization, M.P.; supervision, V.C.; project administration, M.Z. and V.C. All authors have read and agreed to the published version of the manuscript.

Funding: This research received no external funding.

Data Availability Statement: Publicly available datasets were analyzed in this study. These data can be found here: [https://energyplus.net/weather-location/europe_wmo_region_6/ITA//ITA_Rome.162420_IWEC accessed on 2 December 2020] and [https://esgf-node.llnl.gov/search/esgf-llnl/ accessed on 2 December 2020].

Acknowledgments: The authors would like to thank Giovanni Murano for providing excel sheets that apply the methodology of standard EN ISO 15927-4 for generating typical meteorological year (TMY).

Conflicts of Interest: The authors declare no conflict of interest.

References

1. World Meteorological Organization. *WMO Report on The Global Climate in 2015-2019*; World Meteorological Organization: Geneva, Switzerland, 2019.
2. Symon, C. Climate change: Action, trends and implications for business. In *The IPCC's Fifth Assessment Report, Working Group 1*; IPCC: Geneva, Switzerland, 2013.
3. Della-Marta, P.M.; Haylock, M.R.; Luterbacher, J.; Wanner, H. Doubled length of western European summer heat waves since 1880. *J. Geophys. Res. Space Phys.* **2007**, *112*. [CrossRef]
4. National Climate Services Network of Italy, ISPRA. *Future Climate in Italy—An Analysis of Regional Climate Models Projections*; ISPRA: Roma, Italy, 2015; ISBN 978-88-448-0723-8.
5. Muthers, S.; Laschewski, G.; Matzarakis, A. The Summers 2003 and 2015 in South-West Germany: Heat Waves and Heat-Related Mortality in the Context of Climate Change. *Atmosphere* **2017**, *8*, 224. [CrossRef]
6. D'Ippoliti, D.; Michelozzi, P.; Marino, C.; De'Donato, F.; Menne, B.; Katsouyanni, K.; Kirchmayer, U.; Analitis, A.; Medina-Ramón, M.; Paldy, A.; et al. The impact of heat waves on mortality in 9 European cities: Results from the EuroHEAT project. *Environ. Health* **2010**, *9*, 37. [CrossRef] [PubMed]
7. Tan, J.; Zheng, Y.; Tang, X.; Guo, C.; Li, L.; Song, G.; Zhen, X.; Yuan, D.; Kalkstein, A.J.; Li, F.; et al. The urban heat island and its impact on heat waves and human health in Shanghai. *Int. J. Biometeorol.* **2009**, *54*, 75–84. [CrossRef] [PubMed]
8. Salimi, M.; Al-Ghamdi, S.G. Climate change impacts on critical urban infrastructure and urban resiliency strategies for the Middle East. *Sustain. Cities Soc.* **2020**, *54*, 101948. [CrossRef]
9. McEvoy, D.; Ahmed, I.; Mullett, J. The impact of the 2009 heat wave on Melbourne's critical infrastructure. *Local Environ.* **2012**, *17*, 783–796. [CrossRef]
10. Xia, Y.; Li, Y.; Guan, D.; Tinoco, D.M.; Xia, J.; Yan, Z.; Yang, J.; Liu, Q.; Huo, H. Assessment of the economic impacts of heat waves: A case study of Nanjing, China. *J. Clean. Prod.* **2018**, *171*, 811–819. [CrossRef]
11. Herbel, I.; Croitoru, A.-E.; Rus, A.V.; Roşca, C.F.; Harpa, G.V.; Ciupertea, A.-F.; Rus, I. The impact of heat waves on surface urban heat island and local economy in Cluj-Napoca city, Romania. *Theor. Appl. Clim.* **2018**, *133*, 681–695. [CrossRef]
12. Nakicenovic, N.; Alcamo, J.; Davis, G.; Vries, B.D.; Fenhann, J.; Gaffin, S.; Gregory, K.; Grubler, A.; Jung, T.Y.; Kram, T.; et al. *Special Report on Emissions Scenarios*; IPCC: Geneva, Switzerland, 2000.
13. Dimoudi, A.; Tompa, C. Energy and environmental indicators related to construction of office buildings. *Resour. Conserv. Recycl.* **2008**, *53*, 86–95. [CrossRef]
14. Shen, P. Impacts of climate change on U.S. building energy use by using downscaled hourly future weather data. *Energy Build.* **2017**, *134*, 61–70. [CrossRef]
15. Berardi, U.; Jafarpur, P. Assessing the impact of climate change on building heating and cooling energy demand in Canada. *Renew. Sustain. Energy Rev.* **2020**, *121*, 109681. [CrossRef]
16. Flores-Larsen, S.; Filippín, C.; Barea, G. Impact of climate change on energy use and bioclimatic design of residential buildings in the 21st century in Argentina. *Energy Build.* **2019**, *184*, 216–229. [CrossRef]
17. Soutullo, S.; Giancola, E.; Jiménez, M.J.; Ferrer, J.A.; Sánchez, M.N. How Climate Trends Impact on the Thermal Performance of a Typical Residential Building in Madrid. *Energies* **2020**, *13*, 237. [CrossRef]
18. Da Guarda, E.L.A.; Domingos, R.M.A.; Jorge, S.H.M.; Durante, L.C.; Sanches, J.C.M.; Leão, M.; Callejas, I.J.A. The influence of climate change on renewable energy systems designed to achieve zero energy buildings in the present: A case study in the Brazilian Savannah. *Sustain. Cities Soc.* **2020**, *52*, 101843. [CrossRef]
19. Zhai, Z.J.; Helman, J.M. Implications of climate changes to building energy and design. *Sustain. Cities Soc.* **2019**, *44*, 511–519. [CrossRef]
20. Chai, J.; Huang, P.; Sun, Y. Investigations of climate change impacts on net-zero energy building lifecycle performance in typical Chinese climate regions. *Energy* **2019**, *185*, 176–189. [CrossRef]
21. Herrera, M.; Natarajan, S.; Coley, D.A.; Kershaw, T.; Ramallo-González, A.P.; Eames, M.; Fosas, D.; Wood, M. A review of current and future weather data for building simulation. *Build. Serv. Eng. Res. Technol.* **2017**, *38*, 602–627. [CrossRef]
22. Hall, I.J.; Prairie, R.R.; Anderson, H.E.; Boes, E.C. *Generation of a Typical Meteorological Year (No. SAND-78-1096C; CONF-780639-1)*; Sandia Labs: Albuquerque, NM, USA, 1978.
23. Barnaby, C.S.; Crawley, D.B. Weather data for building performance simulation. In *Building Performance Simulation for Design and Operation*; Routledge: London, UK, 2011.
24. Finkelstein, J.M.; Schafer, R.E. Improved goodness-of-fit tests. *Biometrika* **1971**, *58*, 641–645. [CrossRef]

25. Jentsch, M.F.; James, P.A.; Bourikas, L.; Bahaj, A.S. Transforming existing weather data for worldwide locations to enable energy and building performance simulation under future climates. *Renew. Energy* **2013**, *55*, 514–524. [CrossRef]

26. Dias, J.B.; Da Graça, G.C.; Soares, P.M. Comparison of methodologies for generation of future weather data for building thermal energy simulation. *Energy Build.* **2020**, *206*, 109556. [CrossRef]

27. Moazami, A.; Nik, V.M.; Carlucci, S.; Geving, S. Impacts of future weather data typology on building energy performance—Investigating long-term patterns of climate change and extreme weather conditions. *Appl. Energy* **2019**, *238*, 696–720. [CrossRef]

28. U.S. Department of Energy's (DOE). Energy Plus Software, v. 9.0. Available online: https://energyplus.net/ (accessed on 2 September 2020).

29. Köppen, W. Die Wärmezonen der Erde, nach der Dauer der heissen, gemässigten und kalten Zeit und nach der Wirkung der Wärme auf die organische Welt betrachtet. *Meteoro. Z.* **1884**, *1*, 5–226.

30. Hawkins, E.; Sutton, R.T. The Potential to Narrow Uncertainty in Regional Climate Predictions. *Bull. Am. Meteorol. Soc.* **2009**, *90*, 1095–1108. [CrossRef]

31. Hawkins, E.; Sutton, R. The potential to narrow uncertainty in projections of regional precipitation change. *Clim. Dyn.* **2011**, *37*, 407–418. [CrossRef]

32. Ramon, D.; Allacker, K.; Van Lipzig, N.P.M.; De Troyer, F.; Wouters, H. Future Weather Data for Dynamic Building Energy Simulations: Overview of Available Data and Presentation of Newly Derived Data for Belgium. *Energy Environ. Sustain.* **2018**, 111–138. [CrossRef]

33. Uppala, S.M.; Kållberg, P.W.; Simmons, A.J.; Andrae, U.; Bechtold, V.D.C.; Fiorino, M.; Gibson, J.K.; Haseler, J.; Hernandez, A.; Kelly, G.A.; et al. The ERA-40 re-analysis. *Q. J. R. Meteorol. Soc.* **2005**, *131*, 2961–3012. [CrossRef]

34. Laflamme, E.M.; Linder, E.; Pan, Y. Statistical downscaling of regional climate model output to achieve projections of pre-cipitation extremes. *Weather. Clim. Extrem.* **2016**, *12*, 15–23. [CrossRef]

35. Belcher, S.; Hacker, J.; Powell, D. Constructing design weather data for future climates. *Build. Serv. Eng. Res. Technol.* **2005**, *26*, 49–61. [CrossRef]

36. American Meteorological Society. Regional climate model. In *Glossary of Meteorology*; American Meteorological Society: Boston, MA, USA, 2013.

37. Soares, P.M.; Cardoso, R.M.; Miranda, P.M.; de Medeiros, J.; Belo-Pereira, M.; Espirito-Santo, F. WRF high resolution dynam-ical downscaling of ERA-Interim for Portugal. *Clim. Dyn.* **2012**, *39*, 2497–2522. [CrossRef]

38. Van der Linden, P.; Mitchell, J.E. *ENSEMBLES: Climate Change and its Impacts: Summary of Research and Results from the ENSEMBLES Project*; Met Office Hadley Centre: Exeter, UK, 2009.

39. Jacob, D.; Petersen, J.; Eggert, B.; Alias, A.; Christensen, O.B.; Bouwer, L.M.; Braun, A.; Colette, A.; Déqué, M.; Georgievski, G.; et al. EURO-CORDEX: New high-resolution climate change projections for European impact research. *Reg. Environ. Chang.* **2014**, *14*, 563–578. [CrossRef]

40. Giorgi, F. Thirty Years of Regional Climate Modeling: Where Are We and Where Are We Going next? *J. Geophys. Res. Atmos.* **2019**, *124*, 5696–5723. [CrossRef]

41. World Research Climate Program (WRCP) Coordinated Downscaling Experiment—European Domain. Available online: https://www.eurocordex.net/ (accessed on 2 December 2020).

42. EUROCORDEX: Cordex Archive Specifications. Available online: https://is-enes-data.github.io/cordex_archive_specifications.pdf (accessed on 2 December 2020).

43. Jacob, D.; Podzun, R. Sensitivity studies with the regional climate model REMO. *Theor. Appl. Clim.* **1997**, *63*, 119–129. [CrossRef]

44. Jacob, D. A note to the simulation of the annual and inter-annual variability of the water budget over the Baltic Sea drain-age basin. *Meteorol. Atmos. Phys.* **2001**, *77*, 61–73. [CrossRef]

45. Remund, J.; Kunz, S. *METEONORM: Global Meteorological Database for Solar Energy and Applied Climatology*; Meteotest: Bern, Switzerland, 1997.

46. Climate Change 2007: Synthesis Report. Contribution of Working Groups I, II and III to the Fourth Assessment Report of the Intergovernmental Panel on Climate Change. In *Intergovernmental Panel on Climate Change*; Core Writing Team, Pachauri, R.K., Resinger, A., Eds.; IPCC: Geneva, Switzerland, 2007.

47. Remund, J.; Müller, S.C.; Schilter, C.; Rihm, B. The use of Meteonorm weather generator for climate change studies. In Proceedings of the 10th EMS Annual Meeting 2010, Zürich, Switzerland, 13–17 September 2010. EMS2010-417.

48. Met Office HadCM3: Met Office Climate Prediction Model. Available online: https://www.metoffice.gov.uk/research/approach/modelling-systems/unified-model/climate-models/hadcm3 (accessed on 2 December 2020).

49. Jentsch, M.F. Technical Reference Manual for the CCWeatherGen and CCWorldWeatherGen Tools Version 1.2. 2012. Available online: http://blog.soton.ac.uk/serg/files/2013/06/manual_weather_tool1.pdf (accessed on 2 December 2020).

50. Jentsch, M.F.; Bahaj, A.S.; James, P.A. Climate change future proofing of buildings—Generation and assessment of building simulation weather files. *Energy Build.* **2008**, *40*, 2148–2168. [CrossRef]

51. WeatherShift. Available online: http://www.weather-shift.com (accessed on 2 December 2020).

52. Pachauri, R.K.; Reisinger, A. *Climate Change 2007. Synthesis Report. Contribution of Working Groups I, II and III to the fourth Assessment Report*; IPCC: Geneva, Switzerland, 2008.

53. Troup, L.; Fannon, D. Morphing Climate Data to Simulate Building Energy Consumption. In Proceedings of the ASHRAE and IBPSA-USA SimBuild 2016: Building Performance Modeling Conference, Salt Lake City, UT, USA, 12–10 August 2016; Volume 6.

54. CORDEX Data Extractor. Available online: https://agrimetsoft.com/CordexDataExtractor (accessed on 2 December 2020).
55. Flato, G.; Marotzke, J.; Abiodun, B.; Braconnot, P.; Chou, S.C.; Collins, W.; Cox, P.; Driouech, F.; Emori, S.; Eyring, V.; et al. Evaluation of climate models. In *Climate Change 2013: The Physical Science Basis. Contribution of Working Group I to the Fifth Assessment Report of the Intergovernmental Panel on Climate Change*; Cambridge University Press: Cambridge, UK, 2014; pp. 741–866.
56. European Committee for Standardization. *EN ISO 15927-4. Hygrothermal Performance of Buildings Calculation and Presentation of Climatic Data, Part 4: Hourly Data for Assessing the Annual Energy Use for Heating and Cooling*; European Committee for Standardization: Brussels, Belgium, 2005.
57. O. J. of the Italian Republic. *Italian Republic, Interministerial Decree of June 26th, 2015—Calculation Methodologies of the Building Energy Performance and Minimum Energy Performance Requirements (in Italian)*; O. J. of the Italian Republic: Rome, Italy, 2015.
58. European Committee for Standardization. *EN ISO 16798-1. Energy Performance of Buildings—Ventilation for Buildings—Part 1: Indoor Environmental Input Parameters for Design and Assessment of Energy Performance of Buildings Addressing Indoor Air Quality, Thermal Environment, Lighting, and Acoustics—Module M1-6*; European Committee for Standardization: Brussels, Belgium, 2019.
59. Ballarini, I.; Corgnati, S.P.; Corrado, V. Use of reference buildings to assess the energy saving potentials of the residential building stock: The experience of TABULA project. *Energy Policy* **2014**, *68*, 273–284. [CrossRef]
60. Comitato Termotecnico Italiano. *Technical Commission 241, doc. no. 181, Italian National Annex of the EN 16798-1 Technical Standard (Working Draft for Internal Use)*; Comitato Termotecnico Italiano: Milan, Italy, 2020.

Article

Assessment of the Urban Heat Island Impact on Building Energy Performance at District Level with the EUReCA Platform

Pierdonato Romano [1], Enrico Prataviera [1,*], Laura Carnieletto [1], Jacopo Vivian [1], Michele Zinzi [2] and Angelo Zarrella [1]

[1] Department of Industrial Engineering—Applied Physics Section, University of Padova, Via Venezia 1, 35131 Padova, Italy; pierdonato.romano@unipd.it (P.R.); laura.carnieletto@unipd.it (L.C.); jacopo.vivian@unipd.it (J.V.); angelo.zarrella@unipd.it (A.Z.)

[2] Smart Energy Division, Energy Technologies Department, ENEA—Italian National Agency for New Technologies, Energy and Sustainable Economic Development, Via Anguillarese 301, 00123 Rome, Italy; michele.zinzi@enea.it

* Correspondence: enrico.prataviera@phd.unipd.it; Tel.: +39-049-827-6895

Abstract: In recent decades, the cooling energy demand in urban areas is increasing ever faster due to the global warming and the growth of developing economies. In this perspective, the urban building energy modelling community is focusing its research activities on innovative tools and policy actions to improve cities' sustainability. This work aims to present a novel module of the EUReCA (Energy Urban Resistance Capacitance Approach) platform for evaluating the effects of the interaction between district's buildings in the cooling season. EUReCA predicts the urban energy demand using a bottom-up approach and low computational resources. The new module allows us to evaluate the mutual shading between buildings and the urban heat island effects, and it is well integrated with the calculation of the energy demand of buildings. The analysis was carried out considering a real case study in Padua (Italy). Results show that the urban heat island causes an average increase of 2.2 °C in the external air temperature mainly caused by the waste heat rejected from cooling systems. This involves an increase in urban cooling energy and electricity demand, which can be affected between 6 and 8%. The latter is the most affected by the urban heat island (UHI), due to the degradation it causes on the HVAC systems' efficiency.

Keywords: urban modelling; urban heat island; cities; buildings; energy efficiency; decarbonization

Citation: Romano, P.; Prataviera, E.; Carnieletto, L.; Vivian, J.; Zinzi, M.; Zarrella, A. Assessment of the Urban Heat Island Impact on Building Energy Performance at District Level with the EUReCA Platform. *Climate* **2021**, *9*, 48. https://doi.org/10.3390/cli9030048

Received: 19 February 2021
Accepted: 12 March 2021
Published: 16 March 2021

Publisher's Note: MDPI stays neutral with regard to jurisdictional claims in published maps and institutional affiliations.

1. Introduction

The increase in greenhouse gases concentration and average temperature during the past decades is a well-documented phenomenon; it is bound to intensify in the coming decades, with a projected temperature rise between 2.6 and 4.8 °C by the end of the century, with respect to current values [1]. In addition, latest demographical international reports highlight the exponential and fast growth of urban areas and cities [2]. Population moving or living in large cities is constantly increasing, with some projection that states that the global urban population will increase by 2.5 billion people before 2050 [3], with a consequent increase in the associated energy uses.

Urban building energy modelling (UBEM) is trying to tackle this issue focusing on buildings, which account for a significant fraction of urban energy usage and carbon emissions [4]. Research activity in this field grew up intensively in the last 10 years [5], and it is continuously developing novel techniques to: analyze and cluster urban data, simulate and predict buildings' thermal performance and consumptions, model urban climate and user's comfort, and propose new and cutting-edge actions and policies [6]. This research activity started from the pioneer building energy simulation (BES) and developed its methods to the urban scale, focusing on specific challenges that characterize the urban scenarios. Recent literature clarifies the state-of-the-art on several aspects regarding how districts and cities must be modeled in efficient and efficacy way [7]. Several works [8,9] highlight two general approaches in UBEM: (i) top-down, which utilizes aggregated urban data as energy consumptions, population surveys, or building stock databases and (ii) bottom-up methods, which recreate urban energy behavior from the single building modeling transposed to the district or urban scale. Despite the computational efficiency of top-down approaches, bottom-up models are usually preferred due to a multilevel knowledge and a larger flexibility and reliability in recreating possible future scenarios or in predicting the effects of energy conservation policies, which is one of the most considerable goals of UBEM tools [10]. On the contrary, bottom-up methods show off some significant drawbacks, which are still in the UBEM research spotlight. Many of these were listed by Ferrando et al. [11] and consist of a general lack of knowledge on the input data—necessary in bottom-up modeling—efficient data-driven or physical algorithms, and the introduction of urban microclimate and mutual buildings' interactions.

Data on the entire building stock are generally available to large municipalities; however, it is often incomplete and usually not suitable for detailed energy simulation [10]. The most used approach to solve this issue is the creation of archetypes or prototypes, reference buildings where geometry, building envelope, operational schedule, and usage are approximated to average and typical values depending on the age-class or end-use [8]. A city's buildings are usually associated to archetypes by means of their construction year, end-use, or geometry (e.g., their footprint). Several alternatives have been proposed to this method. For instance, Giassi and Mahdavi [12] applied a clustering and re-diversification technique to create realistic scenarios, while Chen et al. [13] used a Monte Carlo calibration on 17 different energy parameters, including both thermal properties and usage profiles. On the last issue, Happle et al. [14] investigated the effect of the building occupant presence models in commercial buildings.

About the calculation model, the most common way to predict the thermal and energy behavior of buildings is through detailed BES tools [11] such as EnergyPlus and TRNSYS, which represent the core of many UBEM platforms, e.g., CityBES [15], UMI [16], and CEA [17]. As pointed out by Nageler et al. [18], physical bottom-up approaches are usually preferable to data-driven methods thanks to the outstanding results they reach without extensive billings data; nonetheless, they require a large computational effort due to the complexity of the physical model. A possible alternative is the utilization of simplified physical equations, as an equivalent electrical network [19,20], being able to capture the dynamic thermal behavior of the building with simple techniques and a lower amount of required input data.

Lastly, additional considerable issues that characterize bottom up UBEM are the mutual interaction of buildings, in terms of both radiant heat transfer and mutual shadowing between city's surfaces and microclimatic urban effects. Freitas et al. [21] described the well-known techniques and tools for a good prediction of the solar potential and shadowing in urban canopy layers, pointing out the trade-off between detailed results and computation effort. For this reason, when considering mutual surfaces shadowing in buildings' thermal balance, simplified view factor techniques are implemented [22,23].

Microclimatic effects consist mainly of the urban heat island (UHI), i.e., the average urban temperature increase with respect to rural data. This phenomenon is well known in literature, and many researches outlined various intensity via measurements depending on the case study and climate region. Meng et al. [24] and [25] reported for Chinese cities an average growth around 2.1–2.2 °C, with a consequent raise of the cooling energy demand. Focusing on Europe, Salvati et al. [26] presented the effect for the city of Barcelona (maximum of 4.3 °C with more than 20% increase on cooling demand), while Zinzi et al. [27,28] pointed out measured results for different neighborhoods of the Italian city of Rome, finding an average UHI effect of 2 °C for summer season. Morabito [29] also provides the evaluation of the UHI effect through satellite data for differ Italian cities, focusing on the low vegetation density of some urban areas. UHI is the result of several physical phenomena occurring in the urban environment. According to Bueno et al. [30], the different morphology of the city canopy layer reduces the average wind speed, and inter-reflections between the surface raise the urban albedo. Moreover, infrastructures and buildings increase the thermal inertia of urban canyon, whose effect increases urban temperature. Some studies highlight how this effect can be mitigated through the novel technologies for green buildings [31,32], but additional research is required to deal with these physical phenomena. Many models have been developed in order to integrate UHI in UBEM researches [33]. In microscale applications (less than ten buildings), computational fluid dynamics (CFD) is usually coupled with detailed building energy simulation [34], which, on the contrary, become unsuitable for the larger scale due to its huge computational effort [10]. In larger scenarios, for district or small cities, urban canopy models (UCMs) simplify the energy balance to bigger control volumes, calculating the average urban temperature without significant computational needs [35]. The heat from traffic, human metabolism, and buildings must be considered in UCMs. In the literature, many different anthropogenic heat models have been tested, both with top-down approaches [36] or with a bottom-up model [37]. Some UCM-BEM coupling has also been presented by Kikegawa et al. [38,39], considering steady-state BEM. More recent work on UCM results on Mediterranean medium cities can be found in Salvati et al. [26,40], who applied the urban weather generator to quantify the importance of the building density in the city of Rome and Barcellona, comparing results with experimental data as well. UCM techniques can be coupled in chain schemes to improve results accuracy [41].

The present work focuses on analyzing the urban heat island effect on the cooling energy consumptions of Padua, a Mediterranean medium city located in north Italy. The case study is also replicable in other typical Italian cities. A real district of the historical center has been simulated through EUReCA (Energy Urban Resistance Capacitance Approach), a UBEM platform based on lumped-parameters thermal networks developed by the authors. For the simulations, the tool has been improved introducing an integrated connection to the urban weather generator, a UCM model developed by Bueno et al. [30,42], solving simultaneously both the urban canyon and the buildings' thermal balances. As a result, the impact of the air conditioning of buildings is directly considered on the urban heat island effect and vice versa. The aim of the present work is the development of a reliable simple UBEM tool, where a functional thermal zone balance is coupled with an efficient UCM, improving energy consumption predictions as well, especially for the cooling season.

2. Models

EUReCA, a new urban building energy modelling tool developed by the authors and proposed in this paper, aims to simulate both single building and entire neighborhoods or cities. To simulate the building thermal behavior, the platform includes two lumped-parameter models: (i) the first one is based on the 5R1C thermal network, proposed by ISO 13790 Standard [43] and widely described in a previous work [20]; (ii) the second model implements the 7R2C thermal network presented in VDI 6007 German Standard [44]. The platform is built in Python with an object-oriented structure, which offers the opportunity to include different additional modules, improving the efficacy of the model in predicting the urban heating and cooling energy demand. Moreover, the platform is able to handle both CityJSON and GeoJSON semantic geo-referenced data, from which buildings geometry is created and processed. In particular, to assess the growth of district cooling demand in urban areas, the platform considers the interactions between buildings evaluating both the mutual shading between them and the urban heat island effects, considering the anthropogenic heat from buildings' systems and traffic.

According to a previous research [45], both 5R1C and 7R2C models are able to evaluate well the buildings' energy demand. Nevertheless, the accuracy of the two-capacitance model (i.e., 7R2C model) in the calculation of the daily peak load is better than that outlined by 5R1C network. For this reason, in this work, the simulations have been carried out using the 7R2C model in the EUReCA platform.

An overview of the method implemented in the EUReCA platform for evaluating the thermal behavior of the buildings is presented in Section 2.1. Moreover, a detailed description of the additional modules and their characteristics is given in Section 2.2, Section 2.3, and Section 2.4.

2.1. Thermal Zone Model

A simplified dynamic approach, such as EUReCA, is suitable in a city scale simulation due to less computational effort and limited input data, while producing a good result both for a single building and entire districts. EUReCA implements lumped-capacitance methods to simulate the building's behavior, using the electrical analogy to solve the heat balance in the unsteady state of the thermal zone.

The method used in this work is based on the 7R2C model presented in the VDI 6007 German Standard [44], where structures are distinguished in adiabatic and non-adiabatic ones. All the thermal characteristics of the building are lumped in seven thermal resistances and two thermal capacitances. Figure 1 shows a simple scheme of the model considering a simplified thermal zone, outlining the boundary conditions in terms of climatic data and heat gains.

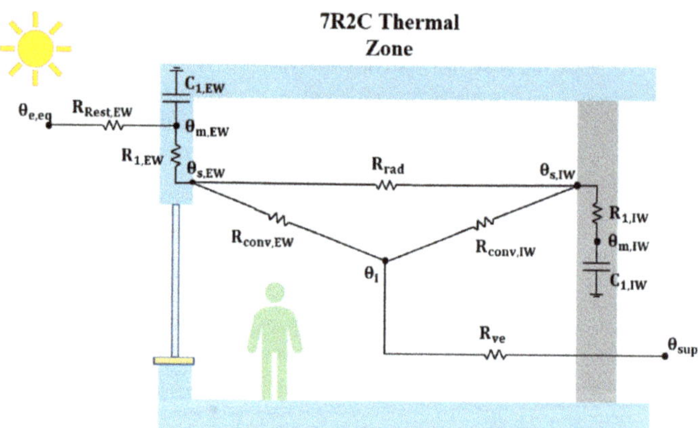

Figure 1. 7R2C thermal network simplified representation.

Adiabatic surfaces are modelled with the thermal resistance $R_{1,IW}$ and thermal capacitance $C_{1,IW}$, while the thermal resistance $R_{1,EW}$ and the thermal capacitance $C_{1,EW}$ refer to the non-adiabatic surfaces. The long-wave radiant heat transfer between these two structures' categories is considered through the thermal resistance R_{rad}. The convective heat transfer between the internal air and the non-adiabatic and adiabatic surfaces is considered using the thermal resistances $R_{conv,EW}$ and $R_{conv,IW}$, respectively. In addition, the thermal mass node of the non-adiabatic surfaces $m_{,EW}$ is connected to the external environment through the thermal resistance $R_{Rest,EW}$. Finally, ventilation and infiltration heat losses are considered via the thermal resistance R_{ve}. The following equations briefly describe the procedure for the calculation of the network's parameters.

According to the VDI 6007 Standard, the unsteady-state behavior of the structures is modelled via characteristic parameters. For each wall's material, a chain matrix A_{BL} is calculated, considering the resistance R_{BL}, i.e., the ratio between the thickness s_{BL} and the thermal conductivity λ_{BL}, and the specific capacity C_{BL}, i.e., the product between the thickness s_{BL}, the density ρ_{BL}, and the specific heat c_{BL} of the material. Each component of the matrix derives from a different function f, listed in detail in the Standard.

$$A_{BL} = \left\| \begin{matrix} a_{11} & a_{12} \\ a_{21} & a_{22} \end{matrix} \right\| \tag{1}$$

$$a_{ii} = f(R_{BL}, C_{BL}) = f(s_{BL}, \rho_{BL}, \lambda_{BL}, c_{BL}) \tag{2}$$

Once the chain matrix of each structure's layer A_{BL} is calculated, the product of them (from inward to outward) gives the element's matrix A_{BS}. Thus, the thermal response of the structure can be modeled through its overall impedance Z_{BS}, which can be expressed with Equation (3):

$$Z_{BS} = R_{BS} + \frac{1}{i\omega_{BS}C_{BS}} \tag{3}$$

where ω_{BS} is the angular frequency, R_{BS} is the thermal resistance, and C_{BS} is the thermal capacitance of each building structure.

When the impedance of each building structure is determined, it is possible to calculate the overall impedance Z_W of both non-adiabatic and adiabatic surfaces:

$$Z_W = \frac{1}{\sum_{i=1}^{m} \frac{1}{Z_{BS,i}}} = R_W + \frac{1}{i\omega_{BS}C_W} \tag{4}$$

where R_W and C_W are the thermal resistance and thermal capacitance of adiabatic or non-adiabatic structures.

The outdoor environment is described through an equivalent external air temperature $\theta_{e,eq}$, which makes it possible to properly consider both convective and radiant heat transfer on the building envelope:

$$\theta_{e,eq} = \theta_e + \Delta\theta_{e,lw} + \Delta\theta_{e,sw} \tag{5}$$

where θ_e is the external air temperature taken from the weather data file, $\Delta\theta_{e,lw}$ is an equivalent temperature difference to model the long-wave radiant heat exchange between surfaces, sky, and ground, and $\Delta\theta_{e,sw}$ is an equivalent temperature difference to model the short-wave radiant heat contribution due to the global solar radiation. In EUReCA, the epw files [46] are used to import the weather data of the location.

On the other hand, solar gains through the glazed components are considered as completely radiant and distributed to the internal and external surface nodes proportionally to their total area. Internal heat gains can be set both as radiant and convective and are distributed to both the internal air node and the non-adiabatic and adiabatic surface nodes. Further details on the entire procedure can be found in [20].

After calculating all parameters, the heat balance equation at each thermal node can be written obtaining a linear system of equations, which can be easily solved either by fixing the internal air node setpoint temperature or by fixing the heating–cooling load.

2.2. Surfaces Mutual Shading Calculation

In an urban context, simulating each building separately without considering the interaction between them can lead to inaccuracies, especially for those cities characterized by a high density or average height of buildings. In the cooling season, the interaction between buildings mostly affects the calculation of solar gains due to the mutual shading between adjacent buildings. For this reason, a module to evaluate the shading between buildings has been implemented in the EUReCA platform. To this purpose, a set of equations is provided to calculate an hourly reduction coefficient of the solar gains through each external surface.

The calculation is performed in two sections: firstly, distance, view factor, and mutual orientation are assessed for each couple of district's surfaces, and only those couples that respect specific tolerances are stored; secondly, the actual shading effect is evaluated. A more detailed description of the procedure is explained below.

At the beginning, all external surfaces of each building are temporarily stored in a list, and the distance d_{ij} between their barycenters is evaluated as follows:

$$d_{ij} = \sqrt[2]{\left(x_{ci} - x_{cj}\right)^2 + \left(y_{ci} - y_{cj}\right)^2} \tag{6}$$

where x_{ci} and x_{cj} are the x-coordinates, and y_{ci} and y_{cj} are the y-coordinates of the surface barycenters under examination. Only couples of surfaces closer than a specified tolerance are considered. In this work, the tolerance has been set to 100 m.

If the tolerance condition is satisfied, the second check consists of verifying if the centroids direction angle CDA_{ij} is included in a second tolerance range (which is set with a maximum value of 80°) with respect to the normal direction of the first surface. The centroids direction angle is calculated with the equation:

$$CDA_{ij} = \cos^{-1}\left(\frac{x_{cj} - x_{ci}}{d_{ij}}\right) \tag{7}$$

Finally, the last check concerns the mutual orientation angle MOA_{ij} of the couple of surfaces. This parameter consists of the angle laying between the two surfaces versors; if this mutual angle falls in a tolerance range the surfaces faces each other, and they can shade mutually. The tolerance for this parameter is set to 80°. A simple scheme of the geometrical model is shown in Figure 2.

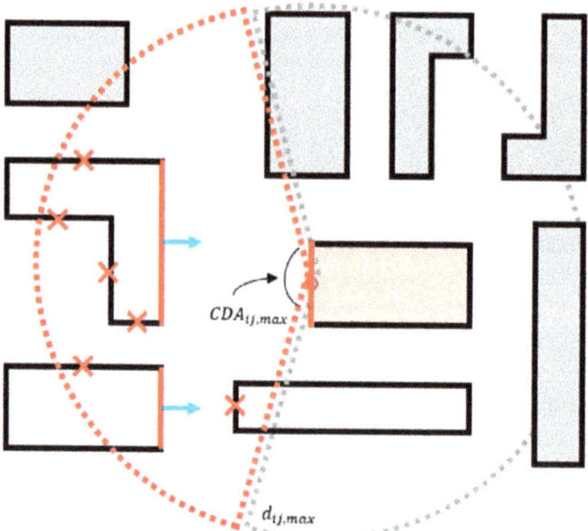

Figure 2. Simplified scheme of the calculation of mutual shading factor.

The red line on the central building is the surface for which the calculation is performed. The dashed circle represents the distance tolerance, while the red zone of the circle shows the second check on the centroids direction angle. Red lines on white buildings represent the surfaces that face and can shade the surface under calculation, while the surfaces marked with a cross represent those that do not respect the third check regarding the mutual orientation angle. Blue arrows indicate the normal direction of the shading surfaces.

Once all possible shading connections between surfaces are detected, the limit solar height and the limit azimuth angles are calculated for each couple of surfaces. The limit solar height LSH_{ij} is evaluated through the following equation:

$$LSH_{ij} = \tan^{-1}\left(\frac{z_{sj} - z_{ci}}{d_{c,ij}}\right) \tag{8}$$

where z_{sj} is the z-coordinate of the highest point of the second surface, and z_{ci} is the z-coordinate of the barycenter of the first surface.

The two limit azimuth angles $LAA_{ij,x}$ are calculated, with the following equation, considering the horizontal length of the second surface with respect to the first barycenter.

$$LAA_{ij,x} = \frac{x_{vx,sj} - x_{ci}}{\sqrt[2]{\left(x_{ci} - x_{vx,sj}\right)^2 + \left(y_{ci} - y_{vx,sj}\right)^2}} \tag{9}$$

where $x_{vx,sj}$ and $y_{vx,sj}$ are the coordinates of one of the vertices of the second surface.

Finally, for each time step, if the solar height is lower than the limit solar height, and the solar azimuth angle is between the limit azimuth angles, the first surface is shaded by the second one. In this case, a reduction coefficient is applied to the calculation of the solar gains. This procedure is repeated for each surface of the district.

2.3. Urban Heat Island Evaluation

Recently, several research studies have point out the impact of the urban heat island (UHI) on the energy performance of buildings [47]. This phenomenon refers to an increase in the external air temperature in urban areas compared to the surrounding rural ones, due to the different morphology, the inter-reflections between building surfaces, and the reduction in the wind speed caused by the higher building density. Moreover, the heat gains due to the anthropogenic sources (e.g., the heat waste of cooling systems and the heat caused by traffic) contribute to increase the effect.

The EUReCA platform includes a module to evaluate the urban heat island effect based on the urban weather generator (UWG) presented by Bueno et al. [30]. The model allows us to estimate the temperature of the urban district considering the interaction between the rural and urban areas. The model involves only the sensible balance, without considering further phenomena connected to evaporation heat transfers. The procedure is divided into four sections, in which rural and urban areas are connected by the rural vertical diffusion and the canopy boundary layer, located over the urban area.

In the first section, the calculation of the rural surface temperature θ_{soil} is performed to take into account the sensible heat flux H_{rur} that flows from the rural area to the boundary layer. To this end, the following equation is used:

$$H_{rur} = h_{conv}\left(\theta_{rur} - \theta_{soil}\right) \tag{10}$$

where h_{conv} is the convective heat transfer coefficient, and θ_{rur} is the rural air temperature derived from the weather data file.

In the second section, the vertical distribution of the rural air temperature θ_{VDM} is calculated with the following equation:

$$\frac{\partial\theta(z)}{\partial z} = -\frac{1}{\rho(z)}\frac{\partial}{\partial z}\left(\rho(z)K_d(z)\frac{\partial\theta(z)}{\partial z}\right) \tag{11}$$

where z is the altitude above the rural area, with a maximum height of 150 meters, beyond which the temperature is considered constant. The diffusion coefficient $K_d(z)$ is calculated as in the original urban weather generator tool, considering the Bougeault and Lacarrere model [48], dependent by the square root of the turbulence kinetic energy (TKE).

In the third section, the boundary layer temperature above the district θ_{UBL} is calculated considering the heat fluxes from the rural and urban areas H_{rur} and H_{urb}, respectively. The thermal balance at the time step τ can be expressed with the following equation:

$$\theta_{UBL}(\tau) - \theta_{UBL}(\tau - 1) = C_{surf} + C_{adv}\theta_{eq} - C_{adv}\theta_{UBL}(\tau) \tag{12}$$

where the surface coefficient C_{surf}, the advection coefficient C_{adv}, and the equivalent temperature θ_{eq} depend on the hour of the day and on the difference between the circulation air velocity into the urban area and the undisturbed wind speed into the rural area. The detailed procedure to calculate C_{surf} and C_{adv} is described in [30].

In the last section, the temperature of the urban canyon θ_{urb} and the specific heat flux exchanged with the boundary layer above the district H_{urb} are calculated. The thermal balance that makes it possible to calculate the temperature θ_{urb} is indicated in the following equation:

$$V_{can}\rho c_v \frac{d\theta_{urb}}{d\tau} = A_w h_w(\theta_w - \theta_{urb}) + A_r h_r(\theta_r - \theta_{urb}) + A_r h_{r,sky}(\theta_{sky} - \theta_{urb}) + A_{win}U_{win}(\theta_{in} - \theta_{urb})$$
$$+\dot{V}_{inf}\rho c_p(\theta_{in} - \theta_{urb}) + A_{can}u_{ex}\rho c_p(\theta_{UBL} - \theta_{urb}) + Q_{waste} + Q_{traffic} \tag{13}$$

where V_{can} is the control volume of the district calculated considering an average height of the buildings; A_w, A_{win}, A_r, and A_{can} are the external walls, windows, road, and canyon areas, respectively; h_w and h_r are the global heat transfer coefficients, and $h_{r,sky}$ is the radiant heat transfer coefficient between the sky and the district; U_{win} is the average thermal transmittance of the building's windows; V_{inf} is the sum of the infiltration and ventilation volumetric flow rate; u_{ex} is the air velocity due to the mass exchange between the district and the boundary layer above it; θ_w, θ_r, θ_{sky}, and θ_{in} are the external walls, road, sky, and buildings' internal air temperatures, and Q_{waste} and $Q_{traffic}$ are the heat fluxes caused by HVAC systems and other anthropogenic sources of heat, respectively.

Buildings' variables θ_w, V_{inf}, and θ_{in} derive from the building model, Section 2.1. In particular, the building wall temperature is calculated considering the network branch $\theta_{e,eq} - \theta_{m,EW}$, with the external convective and radiant resistance R_e:

$$\theta_w = \theta_{e,eq} + (\theta_{m,EW} - \theta_{e,eq})\frac{R_e}{R_{Rest,EW}} \tag{14}$$

Finally, Equation (12) allows us to calculate the specific heat flux H_{urb}:

$$H_{urb} = u_{ex}\rho c_p(\theta_{urb} - \theta_{UBL}) + h_w(\theta_w - \theta_{UBL}) \tag{15}$$

As shown in Equations (13) and (15), the air mass of the canyon is lumped in a single node, considering perfectly mixed air and neglecting fluid dynamics phenomena as convective currents, stratification, etc. Although this represents quite a significant hypothesis for microclimatic models, it is considered suitable in the evaluation of the buildings' consumption. Moreover, such a model preserves the simplicity and efficiency of the entire software.

Figure 3 summarizes the workflow of the model when the UHI calculation is performed. The external air temperature into the district is hourly replaced with the urban canyon temperature θ_{urb}, affecting buildings' thermal balance and HVAC systems' efficiency. The systems' waste heat is then utilized to calculate the urban temperature for the following time step. This leads to the evaluation of the realistic mutual connection between district energy demand, electrical HVAC consumptions, and the heat island within the urban environment, i.e., the modules are perfectly integrated and allow for evaluating the different aspects.

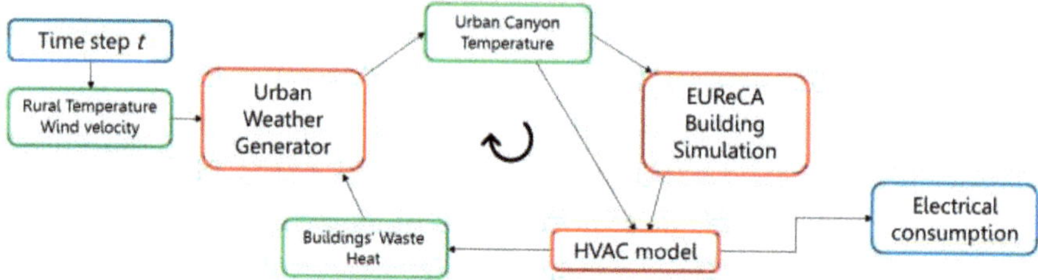

Figure 3. Simplified workflow of the integrated urban heat island model.

2.4. Cooling System Model

In order to better evaluate the interaction between district's buildings, taking into account the waste heat rejected by HVAC systems, a module able to simulate the performance of the energy systems was implemented.

Thanks to the object-oriented structure of the platform, it was possible to implement a "systems" class in which all the systems converge. Exploiting this class, a system object is associated to each building during the creation of the district. To correctly simulate the system's behavior, three steps are fundamental: the first consists of the definition of the system's typology and main characteristics, and it is carried out during the initialization of the building; the second step consists of the setting the size according to the peak thermal load of the building; the third consists of the calculation of the performance parameters of the system and its consumption during the dynamic simulation of the district.

Some typical systems have been modelled in accordance with the UNI-TS-11300 National Standard [49]. EUReCA currently includes two types of heating and cooling systems, even if in this work, only the cooling systems have been used. For the heating season, the *Traditional* and *Condensing Gas Boiler* were taken into consideration. For the cooling season, the *Split Air Conditioner* and the *Air-to-Water Chiller* were considered:

- *Split Air Conditioner*: Italian residential buildings, especially older ones, are usually not equipped with a centralized cooling system; nevertheless, due to the high summer temperature and humidity, it is quite common to use split air conditioners, which are installed for each apartment. This type of system is usually used in the residential buildings.
- *Air-to-Water Chiller*: In non-residential buildings, a centralized cooling system with a higher nominal power than a *Split Air Conditioner* is generally present. In most cases, the system corresponds to an *Air-to-Water Chiller*.

As mentioned above, the model of the building's system previously requires the calculation of the peak load in design conditions. For the heating season, a steady-state calculation is performed, considering a constant external air temperature defined by the Standard [49]. For the cooling season, a dynamic calculation in the two hottest summer weeks based on the test reference year is executed. The design cooling load allows us to estimate the size of the building's system.

Both cooling systems' classes require as input only the values of the energy efficiency ratio (EER) at different part-load factors. Then, the Standard [49] gives several tables to calculate the multiplication factor η_{mf} of the *EER* based on the actual load factor, the external air temperature, and the internal air temperature for the *Split Air Conditioner* and on the actual load factor, the external air temperature, and the supply water temperature for the *Air-to-Water Chiller*. A linear interpolation to reach the right value of the multiplication factor is performed.

Equations (16) and (17) allow us to calculate the electrical power W_{el} required by the cooling system and the waste heat Q_{waste} rejected to the outdoor environment for each timestep, respectively.

$$W_{el} = \frac{\Phi_{cool}}{EER \cdot \eta_{mf}} \tag{16}$$

$$Q_{waste} = \Phi_{cool} + W_{el} \tag{17}$$

where Φ_{cool} is the total cooling load required by the building.

3. The Case Study District

The simulation district considered for this work is located in the city of Padua, a municipality in the northern Italy, coordinates 45.4° N and 11.9° E, with a population of around 210,000 inhabitants. The city can be subdivided in 3 areas: (1) the historical center, (2) a peripherical zone with buildings built between 1950 and today, depending on the area, and (3) the industrial area to the southeast. The case study district, from now on called Belzoni district, is located in the east border of the city center and includes several old residential buildings, apartment blocks, and some large office-didactical buildings belonging to the University of Padua. The spatial extension of the district is around 0.53 km^2. Figure 4a and b represent the map of the Italy and a particular focus on the city of Padua and the Belzoni district. The vegetation density has been estimated considering the real green areas of the district, which are about 24% of the total district area and mainly consist of sports fields and public parks.

UBEM simulations have been carried out by means of the 3D CityJSON model of the district, created within the URBAN GEO BIG DATA project [50,51] and kindly made available by the Department of Land, Environment, Agriculture and Forestry (TESAF) of the University of Padua. The virtual district results in 580 buildings, displayed in Figure 4c.

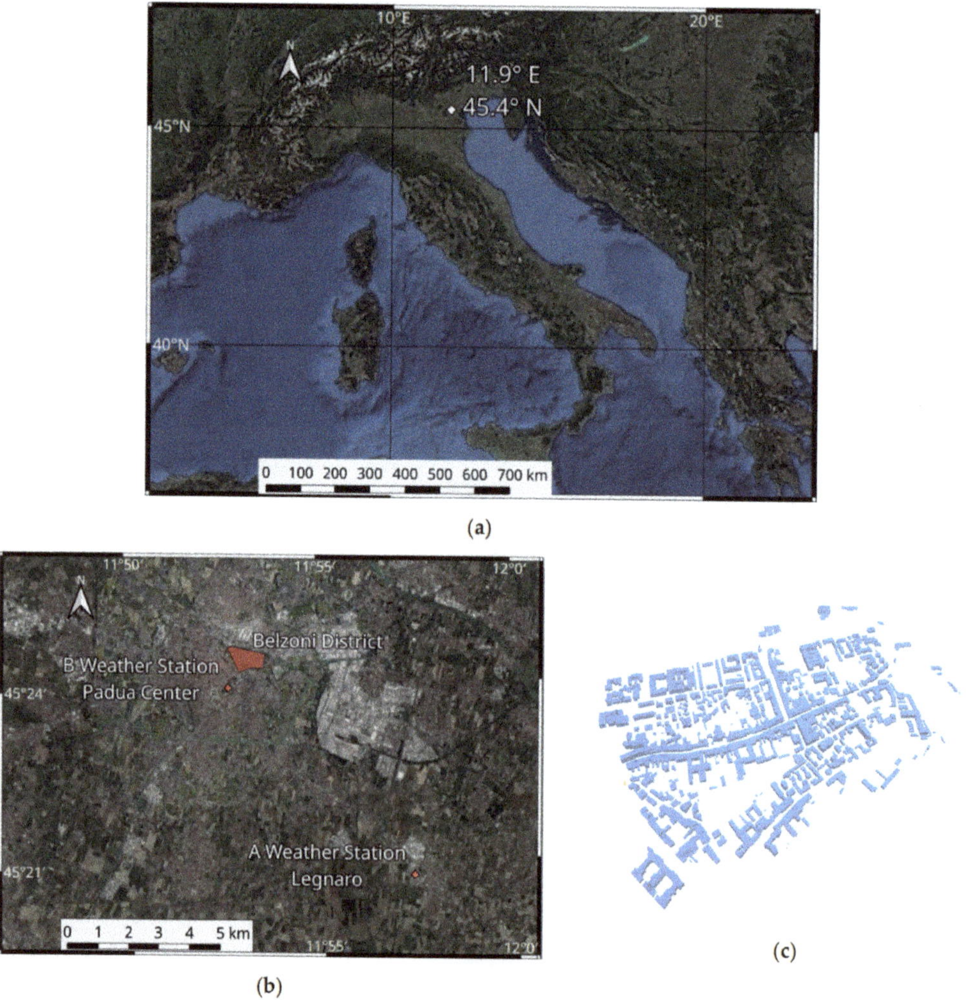

Figure 4. (**a**) Satellite image of Italy; (**b**) a particular of the Belzoni district and the weather stations in Padua; (**c**) virtual image of the Belzoni district.

3.1. Buildings Envelopes

Since no specific information about the age-class was available for the case study, buildings envelopes thermal properties have been set following the well-known archetype method [8]. In particular, the Italian Statistical Institute (ISTAT) buildings' survey [52], released in 2011, includes the national number of buildings for five age-classes, which allowed the extrapolation of the Padua municipality data, listed in Table 1. The percentage of buildings for each age-class has been then utilized to associate an age-class label randomly to each building of the case study, resulting in the distribution listed in Table 1. An envelope archetype has been created for each age-class, following the databases proposed by Carnieletto et al. [53], and is summarized in Table 2. In particular, the oldest age class consists of an external wall with solid bricks and a concrete-slab roof. The external wall from 1960 to 1990 is built-up with hollow bricks with a different thickness of internal air cavity. Bricks are also present in the two most recent age-classes but coupled with different levels of insulation. The concrete slab roof is insulated as well after 1990. These envelopes archetypes refer to typical Italian envelopes and were built up following several national Standards (Carnieletto et al. [53]). The average solar reflection coefficient is set to 0.4 [54]. The combination of building geometry and structures' thermal properties allows the calculation of the model's parameters.

Table 1. Age-classes percentage distribution of the Padua municipality and number of buildings.

Age-Class	Percentage to the Total Building Number (ISTAT 2011)	Percentage to the Total Building Floor Area	Number of Buildings (Belzoni District)	Building Heating System
<1960	24%	15%	134	Traditional boiler
1961–1980	44%	44%	251	Traditional boiler
1981–1990	13%	26%	76	Traditional boiler
1991–2005	15%	10%	97	Condensing boiler
>2005	4%	5%	22	Condensing boiler
Total	100%	100%	580	

Table 2. Age-class attributes of the Belzoni district. U-value of the main structures and solar heat gain coefficient of the windows.

Age-Class	External Wall	Roof	Ground Floor	Windows	
	U-value $[W/(m^2K)]$	U-value $[W/(m^2K)]$	U-value $[W/(m^2K)]$	U-value $[W/(m^2K)]$	SHGC [–]
<1960	1.05	1.34	1.42	4.90	0.82
1961–1980	0.98	1.34	1.42	3.70	0.70
1981–1990	0.67	0.81	1.42	3.70	0.70
1991–2005	0.54	0.52	0.61	3.40	0.70
>2005	0.30	0.33	0.23	2.20	0.27

3.2. Buildings End-Use

The buildings' end-use has been set similarly to the age-class, following the distribution derived from the ISTAT database [52], reported for Padua in Table 3. However, in this case, the association of the end-use to each building was not totally random for two reasons: firstly, non-residential buildings are usually bigger than residential ones, secondly, as previously mentioned, the case study district includes many of the university's large complexes. In order to consider these aspects, the 548 smallest buildings (94%, considering the building footprint) have been labelled as residential buildings, while for the largest 32 buildings (6%), the end-use has been associated randomly between commercial, tertiary, and services, resulting in the distribution of Table 3. For this different association, the percentage to the total building floor area is significantly different from the percentage to the total building number.

With respect to each end-use, a set of internal heat gains and operational schedules has been created considering EN 16798-1 [55] and ISO 18523-1 [56] Standards. The nominal values of the internal heat gains and setpoints are listed in Table 4, while Figure 5 displays the occupancy trend during a weekday and a weekend.

The cooling system typology depends on the end-use as well. Single air-to-water chillers have been coupled with non-residential buildings, while the split air conditioner has been considered as a cooling system in the residential ones. The number of cooling systems' units has been set dividing the peak cooling load of the building by the maximum nominal cooling power of the system, i.e., 200 kW for the air-to-water chiller and 15 kW for the split air conditioner. For instance, a services building with a cooling peak load equal to 1 MW will be provided with 5 air-to-water chillers, while a residential apartment block will be coupled to 9 split air conditioners if its peak cooling load is around 135 kW. This assumption has been taken accordingly to the typical Italian building configurations.

Table 3. End-use percentage distribution of the Padua municipality and number of building.

End-Use	Percentage to the Total Building Number (ISTAT 2011)	Percentage to the Total Building Floor Area	Number of Building (Belzoni District)	Reference Standard	Cooling System
Residential	94%	52%	548	EN 16798 (Residential, apartment)	Split air conditioner
Commercial	3%	21%	16	ISO 18523 (Shop, small store)	Air-water chiller
Tertiary	1%	12%	6	EN 16798 (Office, landscaped)	Air-water chiller
Services	2%	15%	10	EN 16798 (School, classroom)	Air-water chiller
Total	100%	100%	580		

Table 4. End-uses nominal heat gains and setpoint.

	Occupancy	Appliances	Lighting	Heating Setpoint	Cooling Setpoint
	[W/m^2]	[W/m^2]	[W/m^2]	[°C]	[°C]
Residential	4.2	3	3	20	26
Commercial	20	40	30	20	26
Tertiary	7	12	12	20	26
Services	21.7	0	8	20	26

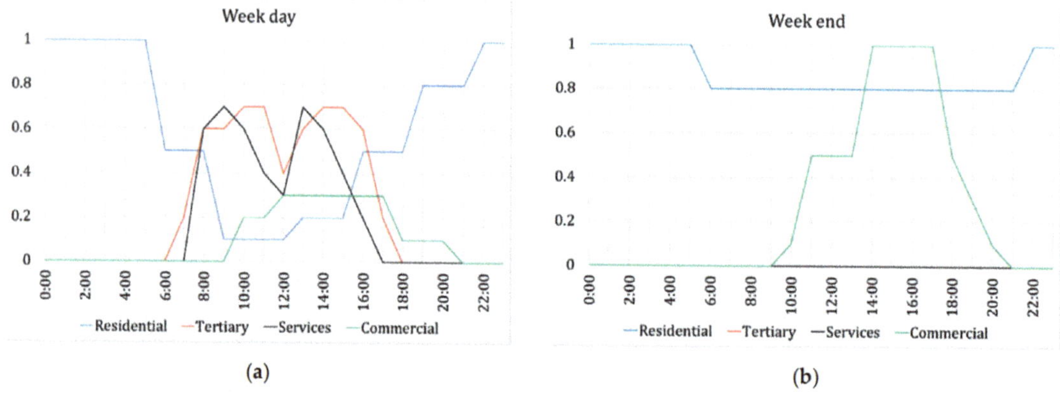

Figure 5. Daily occupancy trend for the considered end-uses: (**a**) week days, (**b**) weekend days, (adapted from [55,56]).

4. Methods

The interaction between the urban district's buildings mainly affects two connected phenomena: a variation of the urban external temperature and, consequently, a variation of the buildings energy demand, especially in the cooling season. To highlight the differences between a single building simulation and an urban one, several simulations have been carried out. The following sections outline the weather input data and the characteristics of the different evaluated simulation scenarios.

4.1. Weather Input Data

The simulations have been performed considering the rural weather data of Legnaro, a little town close to Padua. Weather data were made available by the Regional Environmental Protection Agency of Veneto (ARPAV) [57] and refer to the measured data of summer 2019. The automatic weather station, *A Station* in Figure 4, is located far away from buildings, within agricultural fields, thus not being influenced by any urban effect. Moreover, to evaluate the hourly temperature difference between the measured urban temperature and the estimated value, the weather data of the city center of Padua, *B Station* in Figure 4, have been taken into account, considering those measured by ARPAV in summer 2019 as well. This second weather station is located in a residential area, surrounded by terraced houses (less than 10 m height) and streets. This residential zone and the Belzoni district are similar, although the second one has a slightly higher building density. Despite the fact that this station is not located inside the simulated district, its urban data are shown in the results section in order to give a reference of the measured UHI effect affecting Padua. Moreover, these data represent the unique measurement inside the city center. In any case, for all these reasons, no additional statistical analysis on the measured urban temperature has been carried out.

Both the rural and the urban weather station provides the hourly average air temperature, relative humidity, and global solar radiation.

4.2. Urban Simulation Scenarios

To explain the impact of the urban heat island on the case study district and separately evaluate the different effects, several simulations have been carried out. In particular, three parameters have been chosen as independent variables: (i) the district heat caused by the traffic $Q_{traffic}$, (ii) the waste heat rejected by the cooling systems Q_{waste}, which with the previous one, constitutes the heat caused by anthropogenic sources, and (iii) the district vegetation density VD_d. The list below describes the examined simulation scenarios:

- UHI-WOAH: the first scenario was carried out evaluating the effects of the urban heat island without considering anthropogenic heat sources;
- UHI-ATH: in the second scenario, the anthropogenic heat exclusively caused by the traffic $Q_{traffic}$ was taken into account [58,59];
- UHI-AWH: the third scenario was performed considering only the anthropogenic waste heat rejected by the case study district cooling systems Q_{waste};
- UHI-WAH: in the fourth scenario, both heat fluxes caused by traffic and cooling systems were analyzed;
- HVD: the fifth scenario was executed considering all the effects of the urban heat island and a half district vegetation density VD_d (12% of the total district area);
- DVD: the sixth scenario was performed considering all the effects of the urban heat island and a double district vegetation density VD_d (48% of the total district area).

5. Results

Result section has been split in two subsections. Firstly, urban summer temperatures are exposed, showing the model's performances with respect to real measured data. Then, an aggregated energy analysis has been carried out, investigating the effects of the urban heat island on a small Mediterranean city's consumptions and the potential of simplified bottom-up tools as EUReCA.

5.1. Urban Temperature Analysis

As previously described, the urban weather generator, by Bueno et al. [30], has been fully integrated into EUReCA. Such a method allows the mutual correlation between the UHI and each simplified building model within the UBEM simulation. In this scenario, the hourly consumption of each building is influenced by the corresponding urban temperature θ_{urb}, which, on the contrary, depends on the building systems' waste heat, Figure 3.

As a first result, Figure 6 displays the comparison between the rural hourly temperature, i.e., *A Station—Legnaro*, the urban measured temperature, *B Station—Padua center*, and the model's outcomes for the last days of July. The comparison between the model outcome and the urban temperature data is carried out focusing on their trend, as far as the weather data derive from a single point measurement. Rural temperature ranges between 20 and 34 °C, with a maximum value of 34.5 °C on the 23rd of July. These are typical temperatures trends for a hot summer week in northern Italy. As measured data in *B Station* show (black line), the urban heat island effect is limited for small cities like Padua; nonetheless, it reveals from the daily temperature peak to the central hours of the night, e.g., on the 23rd and 25th of July with a maximum temperature of 36 and 35 °C, respectively. The UWG air node temperature resulting from the simulation (dot red line) follows the measured data trend, with some difference especially in the maximum values. When the daily peak occurs, the model results in a higher value, which starts to decrease later than the measured one. From 22nd and 24th of July, the temperature reaches 36–37 °C. This is an expected trend as far as the buildings' density and height are higher in Belzoni district than around the *B Station*. On the contrary, the model and the measured data are more similar during the daily valleys (night-time), as in the first hours of the 23th, 25th, or 26th of July.

In general, the results of the urban weather generator, coupled with the EUReCA UBEM simulator, agree with the measured urban temperature, capturing the variations during its daily trend coherently with the difference between the Belzoni district and *B Station* morphology.

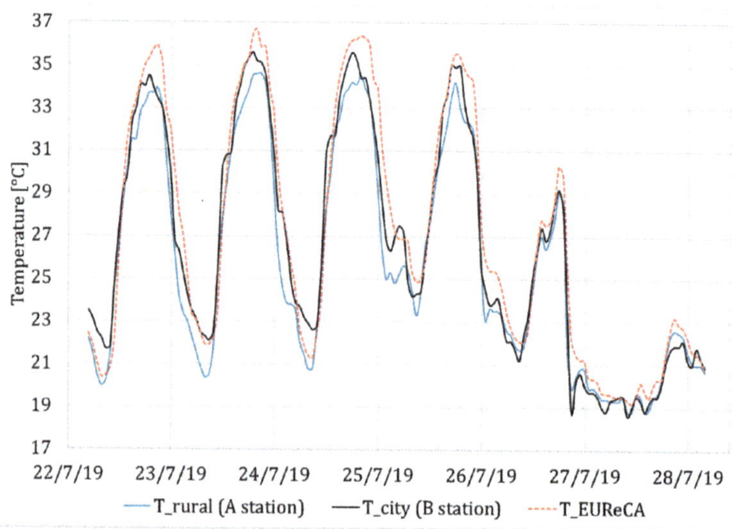

Figure 6. Hourly air temperature in some July days. Comparison between rural measured data (*B station*), city center measured data and simulated temperature (urban weather generator (UWG) air node).

In order to provide quantitative and more general results on the model outcome, the average temperature difference between measured rural temperature and the model estimated urban temperature is shown in Figure 7a, for each summer month. The model accounts for lower values of the urban temperature early in the morning, from 6:00 to 8:00 a.m. On the contrary, the urban canyon temperature overtakes rural data between the late morning and 4:00 a.m., with a slight growth in the evening. The maximum average urban heat island effect appears at 19:00, resulting in an urban temperature 2.2 °C higher than the rural one.

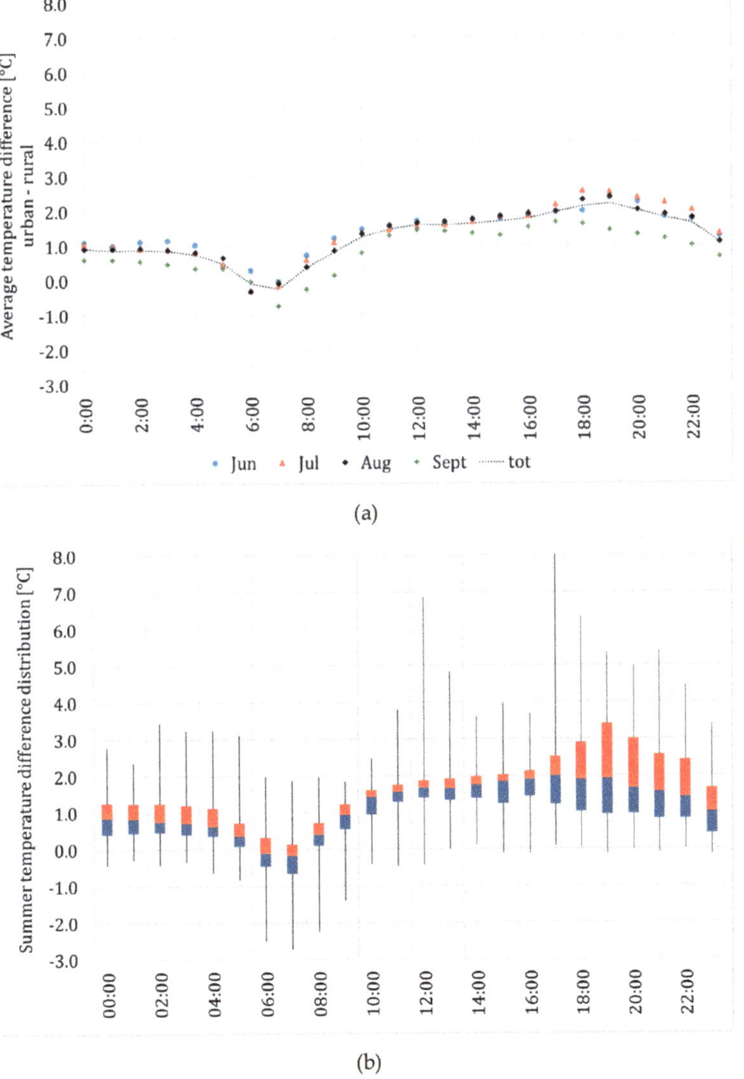

Figure 7. Urban–rural daily temperature difference: (**a**) average monthly trends; (**b**) box plot showing minimum, maximum, and quartiles for the summer season.

The larger UHI occurs during June and July, while the difference is lower in September, when the solar radiation that is absorbed by the urban canyon is lower. Figure 7b shows the statistical distribution of the modeled UHI during summer season, showing the relative quartiles (minimum value, 25, 50, 75%, and maximum). The outcome highlights a lower variability of the UHI effect during the first part of the day, while it starts to increase after 4:00 p.m., reaching differences of up to 2.5 °C in the evening between the 1st and 3rd quartile.

5.2. Urban Heat Island Effects on District Energy Demand

The main effect related to the urban heat island phenomenon is the increase in the external urban air temperature, especially during the summer season. The previous section pointed out the typical trend on a small Mediterranean city like Padua and highlighted the outcome of the UWG model coupled with EUReCA. As further step, the resulting integrated tool has been applied aiming at understanding the effect of the UHI on the cooling energy consumption of the Belzoni district. The method consists of simulating several scenarios, stressing the impact on three different parameters: the traffic anthropogenic heat, the heat load from HVAC systems, and the district vegetation density.

Figure 8 summarizes the district cooling energy demand without considering the urban heat island effects. The first period, from mid-June to mid-July (blue), results in the higher sensible demand, while looking at the latent demand, the second period, from mid-July to mid-August (red), is greater due to high external specific humidity. The period from 15th June to 15th July is the one with the higher total energy demand, resulting in 9751 MWh, followed by the second period, for which the district cooling energy demand is around 9686 MWh. The cooling demand between August and September (gray) is lower because of cooler temperatures.

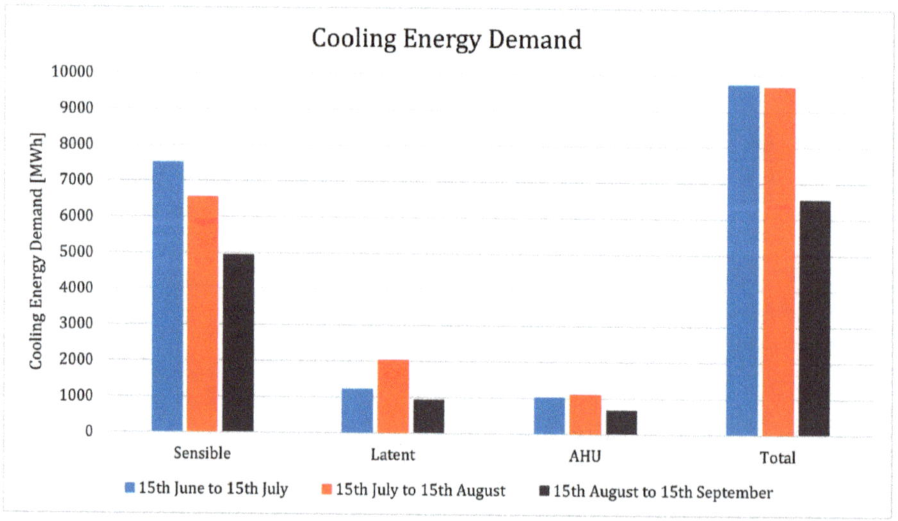

Figure 8. District cooling energy demand without considering the interaction between buildings.

The total cooling demand shown previously is considered as a reference scenario for the following deviations. Figure 9 shows the relative deviation (percentage) of the district cooling energy demand for the different simulation scenarios with respect to the case in which the UHI phenomenon is neglected. In particular, Figure 9a refers to the deviation of the zone demand (sensible and latent without ventilation), Figure 9b shows the deviation considering only the air handling units, while Figure 9c highlights the deviation of the total demand (zone latent, sensible, and Air Handling Unit (AHU)).

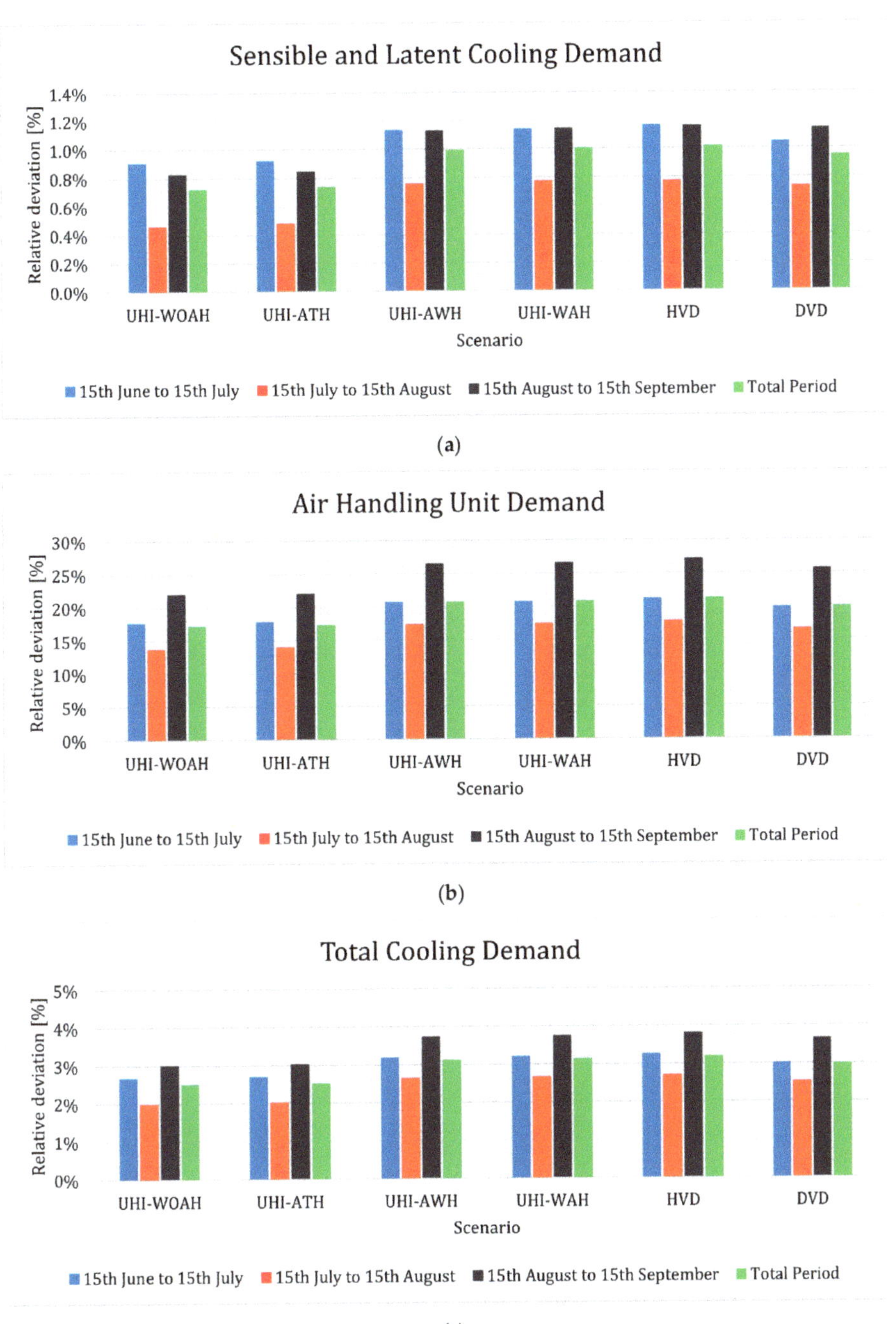

Figure 9. Relative deviation of the district cooling energy demand for different scenarios with respect to the case without considering the urban heat island effect.

Looking at the latter, it shows an increase in cooling energy demand of 2.5% in the case in which the anthropogenic heat is neglected (*UHI-WOAH*), showing the relevance of the urban canyon morphology in the UHI effect and related consumption. The relative deviation grows up to 3.1% if the effects of anthropogenic heat are considered (*UHI-WAH*). The increase in energy demand can be almost entirely attributed to the waste heat rejected from cooling systems, which is about 20 times higher than the heat flux from traffic. This is the reason why considering the traffic anthropogenic heat (*UHI-AWH* and *UHI-WAH*) or not involves a negligible difference.

The last two simulation scenarios refer to the cases in which the vegetation density of the district is halved (*HVD*) and doubled (*DVD*) with respect to the real value (24% of the district area). In the first case, the relative deviation increases up to 3.2%, while in the second case, it decreases up to 3%. Although the deviation is limited, it shows how the increase in the green urban areas can help to reduce the effects of the urban heat island.

The trend shown in Figure 9c also occurs in Figure 9a,b. Nevertheless, Figure 9b shows that the deviation of the AHU demand is considerably greater than the sensible and latent one, ranging from 15 to 25%. This consideration mainly depends on the physical behavior of air handling units, as far as their consumption is directly correlated and more sensitive to the external temperature. However, the total district demand is almost not influenced by the air handling units, as their demand represents a minor fraction of the total. For instance, districts with a higher percentage of commercial buildings, usually equipped with an AHU, can be affected more by the UHI phenomenon.

Figure 10 shows the electrical energy demand of the district's cooling systems for the different simulation scenarios. Even though the district cooling energy demand only increases by 3.1% if all the effects are considered, the electrical energy demand increases by 7.3% under the same conditions (up to 7518 MWh). This outlines that, in an urban context, the urban heat island affects the electrical consumption both by increasing the buildings' cooling demand and reducing the systems' efficiency as well. These two effects, combined together, cause a considerable growth on the electrical energy usage.

Moreover, doubling the vegetation density from 24 to 48% (DVD scenario) seems to compensate partially the consumption increase caused by the cooling systems (UHI-AWH scenario), reducing it by 28 MWh (passing by 7.3 to 6.9%).

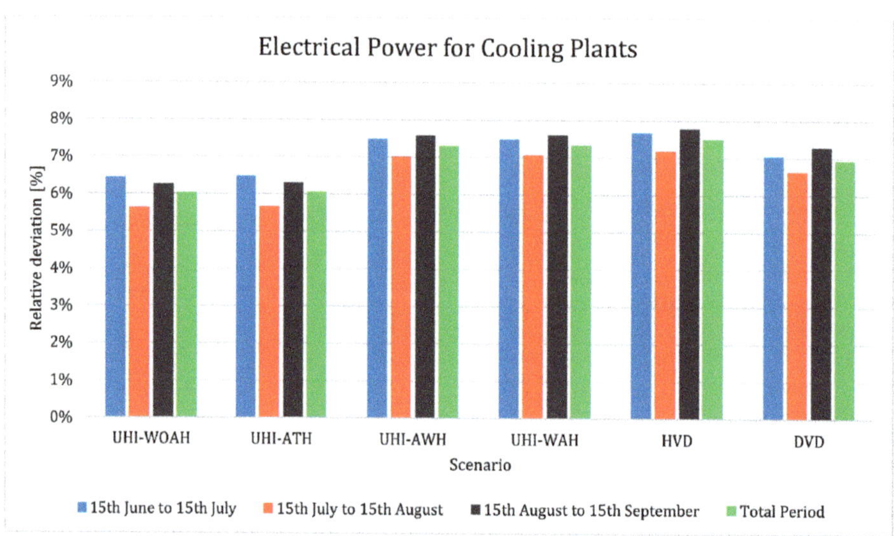

Figure 10. Relative deviation of the district electrical energy demand for different scenarios with respect to the case without considering the urban heat island effect.

6. Discussion

The results presented in the previous section were obtained implementing the urban weather generator tool, a well-known urban canopy model, in EUReCA, a UBEM software based on lumped-capacitance thermal networks to perform energy simulations of building at an urban scale.

The urban canyon temperature predicted by the model has been firstly compared with the measured temperature of an automatic weather station located in a residential part of the city, 500 m close to the simulated Belzoni district. Such a comparison is not intended to be a systematic experimental analysis on the urban heat island of the city, as the measured data are clearly limited for this purpose. A higher distribution of the measured values at the urban scale through the district and their availability for longer time periods are surely necessary to a more extensive statistical evaluation of the UHI effect in the city. However, the comparison was carried out as a reference for the model output. Although the morphology is slightly different, the model output trend is coherent with the measured values, showing the typical urban heat island effect during the late afternoon, evening, and night. Similar results are also reported in other research. In fact, the average trend resulting from the model (Figure 7) is in agreement with experimental values found in similar cities [28,60] and resembles the UHI profile of other studies where the UWG model is utilized [26,40]. In particular, the results outline that, thanks to the implementation of the urban weather generator, the new platform EUReCA captures the typical evening UHI (from 14:00 to midnight), allowing us to carry out a more accurate building energy simulation at the urban scale.

Using the simplified integrated method of EUReCA, a district in the city of Padua has been simulated, aiming at showing the possibility of the software in predicting the district energy consumption and investigating the different components involving UHI. The analysis points out that the buildings' consumption can be affected largely by UHI, especially by the waste heat of the HVAC systems, which represent the most relevant factor from anthropogenic activity. The simulation results show also that the vegetation density can partially compensate the degradation of energy efficiency of cooling systems due to UHI effect; however, the model needs to be improved to consider the evaporation–condensation heat transfer, as the ambient specific humidity can partially affect HVAC systems performances.

As is well known, the building heating and cooling energy demands deeply depend on the outdoor conditions, especially air temperature and solar radiation; at the same time, the urban climate affects the energy efficiency of the systems whose waste heat influences the urban climate in turn. The combination of the two models (urban weather generator and EUReCA) has been carried out, integrating them in a dynamic approach in only one platform, thus capturing the mutual correlation between urban canyon temperature, geometry of the district, and HVAC systems' performances through the hourly simulation: this method represents a novel approach, combining reliable simple models to build up an efficient UBEM software. The cooling demand will certainly increase in the future, and UBEM tools must consider similar contributions in order to properly evaluate the energy consumption and, consequently, to select the appropriate energy conservation measures.

In conclusion, the model is able to perform dynamic integrated simulations of the buildings' behavior, modeling the UHI effect and producing reliable results. Considering the results of this research, future steps will focus on evaluating feasible and convenient energy policies concerning the district's cooling systems.

7. Conclusions

This work investigated the impact of the urban heat island on the energy performance of buildings through a novel module of EUReCA, an urban building energy modeling platform developed by the authors. Global warming will lead to an increase in the cooling demand in buildings, and this will have consequences on the urban climate. To this purpose, the main building's cooling system typologies were modelled to evaluate the waste heat rejected to the urban environment. Then, a case study district in Padua (Italy) consisting of more than 500 buildings was presented in the third section to investigate the results. Several simulation scenarios were analyzed to stress the impact of the UHI effects on the urban district during the cooling season in terms of external temperature, energy demand, and electrical power increase.

The analysis of the simulation results confirms that the UHI involves an average increase in the external air temperature, especially in the late afternoon and night, due to the heat stored during the day. The external air temperature increases by approximately 2 °C, which corresponds to an increase in urban cooling energy demand of around 3.1% and in a growth of the electricity consumption of chillers and split air conditioners of 7.3%. However, the district chosen as a case study is not extensive, and it is characterized by rather low and spaced buildings. This can explain the reason why the UHI leads to a slight increase in the energy demand.

Nevertheless, it can be noted that the increase in the electrical power required by the cooling systems is the most significant effect of the UHI with a more than doubled increase. This makes clear the interrelation between the external air temperature and the performances of the air-to-water chillers and split air conditioners.

Concluding, the urban heat island phenomenon can affect the district, but this mainly depends on the characteristics of the analyzed district and the waste heat rejected by the cooling systems. A higher vegetation density can reduce the effect of the UHI decreasing the external air temperature, while a higher building density and average height can emphasize it. Finally, introducing more efficient cooling systems or considering a system's typologies that reject heat to the ground would help reduce the UHI effect.

Author Contributions: Conceptualization, P.R. and E.P.; methodology, P.R., E.P., L.C., J.V., and A.Z.; software, P.R. and E.P.; formal analysis, L.C. and J.V.; investigation, L.C., J.V., M.Z. and A.Z.; writing—original draft preparation, P.R. and E.P.; writing—review and editing, L.C., J.V., M.Z. and A.Z.; supervision, A.Z. All authors have read and agreed to the published version of the manuscript.

Funding: This research received no external funding.

Data Availability Statement: EUReCA is freely available at https://github.com/BETALAB-team/EUReCA (accessed on 15 March 2021). Publicly available datasets were analyzed in this study. This data can be found here: https://www.istat.it/it/censimenti-permanenti/censimenti-precedenti/popolazione-e-abitazioni/popolazione-2011 (accessed on 30 November 2020).

Acknowledgments: The MS student Nicol Caltranò is gratefully acknowledged. The authors also acknowledge Francesco Pirotti of the Department of Land, Environment, Agriculture and Forestry (TESAF) of the Padua University for providing the information about the case study.

Conflicts of Interest: The authors declare no conflict of interest.

Nomenclature

Subscripts

A	Area (m^2)	BL	Building layer
c_v	Specific heat constant volume (J/(kg K))	BS	Building structure
c_p	Specific heat constant pressure (J/(kg K))	c	Barycenter
C	Heat capacity (J/(m^{2}K))	can	Urban canyon
C_{surf}	Surface coefficient (K)	conv	Convective
C_{adv}	Advection coefficient (-)	cool	Cooling
CDA	Centroids direction angle (-)	e	External
d	Distance (m)	el	Electrical
EER	Energy efficiency ratio (-)	eq	Equivalent
H	Heat transfer coefficient (W/(m^{2}K))	ex	Exchange
H	Heat transfer coefficient (W/K)	EW	Non-adiabatic structures
i	Imaginary unit (-)	in	Internal
K_d	Diffusion coefficient (-)	inf	Infiltration
LAA	Limit azimuth angle (-)	IW	Adiabatic structures
LSH	Limit solar height (-)	lw	Long-wave
Q	Heat flux (W)	mf	Multiplication factor
R	Thermal resistance ((m^2 K)/W)	r	Road
s	Thickness (m)	rad	Radiant
SHGC	Solar Heat Gain Coefficient	Rest	External resistance
U	Velocity (m/s)	rur	Rural
U	U-value (W/m^2 K)	Sky	Sky
V	Volume (m^3)	Soil	Soil
$\dot{V}$	Volumetric flow rate (m^3/s)	sw	Short-wave
W_{el}	Electrical power (W)	traffic	Heat of traffic
x	x-coordinate (m)	UBL	Urban Boundary Layer
y	y-coordinate (m)	Urb	Urban
z	z-coordinate (m)	ve	Ventilation
Z	Thermal impedance ((m^2 K)/W)	vx	x-vertex
		w	Wall group
Greek symbols		waste	Waste heat
λ	Thermal conductivity (W/(m K))	win	Windows
η	efficiency (-)		
θ	Temperature (°C)		
ρ	Density (kg/m^3)		
τ	Simulation timestep (s)		
ϕ	Heat flow rate (W)		
ω	Angular frequency (1/s)		

References

1. Intergovernmental Panel on Climate Change. *The Fifth Assessment Report (AR5)—Climate Change: Actions, Trends and Implications for Business*; IPCC: Cambridge, UK, 2013.
2. EIA. *International Energy Outlook*; EIA: Washington, DC, USA, 2019.
3. United Nations. *World Urbanization Prospects: The 2018 Revision*; UN: New York, NY, USA, 2019.
4. IEA. *Global Status Report for Buildings and Construction 2019*; IEA: Paris, France, 2019.
5. Reinhart, C.F.; Davila, C.C. Urban building energy modeling—A review of a nascent field. *Build. Environ.* **2016**, *97*, 196–202. [CrossRef]
6. Ang, Y.Q.; Berzolla, Z.M.; Reinhart, C.F. From concept to application: A review of use cases in urban building energy modeling. *Appl. Energy* **2020**, *279*, 115738. [CrossRef]
7. Sola, A.; Corchero, C.; Salom, J.; Sanmarti, M. Multi-domain urban-scale energy modelling tools: A review. *Sustain. Cities Soc.* **2019**, 101872. [CrossRef]
8. Swan, L.G.; Ugursal, V.I. Modeling of end-use energy consumption in the residential sector: A review of modeling techniques. *Renew. Sustain. Energy Rev.* **2009**, *13*, 1819–1835. [CrossRef]
9. Li, W.; Zhou, Y.; Cetin, K.; Eom, J.; Wang, Y.; Chen, G.; Zhang, X. Modeling urban building energy use: A review of modeling approaches and procedures. *Energy* **2017**, *141*, 2445–2457. [CrossRef]
10. Hong, T.; Chen, Y.; Luo, X.; Luo, N.; Lee, S.H. Ten questions on urban building energy modeling. *Build. Environ.* **2020**, *168*, 106508. [CrossRef]

11. Ferrando, M.; Causone, F.; Hong, T.; Chen, Y. Urban building energy modeling (UBEM) tools: A state-of-the-art review of bottom-up physics-based approaches. *Sustain. Cities Soc.* **2020**, *62*, 102408. [CrossRef]

12. Giassi, N.; Mahdavi, A. Reductive bottom-up urban energy computing supported by multivariate cluster analysis. *Energy Build.* **2017**, *144*, 372–386. [CrossRef]

13. Chen, Y.; Deng, Z.; Hong, T. Automatic and rapid calibration of urban building energy models by learning from energy performance database. *Appl. Energy* **2020**, *277*, 115584. [CrossRef]

14. Happle, G.; Fonseca, J.A.; Schlueter, A. Impacts of diversity in commercial building occupancy profiles on district energy demand and supply. *Appl. Energy* **2020**, *277*, 115594. [CrossRef]

15. Hong, T.; Chen, Y.; Lee, S.H.; Piette, M.P. CityBES: A web-based platform to support city-scale building energy efficiency. In Proceedings of the 5th International Urban Computing Workshop, San Francisco, CA, USA, 14 August 2016; p. 10.

16. Reinhart, C.; Dogan, T.; Jakubiec, A.; Rakha, T.; Sang, A. UMI—An urban simulation environment for building energy use, daylighting and walkability. In Proceedings of the 13th Conference of International Building Performance Simulation Association, Chambery, France, 25–28 August 2013; pp. 476–483.

17. Fonseca, J.A.; Schlueter, A. Integrated model for characterization of spatiotemporal building energy consumption patterns in neighborhoods and city districts. *Appl. Energy* **2015**, *142*, 247–265. [CrossRef]

18. Nageler, P.; Koch, A.; Mauthner, F.; Leusbrock, I.; Mach, T.; Hochenauer, C.; Heimrath, R. Comparison of dynamic urban building energy models (UBEM): Sigmoid energy signature and physical modelling approach. *Energy Build.* **2018**, *179*, 333–343. [CrossRef]

19. Hedegaard, R.E.; Kristensen, M.H.; Pedersen, T.H.; Brun, A.; Petersen, S. Bottom-up modelling methodology for urban-scale analysis of residential space heating demand response. *Appl. Energy* **2019**, *242*, 181–204. [CrossRef]

20. Zarrella, A.; Prataviera, E.; Romano, P.; Carnieletto, L.; Vivian, J. Analysis and application of a lumped-capacitance model for urban building energy modelling. *Sustain. Cities Soc.* **2020**, *63*, 102450. [CrossRef]

21. Freitas, S.; Catita, C.; Redweik, P.; Brito, M.C. Modelling solar potential in the urban environment: State-of-the-art review. *Renew. Sustain. Energy Rev.* **2015**, *41*, 915–931. [CrossRef]

22. Bhattacharya, S.; Braun, C.; Leopold, U. A Novel 2.5D Shadow Calculation Algorithm for Urban Environment. In Proceedings of the 5th International Conference on Geographical Information Systems Theory, Applications and Management, Heraklion, Greece, 3–5 May 2019; pp. 274–281. [CrossRef]

23. Liao, W.; Heo, Y.; Xu, S. Simplified vector-based model tailored for urban-scale prediction of solar irradiance. *Sol. Energy* **2019**, *183*, 566–586. [CrossRef]

24. Meng, F.; Guo, J.; Ren, G.; Zhang, L.; Zhang, R. Impact of urban heat island on the variation of heating loads in residential and office buildings in Tianjin. *Energy Build.* **2020**, *226*, 110357. [CrossRef]

25. Yang, X.; Peng, L.L.H.; Jiang, Z.; Chen, Y.; Yao, L.; He, Y.; Xu, T. Impact of urban heat island on energy demand in buildings: Local climate zones in Nanjing. *Appl. Energy* **2020**, *260*, 114279. [CrossRef]

26. Salvati, A.; Roura, H.C.; Cecere, C. Assessing the urban heat island and its energy impact on residential buildings in Mediterranean climate: Barcelona case study. *Energy Build.* **2017**, *146*, 38–54. [CrossRef]

27. Zinzi, M.; Carnielo, E.; Mattoni, B. On the relation between urban climate and energy performance of buildings. A three-years experience in Rome, Italy. *Appl. Energy* **2018**, *221*, 148–160. [CrossRef]

28. Zinzi, M.; Carnielo, E. Impact of urban temperatures on energy performance and thermal comfort in residential buildings. The case of Rome, Italy. *Energy Build.* **2017**, *157*, 20–29. [CrossRef]

29. Morabito, M.; Crisci, A.; Guerri, G.; Messeri, A.; Congedo, L.; Munafò, M. Surface urban heat islands in Italian metropolitan cities: Tree cover and impervious surface influences. *Sci. Total Environ.* **2021**, *751*, 142334. [CrossRef] [PubMed]

30. Bueno, B.; Norford, L.; Hidalgo, J.; Pigeon, G. The urban weather generator. *J. Build. Perform. Simul.* **2013**, *6*, 269–281. [CrossRef]

31. Takane, Y.; Kikegawa, Y.; Hara, M.; Grimmond, C.S.B. Urban warming and future air-conditioning use in an Asian megacity: Importance of positive feedback. *NPJ Clim. Atmos. Sci.* **2019**, *2*, 1–11. [CrossRef]

32. He, B.J. Towards the next generation of green building for urban heat island mitigation: Zero UHI impact building. *Sustain. Cities Soc.* **2019**, *50*, 101647. [CrossRef]

33. Frayssinet, L.; Merlier, L.; Kuznik, F.; Hubert, J.L.; Milliez, M.; Roux, J.J. Modeling the heating and cooling energy demand of urban buildings at city scale. *Renew. Sustain. Energy Rev.* **2018**, *81*, 2318–2327. [CrossRef]

34. Adelia, A.S.; Yuan, C.; Liu, L.; Shan, R.Q. Effects of urban morphology on anthropogenic heat dispersion in tropical high-density residential areas. *Energy Build.* **2019**, *186*, 368–383. [CrossRef]

35. Bueno, B.; Norford, L.; Pigeon, G.; Britter, R. A resistance-capacitance network model for the analysis of the interactions between the energy performance of buildings and the urban climate. *Build. Environ.* **2012**, *54*, 116–125. [CrossRef]

36. Sailor, D.J.; Lu, L. A top-down methodology for developing diurnal and seasonal anthropogenic heating profiles for urban areas. *Atmos. Environ.* **2004**, *38*, 2737–2748. [CrossRef]

37. Ichinose, T.; Shimodozono, K.; Hanaki, K. Impact of anthropogenic heat on urban climate in Tokyo. *Atmos. Environ.* **1999**, *33*, 3897–3909. [CrossRef]

38. Kikegawa, Y.; Genchi, Y.; Yoshikado, H.; Kondo, H. Development of a numerical simulation system toward comprehensive assessments of urban warming countermeasures including their impacts upon the urban buildings' energy-demands. *Appl. Energy* **2003**, *76*, 449–466. [CrossRef]

39. Kikegawa, Y.; Genchi, Y.; Kondo, H.; Hanaki, K. Impacts of city-block-scale countermeasures against urban heat-island phenomena upon a building's energy-consumption for air-conditioning. *Appl. Energy* **2006**, *83*, 649–668. [CrossRef]
40. Salvati, A.; Monti, P.; Roura, H.C.; Cecere, C. Climatic performance of urban textures: Analysis tools for a Mediterranean urban context. *Energy Build.* **2019**, *185*, 162–179. [CrossRef]
41. Wong, N.H.; He, Y.; Nguyen, N.S.; Raghavan, S.V.; Martin, M.; Hii, D.J.C.; Yu, Z.; Deng, J. An integrated multiscale urban microclimate model for the urban thermal environment. *Urban Clim.* **2021**, *35*, 100730. [CrossRef]
42. Bueno, B.; Roth, M.; Norford, L.; Li, R. Computationally efficient prediction of canopy level urban air temperature at the neighbourhood scale. *Urban Clim.* **2014**, *9*, 35–53. [CrossRef]
43. International Standard Organisation—ISO. *ISO 13790:2008 Energy Performance of Buildings—Calculation of Energy Use for Space Heating and Cooling*; ISO: Geneva, Switzerland, 2008.
44. German Association of Engineers. *Calculation of Transient Thermal Response of Rooms and Buildings—Modelling of Rooms (VDI 6007-1)*; German Association of Engineers: Düsseldorf, Germany, 2012.
45. Vivian, J.; Zarrella, A.; Emmi, G.; de Carli, M. An evaluation of the suitability of lumped-capacitance models in calculating energy needs and thermal behaviour of buildings. *Energy Build.* **2017**, *150*, 447–465. [CrossRef]
46. Weather Data | EnergyPlus. Available online: https://energyplus.net/weather (accessed on 14 December 2020).
47. Li, X.; Zhou, Y.; Yu, S.; Jia, G.; Li, H.; Li, W. Urban heat island impacts on building energy consumption: A review of approaches and findings. *Energy* **2019**, *174*, 407–419. [CrossRef]
48. Bougeault, P.; Lacarrere, P. Parameterization of orography-induced turbulence in a mesobeta-scale model. *Mon. Weather Rev.* **1989**, *117*, 1872–1890. [CrossRef]
49. UNI Ente Nazionale Italiano di Unificazione. *UNI/TS 11300-3:2010 Prestazioni Energetiche Degli Edifici—Parte 3: Determinazione del Fabbisogno di Energia Primaria e Dei Rendimenti per la Climatizzazione Estiva*; UNI Ente Nazionale Italiano di Unificazione: Milan, Italy, 2010.
50. Kilsedar, C.E.; Fissore, F.; Pirotti, F.; Brovelli, M.A. Extraction and visualization of 3d building models in urban areas for flood simulation. *ISPRS Ann. Photogramm. Remote Sens. Spat. Inf. Sci.* **2019**, *42*, 669–673. [CrossRef]
51. GitHub—Kilsedar/Urban-Geo-Big-Data-3d: Enables the Visualization, Query and Processing of Multidimensional Vector and Raster Geospatial data Related to Urban Areas on the Web. Available online: https://github.com/kilsedar/urban-geo-big-data-3d (accessed on 30 November 2020).
52. Istituto Nazionale di Statistica. *15° Censimento della Popolazione e delle Abitazioni 2011*; Istituto Nazionale di Statistica: Rome, Italy, 2011.
53. Carnieletto, L.; Ferrando, M.; Teso, L.; Sun, K.; Zhang, W.; Causone, F.; Romagnoni, P.; Zarrella, A.; Hong, T. Italian Prototype Building Models for Urban Scale Building Performance Simulation. *Build. Environ.* **2021**. [CrossRef]
54. Dornelles, K.; Roriz, V.; Roriz, M. Determination of the solar absorptance of opaque surfaces. In Proceedings of the PLEA 2007—The 24th Conference on Passive and Low Energy Architecture, Singapore, 22–24 November 2007. [CrossRef]
55. European Committee for Standardization. *EN 16798-1 Energy Performance of Buildings—Ventilation for Buildings*; European Committee for Standardization: Brussels, Belgium, 2019.
56. International Standard Organisation—ISO. *ISO 18523-1:2016 Energy Performance of Buildings—Schedule and Condition of Building, Zone and Space Usage for Energy Calculation*; ISO: Geneva, Switzerland, 2016.
57. Agenzia Regionale per la Prevenzione e Protezione Ambientale del Veneto—APRAV. Available online: https://www.arpa.veneto.it/ (accessed on 30 November 2020).
58. Allen, L.; Lindberg, F.; Grimmond, C.S.B. Global to city scale urban anthropogenic heat flux: Model and variability. *Int. J. Climatol.* **2011**, *31*, 1990–2005. [CrossRef]
59. Gabey, A.M.; Grimmond, C.S.B.; Capel-Timms, I. Anthropogenic heat flux: Advisable spatial resolutions when input data are scarce. *Theor. Appl. Climatol.* **2018**, *135*, 791–807. [CrossRef]
60. Parker, J. The Leeds urban heat island and its implications for energy use and thermal comfort. *Energy Build.* **2020**, 110636. [CrossRef]

MDPI

St. Alban-Anlage 66

4052 Basel

Switzerland

Tel. +41 61 683 77 34

Fax +41 61 302 89 18

www.mdpi.com

Climate Editorial Office

E-mail: climate@mdpi.com

www.mdpi.com/journal/climate

www.ingramcontent.com/pod-product-compliance
Lightning Source LLC
LaVergne TN
LVHW071307170726
843515LV00006BA/1514